Wayne
KF126
v.1   Y0-AKD-926

**Privacy in the digital age
: 21st-century challenges to the**

Privacy in the Digital Age

Privacy in the Digital Age
21st-Century Challenges to the Fourth Amendment

Volume I

Nancy S. Lind and Erik Rankin, Editors

PRAEGER™

An Imprint of ABC-CLIO, LLC
Santa Barbara, California • Denver, Colorado

Copyright © 2015 by ABC-CLIO, LLC

All rights reserved. No part of this publication may be reproduced, stored in a retrieval system, or transmitted, in any form or by any means, electronic, mechanical, photocopying, recording, or otherwise, except for the inclusion of brief quotations in a review, without prior permission in writing from the publisher.

Library of Congress Cataloging-in-Publication Data

Privacy in the digital age : 21st-century challenges to the Fourth Amendment / Nancy S. Lind and Erik Rankin, editors.
   pages cm
  Includes bibliographical references and index.
  ISBN 978-1-4408-2970-3 (print : alk. paper) — ISBN 978-1-4408-2971-0 (e-book)   1. Privacy, Right of—United States.   2. United States. Constitution. 4th Amendment.   3. Searches and seizures—United States.   I. Lind, Nancy S., 1958– editor.   II. Rankin, Erik, editor.
KF1262.P747 2015
345.73'0522—dc23   2014040017

ISBN: 978-1-4408-2970-3
EISBN: 978-1-4408-2971-0

19 18 17 16 15   1 2 3 4 5

This book is also available on the World Wide Web as an eBook.
Visit www.abc-clio.com for details.

Praeger
An Imprint of ABC-CLIO, LLC

ABC-CLIO, LLC
130 Cremona Drive, P.O. Box 1911
Santa Barbara, California 93116-1911

This book is printed on acid-free paper ∞

Manufactured in the United States of America

Contents

Acknowledgments ix

Introduction xi
Nancy S. Lind

VOLUME 1

1. Developments on the Fourth Amendment and Privacy to the 21st Century   1
   Katharine Leigh

2. Wiretaps, Electronic Surveillance, and the Fourth Amendment   35
   Jason Hochstatter

3. Forensic DNA Analysis, the Fourth Amendment, and Personal Privacy   61
   Wendy Watson

4. Biometric Identification as a Requirement for Work Access and Forced Surrendering of Private Information   83
   Pamela LaFeber

5. Employee Expectations of Privacy in the Workplace: Drug Tests, Work Spaces, Computers, and Social Media   109
   R. Craig Curtis

6. The Privacy Rights of Minors in a Digital Age   131
   Gardenia Harris

7. Library Patrons and the National Security State 151
   Chad Kahl

8. Where Is the Suspect? The Potential for the Use
   of Private Location-Tracking Data by Law Enforcement 175
   R. Craig Curtis

9. Drones and Police Practices 199
   John C. Blakeman

10. So Long, Stakeout? GPS Tracking and the Fourth Amendment 221
    Maureen Lowry-Fritz and Artemus Ward

11. Drones, Domestic Surveillance, and Privacy: Legal and
    Statutory Implications 243
    David L. Weiden

12. 21st-Century Developments in Fourth Amendment
    Privacy Law 267
    Timothy O. Lenz

13. The Changing Expectations of Privacy in the Digital Age 307
    Meghan E. Leonard

VOLUME 2

14. Beyond OnStar: The Future and the Trespass-versus-
    Privacy Debate 331
    Ronald L. Nelson

15. Closed Circuit TVs, Videomation, and Privacy 357
    Elizabeth Wheat

16. Airport Scanners and the Fourth Amendment 377
    Cara E. Rabe-Hemp

17. Social Media and the Fourth Amendment Privacy Protections 399
    Renee Prunty and Amanda Swartzendruber

18. Hacking, the Limits to the Fourth Amendment, and
    Challenges to Local Administration in the 21st Century 429
    Eric E. Otenyo

19. Data Mining in the 21st Century *Todd C. Hiestand*	463
20. The Role of Security in Wireless Privacy *Glen Sagers*	495
21. Identity Theft in the 21st Century *John C. Navarro and Cara E. Rabe-Hemp*	519
22. Maintaining the Technological Neutrality of the Fourth Amendment *L. J. Zigerell*	543
23. Developments in Search-and-Seizure Cases in the Post–September 11 Era *Thomas E. McClure*	567
Appendix: Table of Cases	595
Bibliography	605
About the Editors and Contributors	647
Index	653

Acknowledgments

Nancy S. Lind and Erik T. Rankin would like to thank Mr. Chris Farrer and Mr. Brian Madden for serving as research assistants on the book. Additionally, we would like to thank our editors at ABC-CLIO, including Denver Compton and Steve Catalano, and Pete Feely at Amnet Systems. We would also like to extend a thank-you to all the contributors of original chapters to this book.

Introduction

Nancy S. Lind

A public debate is unfolding about the definition and boundaries of privacy in modern American life. Still, privacy has always been a concern in the United States. The framers of the Constitution protected citizens against illegal searches and seizures by the federal government in the Fourth Amendment to the Bill of Rights. Although the word "privacy" was not specifically mentioned, this amendment has provided some privacy protections over the past 200 years.

The history of privacy and the Fourth Amendment in the United States are explored, from Colonial times up to the 21st century, in the Leigh chapter. There is a discussion on various issues, people, and Supreme Court cases that have had a significant impact on the Fourth Amendment. Also discussed are the many inventions that have had an impact on privacy, such as the telephone, surveillance equipment, and the Internet. The concerns over privacy are not a new phenomenon, but rather are everlasting and ever-changing.

Hochstatter's chapter examines the past and present ways that a variety of government organizations have used wiretaps and Internet traffic monitoring under the Fourth Amendment. It examines the ways in which surveillance is performed and the legal restraints put in place to protect people's privacy. Additionally, it examines past and present controversies related to wiretapping and online surveillance

Watson's chapter explains the ways in which state and national law enforcement currently use DNA databases to solve crimes. It then turns to more specific analysis of "confirmation" and "trawling" uses of DNA, their implications for the Fourth Amendment and privacy concerns, and future incursions on those rights through methods of expanding existing databases. The article concludes that current jurisprudential models—which seek to treat DNA testing as comparable to other methods of evidence collection, such as fingerprint analysis—are insufficient to address the growing threat of forensic DNA analysis on constitutional rights.

The use of biometric technology is not new. We have freely handed over our most personal and private information to others for decades. Even when our information is taken from us, it is not without our knowledge, and its use is generally understood. However, as with any technology, biometric technology has become more sophisticated. Today, information gathering, storage, dissemination, and search capability has blurred the line between personal protection and invasion of privacy. LaFeber's chapter broadly describes biometrics, discusses why the use of biometric technology is growing, gives an overview of the application of biometrics by the government and private sector organizations, considers the privacy implications of the use of biometrics, discusses court rulings on the use of biometrics in the workplace, and details a standard set of principles when employing this type of technology.

Curtis notes that the most important fact in determining the privacy rights of an employee in his or her workplace is the type of employer. In the private sector, the employee's privacy rights are determined by the employment contract and any applicable state law. This means that any private-sector employer may drug test or search the workspace of an employee for whatever reason, unless prohibited by state or local law or the terms of any personnel manual, employment contract, or collective bargaining agreement. In the public sector, by contrast, the principle of state action applies. Searches must be justified according to the principles of the Fourth Amendment as they govern administrative searches. Thus, drug testing and searches of premises and electronic devices must be justified under analysis of the jurisprudence of the Supreme Court. This chapter is concerned primarily with the detailed analysis of the extent to which that jurisprudence allows searches of public sector employees and their workspaces.

Harris's chapter reviews the principles underlying Fourth Amendment protections in the digital age and then examines minors' privacy rights within the home and school settings. The physical language used in the Fourth Amendment has limited applicability when digital information that exists in nonphysical form is considered. The long-standing legal doctrines of the "reasonable expectation of privacy" and the "third-party doctrine" are analyzed for their applicability to computerized information. Courts have acknowledged that minors possess privacy rights but have also ruled that under certain contexts, minors have reduced Fourth Amendment rights compared to adults. The courts' justifications for placing limitations on minors' privacy rights are discussed. Finally, congressional concerns about children's information privacy on the Internet prompted the passage of the Children's Online

Privacy Protection Act (COPPA), which limits the amount of information that website operators can collect from minors without the express consent of their parents.

In Kahl's chapter, readers learn how two competing interests, the government and the librarianship profession, differ on the sanctity of library records; one is driven by the ability to expand its surveillance capabilities, and the other is driven by a strong professional ethos. The chapter examines the development of the library profession's shared values; the historical arc of Supreme Court decisions on the Fourth Amendment, especially those related to search and seizure; how governmental efforts to increase information gathering have impacted library patrons; and how libraries responded to governmental intrusions over the past century with a special emphasis on the protection of library patrons' rights.

Curtis emphasizes that it is easy to believe that the founders never anticipated the effects that advances in information technology would have on privacy rights. The word "privacy" is not in the Fourth Amendment. The language itself only requires that searches be reasonable. It has been interpreted to mean that warrantless searches are suspect in most contexts, but, as written, how it applies to the use of a new generation of electronic gadgets is not at all clear. From the earliest decisions of the United States Supreme Court, in which conversations were intercepted via the use of "bugs," the justices have made it clear that the effects of technology on the meaning of the Fourth Amendment are hard for them to address. Originalism, as understood and applied by the more conservative members of the current Court, makes this task even harder, and the avoidance of the task in *United States v. Jones* is a prime example. The frequent requests for data on the location of suspects by local and federal law enforcement agencies, and the revelation that the National Security Agency is monitoring cell phone calls, text messages, and Internet usage, only makes it more urgent that the federal judiciary adapt the meaning of the Fourth Amendment to provide clear guidance to the government agencies and the public about what is private and what is not.

Law enforcement agencies within the United States are using unmanned aerial vehicles, or drones, with increasing frequency. Drones are affordable, easy to deploy, and adaptable to a wide range of public-safety missions. As drones become more prevalent and are used for an array of criminal justice missions such as surveillance and criminal investigations, the procedures that govern their use are ever more critical, given the constitutional constraints on police power imposed by the Fourth Amendment to the U.S. Constitution. Blakeman's chapter reviews the use of drones by police in the United States. It discusses some of the basic types of drones currently available, the incentives that police have to use them, and how drones are used for public-safety purposes. It also examines current Supreme Court case law concerning the Fourth Amendment search-and-seizure clause and how it will regulate the police use of drones. The chapter concludes with an assessment of the ongoing policy debates in Congress on how to best regulate drones as well as laws passed by state legislatures that place clear restrictions on their use.

In their chapter, Lowry-Fritz and Ward address the current state of technology and the Fourth Amendment, with an emphasis on law enforcement officials' use of Global Positioning System (GPS) technology. Case law regarding this matter has evolved over time, establishing and revising standards by which courts assess the constitutionality of law enforcement strategies. Perhaps the two most noteworthy standards have been the "common law trespass" (*Olmstead* 1928) and the "reasonable expectation of privacy" standards (*Katz* 1967). Recently, the United States Supreme Court revisited the question of Fourth Amendment rights in *U.S. v. Jones*, a case involving the warrantless use of GPS technology. In this case, the Court again faced questions regarding the constitutional rights of criminal suspects and reexamined its approach to the protections afforded in the Fourth Amendment. While the Court unanimously held that the GPS-based surveillance did constitute a "search" pursuant to the Fourth Amendment, the justices divided regarding the appropriate standard. In the majority opinion, Justice Scalia emphasized a return to the objective "common law trespass" standard. However, Justices Alito and Sotomayor authored separate concurring opinions that advocated the use of the "reasonable expectation of privacy" standard. This disagreement regarding the appropriate standard leaves a number of questions unanswered. In this chapter, Lowry-Fritz and Ward provide a typology of five potential standards that the Court may apply in future Fourth Amendment cases.

The use of unmanned drone aircraft has been the subject of considerable controversy as the use of these vehicles increases. One area of particular concern is the utilization of drones by law enforcement authorities for the purpose of criminal investigation and domestic surveillance. The Fourth Amendment of the U.S. Constitution requires that a judicial warrant be obtained before a criminal search is conducted, but the ability of drones to peer into nearly every area of society makes the issue of a police search difficult to determine. Furthermore, many governmental agencies are using drones for surveillance and monitoring without obtaining prior judicial approval. Lawmakers across the nation are grappling with the tradeoff between privacy and security interests, and there is no question that judges will be required to confront legal questions regarding the boundaries of the Fourth Amendment as it applies to domestic drone usage and warrantless surveillance. Weiden's chapter examines drone technology, the early state and federal legislative responses to drones, major U.S. Supreme Court cases regarding the Fourth Amendment's search jurisprudence, and how these precedents may apply to domestic drone surveillance. The chapter concludes with an analysis of the possible constitutional approaches that may be used by the Supreme Court on the use of unmanned aerial vehicles for surveillance by law enforcement.

The Fourth Amendment is the most criticized and dynamic area of criminal law. The main theme of Lenz's chapter is that digital technology has renewed old debates about where to strike the right balance between granting the government enough power to fight crime and protect national security while also limiting power to protect civil rights and liberties. The chapter examines three developments related to Fourth Amendment privacy rights: first, the exceptions to the rule that the Fourth Amendment requires specific warrants for a search to be reasonable; second, digital technology's impact on law enforcement methods and missions—particularly intelligence-based preventive policing; third, the impact of counterterrorism policies that integrate the domestic law enforcement and national security missions of law enforcement agencies and link surveillance programs. These developments are challenging the current legal regime of Fourth Amendment privacy rights, which provides limited, or even no, protection for electronic data.

Leonard's chapter explores how changing technologies have revised our expectations of privacy in the digital age. The chapter begins by reviewing the historical roots of our right to privacy, tracing this right to *Griswold v. United States* (1965) and *Katz v. United States* (1967), in which the Supreme Court argued that citizens have both a right to privacy and a reasonable expectation of privacy. Given this, Leonard considers how technology has made invasions of privacy easier and maybe even more accepted by individuals. Yet, problematically, the law and legal rulings have not been able to keep up with this ever-changing world. Using the examples of Facebook, smartphone location-based applications, and digital evidence, the chapter discusses how these new technologies beget new questions and how the current law is incapable of addressing these very questions.

As more and more advanced surveillance technologies become part of our modern society, the relevance and application of the Fourth Amendment protections of individual liberties come increasingly into question. In particular, the traditional tests employed by the courts in Fourth Amendment analysis require updating if the Amendment is to fit the demands of the times. The development of new surveillance technologies as well as acquisition and analysis techniques raise increased tensions between the needs of law enforcement and the privacy interests of the individual. This tension calls for a response from the American judiciary that modernizes the place of the Fourth Amendment in contemporary jurisprudence. Over the years, the courts have addressed the arrival of various new technologies with the development of different means of analysis. The traditional test of trespass and the subsequent use of an analysis based on the expectation of privacy have been and are currently used to apply the Fourth Amendment to the search and seizure issues of the day. These tests are not now able to adequately address Fourth Amendment issues in a world of On Star tracking and drone surveillance. In his chapter, Nelson notes that a new approach is needed—an approach that recognizes that these new technologies should be viewed with the presumption that they are violative of privacy and that they bear the burden to prove otherwise.

The Fourth Amendment prohibits the government from conducting unreasonable searches and seizures and requires probable cause for a search warrant. The Supreme Court has consistently held there is no reasonable expectation of privacy in public, argues Wheat. Closed-circuit television (CCTV) and surveillance footage technology, such as what the Boston Police Department used to identify and apprehend the Boston Marathon bombing suspect Dzhokhar Tsarnaev, challenge Fourth Amendment protections to privacy by enabling police to obtain enhanced footage from public areas. In a poll conducted by CBS and the *New York Times* after the marathon, respondents indicated overwhelming support for surveillance cameras in public areas. However, a separate poll conducted by CNN and Time Inc. suggested growing concerns over how the government and law enforcement might expand such surveillance and anti-terrorism policies, worrying civil rights advocates. These responses suggest the inherent tension in the United States over the tradeoff between privacy and security.

Washington, D.C., and New York City are considered model programs, using public surveillance systems as methods of crime deterrence and detection. Additional CCTV systems are used in Baltimore, Chicago, Dallas, Denver, and the Meadowlands Stadium in New Jersey for NFL games. Internationally, the United Kingdom is the most supervised country in the world, with 4.5 million cameras. As CCTV programs rapidly expand and private-sector companies develop new facial recognition technologies (FRTs) offering promise, police departments must address this privacy and security tradeoff, making the case for CCTV as an important tool in crime deterrence and detection while also persuading the public its Fourth Amendment rights remain protected. Wheat's chapter discusses the constitutional concerns of CCTV, successful implementation of these systems, and recommendations on how a constitutional surveillance system can be designed.

Following the December 25, 2009, thwarted attack of an Al Qaeda operative upon Flight 253, the Transportation Security Administration (TSA) sought technologies that could better detect nonmetallic biological, radiological, and chemical weapons. One such technology, the Advanced Imaging Technology (AIT) unit, has been constitutionally challenged due to its ability to see through clothes to produce an image of the human body and weapons hidden underneath. Rabe-Hemp's chapter explains how the AIT technology works; provides an historical analysis of airport security, highlighting the vulnerability of air travel to terrorist attack; and establishes the relevance of the AIT machines and airport security to the Fourth Amendment. Rabe-Hemp's chapter concludes with a few lingering questions about the constitutionality of the expanded use of the full-body scanners in commercial airports throughout the United States.

The Fourth Amendment is one of the shortest, yet most complicated Amendments to the Constitution. Advancing technology has further complicated both the issue of when a search has taken place and when a search warrant is required. The new technology and usage also means the government can spy into people's lives without a physical intrusion. The *Katz* test, created to determine whether or not a search has occurred when there is no physical intrusion, is dependent upon the subjective belief of the individual and objective belief of society in a reasonable expectation of privacy. The question being addressed by Prunty and Swartzendruber is whether the current use of social media has changed either the subjective or objective expectation of privacy. What is the expectation of privacy for an individual who shares personal information on Twitter or Facebook? The Supreme Court has yet to accept an Internet case involving the Fourth Amendment right to privacy; therefore, individuals using these technologies have to look to precedent in other areas, rely on outdated legislation, or hope for the passage of current social media legislation.

In his chapter, Otenyo discusses the interactions between the Fourth Amendment and hacking. Research on hacking and breaches to Internet security has grown beyond casual front-page news commentary to nuanced reporting of its implication on the limits of Fourth Amendment privacy rights. This chapter discusses the nature and scope of the problems associated with hacking and attempts a framework for security management of local government websites. In the end, Otenyo reflects on the nexus between hacking and infringement of rights to privacy in a democracy.

Data mining may be a valuable tool to prevent terrorism, but it may also be an example of governmental overreaching that profoundly diminishes individual privacy. Hiestand's chapter discusses definitions of data mining and provides an overview of confirmed and suspected governmental data mining programs. Additionally, Fourth Amendment cases relevant to data mining and issues related to the use of data mining in counterterrorism efforts are addressed.

The extent to which electronic surveillance can be conducted depends to some degree on the technologies used, but generally, most wireless technologies are subject to electronic surveillance. Common technologies in use today, such as cell phones, various handheld radio systems, and the ubiquitous wireless Internet used in our daily lives are discussed, especially from the standpoint of ease of eavesdropping. Sagers's chapter concludes with a discussion of various countermeasures the average citizen can take to minimize or completely thwart most monitoring. We are being monitored, but there are both technical and political steps that can be taken to insure it's not being done illegally.

With the emergence of the Internet, an increasing number of consumer purchases occur online, leaving Americans susceptible to identity theft. The U.S. government has responded with a progression of laws and strategies designed to protect citizens, but in doing so, they have been accused of being overzealous. Navarro and Rabe-Hemp's chapter explores the difficult balance between protecting citizens' Fourth Amendment rights with the need to protect them against victimization. The prevalence, definition, forms, and offender and victims demographics are discussed as well as a history of U.S. legislation. A review of important U.S. Supreme Court cases illustrates the government's Fourth Amendment legal boundaries. The chapter concludes with strategies and policies that encourage government involvement in the investigation and persecution of identity theft while defending American's rights to privacy.

L. J. Zigerell notes that individuals forfeit Fourth Amendment rights in information voluntarily provided to a third party, according to the third-party doctrine. Some scholars have proposed that this doctrine maintains the technological neutrality of the Fourth Amendment because the doctrine prevents suspects from using technology such as a phone or e-mail to hide from law enforcement actions that law enforcement could have observed in the absence of such technology. The purpose of this chapter is to explain why this third-party doctrine does not maintain the technological neutrality of the Fourth Amendment, to propose a modified rule regarding law enforcement access to and use of third-party records, and to illustrate that this modified rule maintains the technological neutrality of the Fourth Amendment better than the third-party doctrine. The chapter also explains how this modified rule can be applied to assess the constitutionality of enhancements to law enforcement's ability to search.

McClure's chapter discusses noteworthy Fourth Amendment decisions rendered by the United States Supreme Court between September 11, 2001, and June 30, 2014. The cases reflect trends in the interpretation of traditional Fourth Amendment issues, such as execution of warrants, exceptions to the warrant requirement, consensual searches, the exclusionary rule, stop and frisk, causes of action for Fourth Amendment violations, and motor vehicle searches. It discusses developments concerning the use of technology, scientific testing, and police dogs employed in investigating crime. Additionally, this chapter discusses how the Court has applied the Fourth Amendment to specific classes of individuals. Finally, the argument is made that the War on Terror initiated following the September 11 attacks does not appear to have impacted search-and-seizure jurisprudence.

1

Developments on the Fourth Amendment and Privacy to the 21st Century

Katharine Leigh

> The fantastic advances in the field of electronic communication constitute a greater danger to the privacy of the individual.
> —Earl Warren[1]
>
> You already have zero privacy. Get over it!
> —Scott McNealy, CEO Sun Microsystems, 1999[2]

INTRODUCTION

The definition of the right to privacy can be found in many places. *Merriam-Webster Online* defines the right to privacy as "the qualified legal right of a person to have reasonable privacy in not having his private affairs made known or his likeness exhibited to the public having regard to his habits, mode of living, and occupation."[3] The right to privacy is considered to be a human right, designed to restrain both governments and private parties from violating a person's privacy. This right developed in North and South America over centuries. In Colonial times, concerns over privacy issues affected the burgeoning Postal Service. The issues could also be seen in the search and seizures conducted by British government officials.

Concerns over the right to privacy can been seen within the first ten amendments, also known as the Bill of Rights. The First Amendment protects the freedom of beliefs, including religion, free speech, the right to assemble, and the right to petition the government for a redress of grievances. The Third Amendment protects the privacy of the home, stating that a soldier cannot be quartered in any house without the consent of the owner. The Fourth

Amendment protects the privacy of the person and possessions, stating that the government cannot engage in unreasonable searches and seizures and that warrants can only be issued with probable cause. The Ninth Amendment could provide a more general protection of privacy, since it states that "The enumeration in the Constitution, of certain rights, shall not be construed to deny or disparage others retained by the people."[4] The amendments found in the Bill of Rights only applied to the federal level until the adoption of the Fourteenth Amendment in 1868, with its due process clause. The aforementioned clause made most of the Bill of Rights applicable to state and local governments as well.

The Fourth Amendment has protected the privacy of U.S. citizens and some others from unreasonable searches and seizures by its government for over 200 years. This book will explore privacy and the Fourth Amendment in the 21st century; however, it is important to look at the developments leading up to the current century.

COLONIAL TIMES AND LATE 18TH CENTURY

Postal Service

It is possible that the only truly private communications we have are the thoughts in our heads. This has been true for hundreds of years; even in colonial times, people had privacy concerns in regard to their communications. In Colonial America, one method of communication was sending messages through the mail. For many years, mail was delivered via private messengers. A postal system was established in the late 1600s when Thomas Neale was appointed postmaster general of America by the British government. His charge was to establish post offices that were responsible for sending, receiving, and delivering letters in the colonies. Despite being postmaster general, Thomas Neale never traveled to the United States. The nascent postal service was effectively run by the deputy postmaster, Alexander Hamilton.

A respect for privacy can be seen in the Post Office Act 1710. This act, also known as Queen Anne's Act, was passed by the British Parliament and took effect in North America on June 1, 1711, and remained in effect until 1789. It established fixed rates for transporting letters and created a deputy postmaster general for the colonies. Additionally, it included wording concerning privacy, stating that "No Person or Persons shall presume wittingly, willingly, or knowingly, to open, detain, or delay, or cause, procure, permit, or suffer to be opened, detained, or delayed, any Letter or Letters, Packet, or Packets."[5] Postal employees were required to take an oath swearing to the above, and violators faced a fine of up to 25 pounds.

Benjamin Franklin, one of the Founding Fathers, had a long relationship with the postal system. In 1737 Franklin, then a printer, was declared to be the operator of "the Post-Office of Philadelphia." This helped his printing business, as he was now in charge of both creating and distributing his materials. In 1753 the British government named him a deputy postmaster general of the colonies (along with William Hunter of Alexandria, Virginia). During his time with the postal service, Franklin recommended many improvements, such as surveys for new post roads, the installation of mile markers, the rerouting of some postal routes to shorten delivery times, and the use of lanterns to enable nighttime deliveries.[6]

Franklin and Hunter addressed privacy concerns during their tenure. They followed the 1710 Post Office Act, and they added other requirements to ensure privacy. Their requirements included the following: postmasters were required to keep the post office separate from their personal residence, only authorized persons were allowed to handle the mail, mail bags were only allowed to be opened once they had already arrived at their city of destination, and postmasters were required to ask for proof of identification when a person came to retrieve a message.[7]

Benjamin Franklin lost his position as postmaster when he appeared to be sympathetic to the revolutionaries in the American colonies. However, he returned to his position in 1775 when the Second Continental Congress established the United States Post Office and named Franklin its first United States Postmaster General. Franklin remained in this post for 15 months, but the concerns over privacy within the mail system continued to be a contentious issue in the coming centuries.

Writs of Assistance

Over the years, many historians have credited the writs of assistance with being a spark for the American Revolution and a significant contributing factor in the adoption of the Fourth Amendment to the U.S. Constitution. A writ of assistance is a written order issued by a court that provides instructions to a law enforcement office to perform a particular task, similar to a general warrant. The writ was used by the British in Colonial times as a way to limit the colonies from trading with non-English industry. Customs officers were allowed to enter any house or business to search for illegal imports, and they did not need a warrant to do so. "The only practical limitation on the writ was the life-span of the monarch in whose name the writ was issued; under the terms of the statute adopted by Parliament, all writs of assistance automatically expired six months after the death of the British king or queen."[8] In addition, after a judge issued a writ, there was no further requirement for judicial approval. An applicant need not provide what search the writ was needed for, and "it could be transferred from one person to another without judicial approval, and there was no requirement that the writs be returned to a court or magistrate with a report of the results of the search."[9] The incredible amount of power in a general writ was widely regarded as abusive, and it inflamed the colonists against the British.

Perhaps the most famous case in Colonial America in regard to the writs of assistance was the 1761 case involving Thomas Lechmere, surveyor of the customs in America, applying for the continuation of his writ of assistance to the Superior Court of Massachusetts. Sixty-three Boston merchants hired lawyers Oxenbridge Thatcher and James Otis to oppose the writ in court. John Adams, the future second president of the United States, was in court on the day James Otis made an impassioned speech against the writs. "According to Adams' account written many years later, Otis, in a rousing speech that lasted over four hours, denounced the writs on the grounds that they were unenforceable and void because they were in conflict with the Magna Carta."[10] In the end, the judge postponed deciding on whether to grant an extension to Lechmere's writ and instead decided to ask for advice from the provincial agent in England, William Bollan. Not surprisingly, Bollan's opinion was that the writs were perfectly legal, and Chief Justice Hutchinson ruled in favor of reissuing the writ.

Otis may have lost that battle, but the case was later cited by Adams as a factor in American independence. Some historians have debated the importance of the case in the revolution, but it is clear that the writs of assistance were considered a factor. Richard B. Morris stated in his book *The Era of the American Revolution* that writs of assistance were rarely issued in the colonies. When there were applications for them, Morris cites defeat after defeat, wherein judges from the various colonies did not issue writs of assistance. Morris believes that even though few writs were granted, the fact that they were so universally fought showed what the Americans thought of them. "It took courage for judges to refuse writs of assistance when demanded by the customs officers, since they held their commissions at the will of the Crown and were dependent for their salaries upon the revenues collected by the Customs Commissioners."[11]

There were other pieces of legislation that similarly inflamed the colonists, and many of them had to do with privacy. The Stamp Act of 1765 was "a tax on legal documents, almanacs, newspapers, and nearly every form of paper used in the colonies."[12] It was a direct tax imposed by the British Parliament, requiring that printed materials be produced on stamped paper created in London and displaying an embossed revenue stamp. The tax could not be paid in colonial paper money, only in valid British currency. The colonists did not have representation in Parliament, so they had no control or say over the taxes or money was to be used.

The original intent of the tax was to help the British pay for troops in North America after the Seven Years' War. Despite the intent, the act was viewed by the colonists as taking money out of their pockets. It was also seen as censorship, because it made communication even more financially difficult.

The Quartering Act of 1765 was passed by the British Parliament just a few months after the Stamp Act. It was designed to provide provisions for British troops that had just finished fighting the French and Indian War (the North American part of the Seven Years' War), who were continuing their stay in North America. This act stated that British troops must be lodged somehow by the colonists, although not necessarily in their homes (as is commonly believed). If there was not enough room in public houses and barracks for the soldiers, then a town official was "required to take, hire and make fit for the reception of his Majesty's forces, such and so many uninhabited houses, outhouses, barns, or other buildings, as shall be necessary, to quarter therein the residue of such officers and soldiers for whom there should not be room in such barracks and publick houses as aforesaid."[13] Additionally, Colonial authorities were required to pay the cost of both housing and feeding the British troops. This did not please the colonists, as they viewed this to be yet another seizure of their property by the government.

The Townshend Acts were passed two years after the Quartering Act. They were a series of acts named after the Chancellor of the Exchequer, Charles Townshend, and included: the Revenue Act of 1767, the Indemnity Act, the Commissioners of Customs Act, the Vice Admiralty Court Act, and the New York Restraining Act. The purpose of these acts was to raise money to pay the salaries of judges and governors, to create a more effective way of enforcing compliance with trade regulations, to punish New York for failing to comply with the Quartering Act, and to establish the precedent that the colonies could be taxed by the British Parliament. The acts taxed commonly used items such as paper, glass, and tea. These taxes led to a boycott of those products, further angering the British government. Resistance to the taxes led to British troops in Boston and ultimately to the Boston Massacre in 1770. The British government eventually repealed most of the taxes, but not the tax on tea.

The framers of the Constitution wrote the First, Third, and Fourth Amendments to address the abuses of the British government over the American colonies. The Stamp Act, the Quartering Act, and the Townshend Acts were perceived as violations of American colonists' privacy and showed the need for the framers to include protections against these types of violations by the federal government in the supreme law of the new country.

Fourth Amendment and the Constitution

The Articles of Confederation was the first document to rule the fledgling United States. Created on November 15, 1777, and ratified on March 1, 1781, it was mostly a vague document that barely held the thirteen founding states together. It was framed so that the states continued to have most of the power; the federal government had power predominantly over foreign affairs. The only branch of the federal government it created was Congress, and it had no Bill of Rights. This document worked during the Revolutionary War, but it was soon clear that a new supreme law would be needed for the United States.

On May 25, 1787, the Constitutional Convention began in Philadelphia. The new document gave greater prominence to a federal government, creating the executive and judicial branches to go along with the Congress to establish a balance of power. Still, the non-Nationalists were worried about the power the more prominent federal government would have over states and individuals. In order for the U.S. Constitution to receive the required three-fourths vote needed by the states, a Bill of Rights would have to be approved. The search-and-seizure clause of the Fourth Amendment was considered important because of worries by anti-Federalists about the federal government having too much power, which could lead to intrusions into the business and personal lives of citizens. James Madison was the leader of the movement in the first Congress to add the Bill of Rights to the first Constitution. He also wrote the original version of the Fourth Amendment. It stated:

> The rights of the people to be secured in their persons, their houses, their papers, and their other property from all unreasonable searches and seizures, shall not be violated by warrants issued without probable cause, supported by oath or affirmation, or not particularly describing the places to be searched, or the persons or things to be seized. (Quoted in Davies 1999, 697)[14]

Many of the framers considered this wording to be weak because it appeared to ban just general warrants and not warrantless intrusions. To solve this issue, the framers added the phrase "and no warrant shall issue, but upon probable cause," creating the language citizens see today. The more strongly worded amendment then read, it is important that a search be "reasonable," regardless of whether there is a warrant.

> The right of the people to be secure in their persons, houses, papers, and effects, against unreasonable searches and seizures, shall not be violated,

and no Warrants shall issue, but upon probable cause, supported by oath or affirmation, and particularly describing the place to be searched, and the persons or things to be seized.[15]

Twelve amendments were eventually submitted to the states for approval, with ten receiving the required amount of votes to be a part of the Constitution. Under the original numbering, what we know as the Fourth Amendment would have been the Sixth Amendment. The Fourth Amendment and others in the Bill of Rights were ratified on December 15, 1791.

19TH CENTURY

The 19th century saw significant advancements in technology and communications. The first permanent photograph was made in 1822 by Joseph Nicéphore Niépce, a French inventor. It was later expanded upon by Niépce and Louis-Jacques-Mandé Daguerre, which led to the well-known daguerreotype. Later in the century, the first telegraph message was sent by Samuel Morse. It was sent in 1844 from Washington, D.C., to Baltimore, and it said, "What hath God wrought?" Both inventions have had a significant impact on personal privacy.

The 19th century was not a very active time for the Fourth Amendment in the court system; there were only two Supreme Court decisions that dealt directly with the amendment, *Ex parte Burford* (1806) and *Boyd v. United States* (1886). During this century, there were many instances that could be seen as privacy infringements by the government, most notably the changing nature of the United States Census. During the Civil War, the U.S. government exercised a great deal of power that it formerly had not used. After the war was over, the federal government kept much of the power it had gained during the past four years. A major result was the passage and adoption of the Fourteenth Amendment, which had significant impact on the Fourth Amendment. The most significant development in regard to privacy happened near the end of the century, when a law review article changed the way the courts viewed the right to privacy from that time forward.

U.S. Census

The United States conducts a census every 10 years, asking many sensitive questions of its citizens in the process. This potential invasion of privacy by the federal government is considered outside the Fourth Amendment's purview because it is explicitly required in article 1, section 2 of the Constitution. A census is needed to count the United States' population, and that count is used to determine representation in Congress. However, the use of the census has expanded over the years, and the federal government has continually demanded more (and more personal) information from its citizens.

The first census was conducted in 1790 under the direction of Secretary of State Thomas Jefferson. It asked just six questions. It wanted the name of the head of the family and the number of persons in each household matching a certain description. During the 19th century, the census evolved to include information regarding economics, education, marital status, and medical issues. As the information that was collected grew, so did the concerns over personal privacy. Between the 1840 and 1850 censuses, Congress created the Census Board, now called the U.S. Census Bureau, a central governmental bureau for planning, implementing, and tabulating the results from the census. The Census Board also increased the amount of information asked of U.S. citizens. To allay fears about privacy invasions, the U.S. Secretary of the Interior admonished his employees to not gather inappropriate information and to not share any information gathered with others. By the 1880 census, the questions asked had increased from 6 to 24, and if people did not want to answer these questions, they could be fined up to $100. In response to privacy concerns, census takers were required to take an oath affirming that they would not misuse the personal information gathered. If they did, they were subject to a fine of up to $500.[16]

The data generated by the first ten U.S. censuses had to be counted manually. The sheer volume of the data could take years to process. The 1880 census, for example, took eight years to tabulate. Technological advances helped with this problem. A man named Herman Hollerith, a statistician and inventor, designed and constructed a new system for tabulating the data from the census. Census takers would record the initial information by hand, and then census data clerks would transfer the information onto standardized manila cards. The clerks used a device that punched small holes in the cards, corresponding to the information. The finished manila cards would then be placed into a mechanical tabulating machine that automatically compiled the data. The use of mechanical tabulation was a resounding success, with the 1890 census being tabulated in only one year. The government would now be able to compile even more information about its citizenry. Herman Hollerith went on to found the Tabulating Machine Company, which later merged with three other corporations to form the Computing Tabulating Recording Company (CTR). In 1924 CTR was renamed International Business Machines Corporation, or as it is commonly known today, IBM.

Ex parte Burford (1806)

Ex parte Burford was the first case before the United States Supreme Court that examined closely the warrants clause of the Fourth Amendment. John A. Burford was a District of Columbia shopkeeper who was jailed after failing to post $4,000 meant to ensure his good behavior. He was asked to pay this by Alexandria County, D.C., judges who had received information that he was not of good name and fame and was an "evil doer." "The circuit court for the District of Columbia had reduced the length of Burford's confinement to one thousand dollars, but Burford filed for a writ of habeas corpus on the basis that his arrest warrant '[did] not state a cause certain, supported by oath.'"[17]

In a unanimous decision, the Supreme Court justices overturned the conviction because the warrant did not state that a crime had actually been committed, and the Fourth Amendment was "designed to protect against imprisonment without a criminal conviction as well as arbitrary searches." The court did find Burford to be of "ill fame" but felt "that the warrant of commitment was illegal for want of stating some good cause certain, supported by oath."[18]

The Antebellum Period and the Civil War

The most controversial political issues of the 19th century in the United States were slavery and secession. The concerns over these issues led those in government, at both the national and local levels, to suggest actions that would seem to violate the privacy protections in the Fourth Amendment.

In the early 19th century, there was great concern in the South that the federal government was going to end slavery. To combat this, the South attempted to control the flow of antislavery ideas coming into their region.[19] Many southern states passed laws that made antislavery materials illegal entirely. Antislavery papers and pamphlets would not always make it to their intended recipients; instead, they would be detained at local post offices. In essence, the Confederacy began seizing private property (antislavery materials) to try to prevent the seizure of their "property" (their slaves). In 1836, Senator John C. Calhoun introduced to Congress a "Gag Bill" that would have made it illegal for the postmaster general to deliver mail that was considered to be illegal. This bill was defeated due to concerns about potential violations of the First and the Fourth Amendments. Antislavery mail continued to be blocked by local officials, however. When Abraham Lincoln was elected, Southerners worried that he would change the system of seizing illegal materials, by appointing Republican postmasters who did not follow the practice of detaining messages that were considered illegal. They feared this might lead to the building of an antislavery force in the South.

Lincoln did not appoint Republican postmasters to the South, but he did seize private communications. The Fourth Amendment did not protect people from having their mail and telegrams seized during the Civil War. In fact, telegrams were significantly easier to intercept. Letters sent through the mail were sealed, either with wax or inside envelopes; with telegrams, the contents of the message were disclosed to the telegraph company and its operator. "Over time, however, a substantial body of case law developed making it clear that despite knowing the contents of every unciphered message presented for transmission, telegraph companies had a duty of confidentiality toward their customers."[20]

Both sides extensively wiretapped telegraph lines during the Civil War. "The American military even set up a parallel telegraph system in an effort to secure its communications, and telegrams became such an integral part of the conduct of the war that President Lincoln often spent hours or even days in the Army Telegraph Office in Washington."[21] In 1861, "Lincoln ordered federal marshals to visit every major Northern telegraph office and seize whatever copies of telegrams they could lay their hands on."[22] This was done because the rails and telegraph lines in Washington, D.C., had been compromised, and Lincoln was desperate for information. Despite the blatant violation of privacy, President Lincoln was given the benefit of the doubt by the Union public because the Union was actively at war.

Reconstruction and the Fourteenth Amendment

The U.S. Constitution needed several amendments to abolish slavery and to fully guarantee rights to African Americans after the Civil War. Three amendments were written in the hopes of achieving racial equality: the Thirteenth, Fourteenth, and Fifteenth Amendments, also known as the Reconstruction Amendments. The Thirteenth Amendment abolished slavery. The Fifteenth Amendment granted voting rights regardless of "race, color, or previous condition of servitude."[23] The Fourteenth Amendment, with its due process clause, made most of the Bill of Rights applicable to the states. This meant that the Fourth Amendment was now applicable to each individual state, whereas before, it was only applicable to the federal government.

Before the Fourteenth Amendment, there was the privileges and immunities clause in the Constitution, also known as the comity clause. It stated that "the Citizens of each State shall be entitled to all Privileges and Immunities of Citizens in the several States,"[24] essentially granting citizens travelling in other states the same rights as natives of those states (example: a New Yorker on business in Georgia would have the same rights as a resident of Georgia). This often did not extend to African Americans in the South, because African Americans were not considered to be equals with their white counterparts.

The Fourteenth Amendment was intended to end this inequality, giving African American males the same protections as white male citizens in all states.

An example of the disparity in African American rights granted by the states is the Black Codes. The Black Codes were laws passed after the Civil War that sought to limit the civil rights and civil liberties of blacks. These were prevalent in Southern states and were meant to control the labor and movement of the newly freed slaves. "Everything from unjustified arrests, mandated passes to move about the countryside, beatings by state officials, legally authorized whippings, banishment, revived patrols, and invasions of homes encroached on fundamental rights to unimpeded locomotion, privacy, and possession and use of property, absent adequate justification, such as because of probable cause or involvement in a crime."[25] In addition, freed people's firearms were often confiscated, leaving them without protection. Even though the Civil War led to slavery being abolished, Southerners used the Black Codes as a way to keep blacks in an inferior position.

There are five sections to the Fourteenth Amendment. The first section contains the citizenship clause, the due process clause, and the equal protection clause. The first section reads, "All persons born or naturalized in the United States, and subject to the jurisdiction thereof, are citizens of the United States and of the State wherein they reside. No State shall make or enforce any law which shall abridge the privileges or immunities of citizens of the United States; nor shall any State deprive any person of life, liberty, or property, without due process of law; nor deny to any person within its jurisdiction the equal protection of the laws."[26] This gave the Fourth Amendment teeth by protecting against unreasonable searches and seizures at the state level. The first section is the most litigated of all the sections. Over the years, the Supreme Court has held that the Fourteenth Amendment's due process clause incorporates all the substantive protections of the First, Second, Fourth, Fifth, and Sixth Amendments. The due process clause also became the foundation of a constitutional right to privacy.

It took two years for the Fourteenth Amendment to be ratified. Several southern states initially rejected it, but they were forced to ratify it in order to regain representation in the United States Congress. The amendment passed and was ratified on July 9, 1868. The Fourth Amendment could now become much more powerful in protecting personal privacy.

Privacy versus Morality (Late 1800s)

In the early days of the postal service, postal employees took an oath to not violate privacy while performing their duties. However, this level of concern did not last forever. In the late 1800s, Anthony Comstock became a crusader for Victorian morality. Comstock was a former soldier, who fought in the Civil War from 1863 to 1865. His campaign against immorality began when he wrote a letter to the New York YMCA, asking for funds for his campaign against the sellers of obscene books. He received the funds and was eventually put on the payroll as a member of the YMCA Committee on the Suppression of Vice. In 1873 he founded the New York Society for the Suppression of Vice (NYSSV), which absorbed the YMCA's committee. The NYSSV was committed to overseeing the public's morality.

Comstock was instrumental in Congress's passage of an obscenity law, also referred to as the Comstock Law, in 1873. The Comstock Law made illegal the delivery or transportation of "obscene, lewd, or lascivious" materials, and it also disallowed for the delivery of materials on birth control and venereal disease. After the passage of the law, Comstock was appointed a special agent of the post office, which gave him the authority to implement parts of the new law.[27] He used his role as a U.S. postal inspector to target the mail by lobbying. Comstock was even known to disallow delivery of some anatomy textbooks to medical students due to their graphic nature. "He was, in short, the very embodiment of the type of government abuse that the Framers intended to prevent thought the adoption of the Bill of Rights in general and the Fourth Amendment in particular."[28] Persons targeted by Comstock included: anarchist Emma Goldman, birth control activist Margaret Sanger, and women's suffragette Victoria Woodhall.

Comstock initially received widespread support for his actions, but over time, people began to worry about the constitutionality of his methods. Several media outlets, including the *Washington Post*, began criticizing him. Additionally, Comstock was indicted over opening mail while conducting searches, although the charges were eventually dropped. Despite the growing opposition, he stayed on the job as a U.S. postal inspector for 42 years. Anthony Comstock died in 1915, but his legacy lives on. The Comstock Act is still law, and today it is applied mostly to emerging technologies (including the Internet).

Boyd v. United States (1886)

George and Edward Boyd, both New York merchants, were asked to produce an invoice for plate glass that would later be used against them during forfeiture proceedings brought by the federal government. The Boyds had contracted with the government to import glass plates without having to pay taxes on them, which they otherwise would have had to pay due to federal revenue laws. This was due to the Federal Customs Act, which required "a person to produce his business papers in court when his goods had been seized as contraband or else have the charges of fraudulent importing taken and confessed, forfeiting the goods."[29] If the Boyds did not produce the invoice, they were told that would be as good as a confession of guilt. After providing the invoice, the Boyds lost their case and forfeited their goods. "In subsequent proceedings, the Boyds raised important Fourth and Fifth Amendment questions about the constitutionality of the statute that enabled the judge to order them to produce invoices."[30]

Boyd v. United States (1886) is considered the first far-reaching dissection of the meaning and breadth of the Fourth Amendment to the United States Constitution. The Supreme Court ruled that the forced production of the invoice was a search and a seizure. They also ruled that the search and seizure was not valid under the law and that the evidence, therefore, could not be used against the defendants in a trial. Justice Bradley's opinion in *Boyd* likened the laws that compelled the Boyds to produce the invoice to the writs of assistance that had incited Americans to revolution. "The act of 1863 was the first act in this country, and, we might say, either in this country or in England, so far as we have been able to ascertain, which authorized the search and seizure of a man's private papers, or the compulsory production of them, for the purpose of using them in evidence against him in a criminal case, or in a proceeding to enforce the forfeiture of his property. Even the act under which the obnoxious writs of assistance were issued did not go as far as this."[31] *Boyd* linked the Fourth and the Fifth Amendments, because producing the invoice was seen as self-incriminating in addition to being an unreasonable search and seizure. The Supreme Court also suggested that illegally seized evidence should be excluded at trial, but they did not require exclusion of the evidence. The exclusionary rule would have to wait another thirty years for *Weeks v. United States* (1914).

The Right to Privacy

In 1890, Louis D. Brandeis and Samuel D. Warren Jr. published "The Right to Privacy" in the *Harvard Law Review*. In this article, they laid out a case for the right to privacy, arguing that privacy should be protected with tort law instead of waiting for Congress to pass laws protecting it. "The twenty-eight pages of 'The Right to Privacy,' suggesting that privacy be protected through tort law, may be the most famous law review pages in history: the ideas within them are often credited for laying the foundation for all privacy law in the United States."[32]

Brandeis and Warren met when they were classmates at Harvard University. They were from very different backgrounds. Brandeis was a poor student from a Jewish merchant family in Louisville, Kentucky, and Warren was a member of a wealthy New England family. Their friendship and professional involvement grew over the years, from roommates to partners. In 1879, Brandeis moved to Boston to start a new law firm with Warren. The law firm, Warren & Brandeis, was very successful until 1888, when Warren left to run his family business (S.D. Warren Paper Mill Company) after his father's death. Still, the partners went on to write three articles together for the *Harvard Law Review*: "The Watuppa Pond Case" (December 1888), "The Law of Ponds" (April 1889), and "The Right to Privacy" (December 1890).

Brandeis and Warren defined privacy as the "right to be let alone."[33] They argued that the courts' responses to privacy cases had not been intellectually consistent and that new technologies, including the excessive use of the new technologies by the press, were continually providing more opportunities for invasions of privacy. Laws against slander and libel as well as intellectual property laws were not providing enough protection for personal privacy. In short, the time had come for the courts to recognize a common-law right to privacy. "Instantaneous photographs and newspaper enterprise have invaded the sacred precincts of private and domestic life; and numerous mechanical devices threaten to make good the prediction that 'what is whispered in the closet shall be proclaimed from the house-tops.'"[34]

Warren had some personal experience with newspapers invading personal privacy. Law professor Amy Gajda[35] discovered over 60 newspaper articles, many prominent, published about the Warren family from 1882 to 1890. Several of these articles were about the death of Warren's mother and sister and his marriage to Mabel Bayard, daughter of U.S. Senator Thomas F. Bayard. Warren was known to be upset about this intrusion on his privacy. The two authors recognized the right to privacy was not absolute, and they ended their article with suggestions, based on laws from other countries, about where the right to privacy should end. The right to privacy should not prevent the publication of information in the public or general interest; it should not prevent privileged communication according to the law of slander and libel; the right to privacy ceases if an individual, or someone by consent of the individual, makes public the information themselves. "Making public the deliberations of Congress was a public good; making public the names of mourners at Mrs. Warren's mother's funeral was not."[36]

Brandeis and Warren were hoping the courts would recognize the right to privacy and step in to protect privacy, even without the legislation from Congress and other law-making bodies. "The Right to Privacy" did not become immediately popular in the mainstream press, but it did become very influential in case law due to its excellent arguments. Samuel Warren ran his family's company until his death in 1910. Louis Brandeis went on to become a Supreme Court justice for 23 years, where he was known as a passionate defender of personal privacy. One of his most famous writings was a vehement defense of privacy in his dissent in *Olmstead v. United States* (1928).

19th-Century Conclusion

The 19th century in the United States began quietly in regard to the Fourth Amendment until *Ex Parte Buford* was argued before the Supreme Court. The court decided that searches and seizures could not happen before a crime was actually committed, even if the warrant was granted because it was reasonable. The Civil War and subsequent Reconstruction period had a significant impact on the Fourth Amendment. In times of war, governments and their military often find themselves at odds with the laws they are supposed to protect and enforce. Seizures of personal information were common by both sides in the Civil War and seen as necessary to win the fight. The Fourteenth Amendment was passed during the Reconstruction period, with the due process clause and Equal Protection law impacting several amendments in the Bill of Rights. Now the Fourth Amendment applied to the state level as well as the federal level.

As the century closed out, the Supreme Court protected privacy in *Boyd v. United States* (1886) by overturning a verdict because the forced production of an invoice violated the defendant's Fifth Amendment rights against self-recrimination as well as being an illegal search and seizure. The Supreme Court also gave the recommendation, although not the requirement, to use the Exclusionary Rule for evidence seized illegally. Finally, in 1890, the foundation for modern tort protection for personal privacy was laid with the publication of "The Right to Privacy" by Louis Brandeis and Samuel Warren. The path had been paved for an active 20th century in regard to privacy and the Fourth Amendment.

20TH CENTURY

The 20th century was a much more active time for the Fourth Amendment. During the 1900s, multiple new technologies changed the way U.S. citizens and the court system viewed privacy. Technological developments also impacted the way law enforcement performed duties. Law enforcement officials were able to use various devices to perform surveillance and gather evidence on suspected criminals. The Supreme Court decided several important cases, which had impacts on the exclusionary rule and warrantless searches, and they defined what constituted an unreasonable search.

The Warren Court (1953–1969) was very active in protecting criminal suspects' rights, but later in the century, the Burger Court (1969–1986) and the Rehnquist Court (1986–2005) diluted some of the Warren Court's protections in this area.

This section begins with an examination of the key concepts from the Fourth Amendment that were debated by the Supreme Court and government officials during the 20th century, followed by an examination of important Supreme Court cases that dealt with the exclusionary rule, reasonable expectations of privacy, and the right to privacy held by the press. It will end with a discussion on the impact of government officials' use of technology on personal privacy.

Definitions "of the People," Search, Seizure, Unreasonable, Probable Cause

The Supreme Court spent much of the 20th century debating the exact meaning of the concepts in the Fourth Amendment. There are five concepts: "of the people," search, seizure, unreasonable, and probable cause. The Supreme Court has led the way over the past three centuries in defining these terms, with many of the definitions coming from famous Supreme Court decisions, including *Weeks v. United States*, *Olmstead v. United States*, and *Katz v. United States*. We will examine each concept individually, with a more in-depth discussion of the cases in following sections.

The Fourth Amendment starts with "The right of the people to be . . ." Who is "of the people"? "This protected community includes U.S. citizens (even those who are stationed or traveling abroad) and aliens who have voluntarily entered the United States (or its territory) and have developed substantial connections with this country."[37]

What is the definition of a search? A search can be a search of a person, his or her home, or that person's papers and effects. "Searches of 'persons' include, for example, any physical touching of an individual's body or clothing that causes hidden objects to be revealed; demanding that an individual disclose a concealed object; and extracting an individual's blood or urine for traces of alcohol or drugs."[38] A search can also involve police surveillance methods such as wiretaps and beepers. Seeing observable characteristics of an individual is not considered a search. For example, noting that a person has blond hair and blue eyes is not a search. The Fourth Amendment only applies if a search has occurred.

The searching of houses includes private premises, such as apartments, garages, business offices, stores, and curtilages (the surrounding yards). Seeing observable characteristics of private premises is not a search if not assisted by sophisticated technology that is not available to the public for use.

Searching of "papers" and "effects" is a little trickier to define. It can include a "letter, parcel, backpack, suitcase, trash can, and automobile—that attempts to disclose its contents, unless such object has been abandoned."[39] Seeing observable characteristics of an object is not considered a search. Also, it may not be considered a search if an officer receives an item from a third party.

Over time, the concept of a "constitutionally protected area" has expanded. This expansion began with *Katz v. United States* (1967) when the "reasonable expectation of privacy" standard was implemented. "This requires, first, that a person manifest a subjective expectation of privacy in the object of the challenged search and, second, that the expectation be one that society is prepared to recognize."[40]

Seizures can include arrests, investigatory detentions, or certain other confinements that are against an individual's will. Seizures can also apply to a person's home or personal property. The Supreme Court ruled in *Terry v. Ohio* (1968) that a search can be said to occur whenever an officer accosts an individual and restrains his or her freedom to walk away. If the police do not tell a person he or she is required to answer their questions, then the questioning is not considered a seizure.

An unreasonable search is harder to define, because the word "unreasonable" is inherently subjective. It has usually been interpreted that for a search to be reasonable, a law enforcement officer needs to obtain a warrant from a judge prior to conducting the search. To obtain a warrant, a law enforcement officer must convince the judge that the search is warranted due to probable cause—that is, that the search will most likely uncover criminal activity or contraband. Probable cause exists "where the facts and circumstances within [the government's] knowledge . . . [are] sufficient in themselves to warrant a man of reasonable caution in the belief that an offense has been or is being committed."[41]

It is possible to have a reasonable search without a warrant, as there are many exceptions to the rule. Search warrants are supposed to be issued upon finding probable cause, however, and any warrantless searches need to be the result of probable cause if they are to be admissible in court. Some exceptions to the warrant requirement are: border searches, searches of motor vehicles, searches that a party gives consent for, searches of objects in plain view or in open fields, exigent circumstances, and searches incident to a lawful arrest.

Weeks v. United States (1914) and the Exclusionary Rule

The Fourth Amendment states that warrantless searches and seizures by the government are illegal, but it does not provide any measures for enforcing itself. In answer to this, the exclusionary rule has been created by the Supreme Court to enforce the Fourth Amendment. This rule is a legal principle holding that evidence gathered or used in violation of a person's constitutional rights cannot be used in a court against the offended person or party. This rule applies to not only the Fourth Amendment but also to any situation in which evidence was gathered in a way that violated a party's rights. Furthermore, the rule can be applied to both physical and nonphysical evidence. A defendant will often ask for evidence to be excluded before a trial begins if that person or entity believes rights were violated.

The exclusionary rule was first suggested by Justice Bradley in his opinion on *Boyd v. United States* (1886), but it did not become a requirement until 30 years later with the *Weeks v. United States* (1914) decision. In 1911, Fremont Weeks, an employee of an express company, was arrested for transporting lottery tickets through the mail. The evidence against him was seized by police officers during a warrantless search of his home, and was later turned over to a U.S. marshal. The marshal returned later the same day and performed another warrantless search, seizing additional evidence. Weeks petitioned to have his material returned to him, citing violation of the Fourth and Fifth Amendments, but the petition was denied, and he was subsequently convicted.

The Supreme Court later reversed Weeks's conviction, ruling unanimously that materials seized in violation of the Fourth Amendment cannot be used against a defendant at trial. "The efforts of the courts and their officials to bring the guilty to punishment, praiseworthy as they are, are not to be aided by the sacrifice of those great principles established by years of endeavor and suffering which have resulted in their embodiment in the fundamental law of the land."[42] It is important to note that the exclusionary rule after *Weeks* only applied to federal officers, not state officers. Additionally, the exclusionary rule has been viewed differently by the justices over the years. Some justices believed it was implicit from the Fourth Amendment, while others believed it was needed to preserve judicial integrity. Others believed that it is needed to deter police misconduct.[43] Decades after establishing the exclusionary rule in the *Weeks* decision, in *Mapp v. Ohio*, the Supreme Court would need to decide whether the rule also applied to unconstitutionally gathered evidence at the state level.

Extending the Fourth: *Mapp v. Ohio* (1961)

Earl Warren was chief justice of the United States Supreme Court from 1953 to 1969. During his tenure, the Supreme Court expanded civil rights, civil liberties, judicial power, and the federal power in dramatic ways. From 1961 to 1969, the Warren Court examined many areas of the criminal justice system in the United States, citing the Fourteenth Amendment to extend constitutional protections to all courts in every state. The first major decision during this time was in *Mapp v. Ohio* (1961). That case revolved around the question of whether the exclusionary rule was to be used when a person's Fourth Amendment rights were violated by others below the federal level.

The case began on May 23, 1957, in Cleveland, Ohio, when officers went to the home of Dollree Mapp, whom they thought might be harboring a bombing suspect and housing illegal betting equipment. Initially, Dollree refused to admit the officers without a search warrant. Instead of a search warrant, the officers provided a piece of paper that was later taken from her. She was handcuffed for being belligerent. The officers did not find any evidence of her housing a bombing suspect or gambling equipment, but they did find some pornography. Ms. Mapp was arrested, prosecuted, and found guilty for possession of pornographic materials.

The Supreme Court ruled 6–3 in favor of Mapp, extending the federal exclusionary rule to state criminal prosecutions. This decision overruled *Wolf v. Colorado* (1949), which stated, "In a prosecution in a State court for a State crime, the Fourteenth Amendment does not forbid the admission of evidence obtained by an unreasonable search and seizure."[44] Now, the justices in the majority felt that without the exclusionary rule, there would be no way to ensure the compliance of law enforcement offices to the Fourth Amendment. "Having once recognized that the right to privacy embodied in the Fourth Amendment is enforceable against the States, and that the right to be secure against rude invasions of privacy by state officers is, therefore, constitutional in origin, we can no longer permit that right to remain an empty promise."[45]

Exceptions to Exclusionary Rule: Prohibition Rears Its Ugly Head

The Fourth Amendment to the U.S. Constitution requires that warrants be issued by a judge before any search by federal officials of a person's home and that there be probable cause for such searches. However, the framers who wrote this could not possibly have envisioned the technological advances that would be examined using the Fourth Amendment. One of those technological advances was the automobile. The Supreme Court examined the issue of warrants in regard to automobile searches in *Carroll v. United States* (1925).

George Carroll and John Kiro were arrested for the transportation of alcohol in violation of the Eighteenth Amendment (prohibition). The police had suspected George Carroll of selling liquor and therefore had him under investigation. The police hoped to set up an undercover purchase of liquor from Carroll but this was not completed. Later, Carroll and Kiro had been driving in Michigan when they were pulled over by police officers who found illegal liquor behind the rear seat. There was no warrant for the search, but the police officers considered the search valid because the National Prohibition Act, also known as the Volstead Act (1919), allowed officers to conduct warrantless searches of vehicles, boats, or airplanes if there was reason to believe that illegal liquor was being transported.

The Supreme Court upheld the possibility of warrantless searches of automobiles, now known as the automobile exception, but they did not say it was applicable in all cases. "Search without a warrant of an automobile, and seizure therein of liquor subject to seizure and destruction under the Prohibition Act, do not violate the Amendment, if made upon probable cause, *i.e.*, upon a belief, reasonably arising out of circumstances known to the officer, that the vehicle contains such contraband liquor."[46] The Court felt that it was important for people not suspected of transporting illegal alcohol that they not be stopped and searched. They went further in that police officers needed to secure a warrant if practical. Automobiles can move and disappear, unlike buildings, but that did not mean a warrant was never justified. Thus the *Carroll* doctrine was brought into existence. Essentially, a vehicle could be searched without a search warrant if there was probable cause to believe that evidence was present in the vehicle and police believed that the vehicle could be moved to a new area before a warrant could be obtained.

Communications and Surveillance in the 20th Century

In the 20th century, several new inventions changed the way people communicate. The two most important new communication technologies, at least for our purposes, were the telephone and the Internet. The telephone was first patented in 1876 by Alexander Graham Bell and subsequently developed by others. A telephone lets two or more people to talk to each other over some distance. Despite the increasing popularity of this technology, people were often worried about privacy issues in regard to telephones. "One common complaint in the nineteenth century was that the telephone permitted intrusion into the domestic circle by solicitors, purveyors of inferior music, eavesdropping operators, and even wire-transmitted germs."[47] In the 20th century, privacy concerns continued with the popularity of party lines, where multiple people could listen in on private phone conversations.

The Internet was initially intended as a tool for rapidly sharing information between far-flung researchers, computer programmers, and the military. Development continued over the next several decades, and now millions of people use it daily for a variety of purposes. "Internet growth exploded in the 1990s, from a quarter million hosts and less than 10,000 domains to more than 40 million domains."[48] In the past decade, the Internet has had a significant impact on our culture and on how we do business. It has brought us electronic mail, two-way interactive video calls, instant messaging, and the World Wide Web. The World Wide Web provides blogs, social networking, and online shopping.

Every new communications technology will eventually be used by law enforcement and government officials. Because over the past century people were communicating in different ways, law enforcement needed to develop different surveillance technologies to monitor communications. Searches or surveillance can be of many types: acoustic, electromagnetic, chemical, biological, and so forth. Beepers, wiretapping, the Global Positioning System (GPS), and other methods have all been used. Wiretapping is the monitoring of telephone and Internet connections by a third party. It received its name because, in the past, the monitoring connection was an actual electrical tap on the telephone line. "The beeper is a miniature, battery-powered radio transmitter that emits recurrent signals at a set frequency. By covertly attaching the beeper to a subject's property and monitoring its signals with a separate receiver, the police can electronically track the property, and often the subject, for distances of several miles and for as longs as several weeks."[49] GPS has mostly replaced beepers.

These new surveillance devices have had a variety of side effects for law enforcement officials, including a higher bar to obtain general warrants and Supreme Court challenges to the new surveillance devices. Currently, law enforcement agents must go through a process more complicated than obtaining a general search warrant in order to tap communications. These include getting special permission from the attorney general or other superior entities, providing additional information in the wiretap request, and agreeing to turn off monitoring if there are lengthy conversations on topics unrelated to the investigation. "All these precautions are in place because the tapping of a phone intrudes not just on the privacy of the suspect, who may be discussing issues unrelated to the suspected crime (and who is innocent until proven guilty), but also invades the privacy of the other people with whom the suspect is conversing, who may be innocent of any wrongdoing."[50] However, these extra efforts were not always required. In fact, the Supreme Court initially did not see wiretapping as a potential violation of the Fourth Amendment at all, as in *Olmstead v. United States*.

Throughout the 20th century, the Supreme Court justices attempted to decide if the Fourth Amendment covered each new device or technology, even though no device or technology is specifically mentioned in the Fourth Amendment. The writers of the Constitution could not have known that the right of the people to be secure in their "papers" would one day come in the form of e-mail.

Wiretapping and *Olmstead v. United States* (1928)

In *Boyd v. United States* (1886), the Supreme Court took a broader view of the Fourth Amendment and applied it to all invasions by the government of the privacy of a person's home, including forcing a person to produce papers that could be used as evidence against him or her. In *Olmstead v. United States* (1928), the Supreme Court had to decide whether electronic surveillance by government officials was a violation of the Fourth Amendment. Initially, Roy Olmstead was convicted in a federal court of transporting and selling liquor in violation of the National Prohibition Act. The evidence used against him at trial was obtained via a warrantless wiretap used by federal agents on telephone lines between his home and office. Olmstead appealed the conviction, arguing that this wiretap was a violation of the Fourth Amendment (due to it being an illegal search and seizure) and was a violation of the Fifth Amendment (because the wiretaps amounted to the defendants testifying against themselves). The Supreme Court narrowly decided against Olmstead with a 5–4 result. They did not consider the wiretap a trespass, because a telephone conversation was not considered to have occurred on private property. This decision established the trespass doctrine. The trespass doctrine states that "for a search to occur, government officials must trespass on private property."[51]

Several justices were unhappy with this interpretation of the Constitution. Justice Louis D. Brandeis (coauthor of "The Right to Privacy") stated that in the past, the Court had understood that the language of the Constitution should be interpreted to include the needs of today like it did in *Boyd*. Therefore, it did not matter if the illegal search and seizure was outside of the plaintiff's home. Brandeis claimed that since *McCulloch v. Maryland* (1819), "this Court has repeatedly sustained the exercise of power by Congress, under various clauses of the instrument, over objects of which the Fathers could not have dreamed."[52] Citing *Boyd*, Justice Butler also dissented, saying, "Under the principles established and applied by this Court, the Fourth Amendment safeguards against all evils that are like and equivalent to those embraced within the ordinary meaning of its words."[53] *Olmstead*, with its narrow interpretation of the sphere of privacy, would eventually be overruled by the Supreme Court in *Katz v. United States* (1967) four decades later.

Reasonable Expectation of Privacy

"Because the Fourth Amendment protects people, rather than places."[54] Charles Katz was convicted in a federal court of placing illegal bets and wagers. He used a public telephone booth to assist in placing these wagers. The Federal Bureau of Investigation recorded conversations using an electronic eavesdropping device attached to the exterior of the phone booth. They did not have a warrant. The recordings were the main evidence used against Katz and later led to his conviction. Katz challenged the conviction, arguing that the recordings were obtained in violation of his Fourth Amendment rights.

Proponents of the right to privacy scored a major victory when the Supreme Court decision in *Katz v. United States* (1967) was handed down. "Inarguably the most important Fourth Amendment case of the last half-century, the *Katz* majority rejected *Olmstead's* trespass doctrine, deciding that a search could occur without physical intrusion into a 'constitutionally protected area.'"[55] The Court reversed Katz's conviction because the search was warrantless and because Katz had a reasonable expectation of privacy. Previously, the Court had decided that it was acceptable for law enforcement officials to bug a telephone conversation without a warrant because the conversation was not considered to be private. With this decision in 1967, the justices essentially established that the Fourth Amendment could only be violated if the search happened on private property.

The *Katz* decision definitively overruled *Olmstead*, stating that a person's Fourth Amendment rights can be violated if a person has the reasonable expectation of privacy. In Katz's situation, it was true that the phone booth was public and not on private property; however, he had a reasonable expectation of privacy when he closed the phone booth's door, as he was clearly trying to prevent his conversations from being overheard. In his concurring opinion, Justice Harlan outlined what a reasonable expectation of privacy meant in two parts. First, the individual has exhibited a subjective expectation of privacy. Second, society is prepared to recognize that expectation as reasonable.[56] This test is still being used today as the governing standard in regard to expectations of privacy. "Consistent with this view, the Court has demonstrated a willingness to consider hotel rooms, garages, offices, automobiles, sealed letters, suitcases, and other closed containers as protected by the Fourth Amendment."[57]

This decision was not a unanimous agreement. The majority opinion states, "The Fourth Amendment protects people not places."[58] Justice Hugo Black felt the exact opposite. In his dissent, Justice Black argued that the Fourth Amendment was meant to protect things from physical search and seizure; it was never meant to protect personal privacy. The evidence he cites is that wiretapping is analogous to eavesdropping, which existed during the time the Fourth Amendment was written and adopted. Black argues that if the framers had wanted to protect against eavesdropping, they would have added language about it to the amendment.

Despite the *Katz* ruling, the Supreme Court has made it clear that a reasonable expectation of privacy is not absolute. In 1963, three men, including

John Terry, were observed by a Cleveland, Ohio, police officer for acting in a suspicious way. Based on their behavior, the police officer concluded that the three men were planning a robbery. The officer confronted the men and eventually patted down their outer clothing, in which two pistols were found. The officer removed the pistols and arrested the men. The officer confronted the men and eventually patted down their outer clothing, in which a revolver was found on two of the men. The officer removed the revolvers and arrested the men. The two men with the revolvers were eventually charged with carrying a concealed weapon and later convicted.

In *Terry v. Ohio* (1969), the Supreme Court wrestled with several questions: When is a person considered seized, what constitutes a search, and what is considered reasonable? The Court ruled that the pat-down, or frisk, was a search as defined by the Fourth Amendment and that the frisk was reasonable based on the defendants' suspicious behavior. Additionally, since the police officer suspected that a robbery might take place, the frisk (search) for weapons was a reasonable thing to do. "Our evaluation of the proper balance that has to be struck in this type of case leads us to conclude that there must be a narrowly drawn authority to permit a reasonable search for weapons for the protection of the police officer, where he has reason to believe that he is dealing with an armed and dangerous individual, regardless of whether he has probable cause to arrest the individual for a crime."[59] Because the search had been declared to be reasonable, this meant the seizure of the weapons could not be thrown out as evidence under the exclusionary rule. The *Terry* decision set a precedent for many other Fourth Amendment cases regarding when it is acceptable for a police officer to stop and frisk people to search for weapons in various situations, such as at a traffic stop.

The concept of a reasonable expectation of privacy established in *Katz* also was examined in regards to Supreme Court cases involving surveillance. In *California v. Ciraolo* (1986),[60] the Supreme Court ruled that warrantless aerial observation of a person's backyard did not violate the Fourth Amendment. Dante Ciraolo was arrested for growing marijuana plants in his backyard that were seen by police officers. Ciraolo had two fences to shield the plants from view; however, Santa Clara police officers in a private airplane flew over the backyard and photographed the house. The officers were granted a search warrant to photograph the plants based on an officer's naked-eye observation of the yard. At his trial, Ciraolo requested that the Court suppress the photographs, citing the exclusionary rule. His request was denied, and he pled guilty. The California Court of Appeals reversed the decision, stating that the aerial observation violated Ciraolo's Fourth Amendment rights because the curtilage of his home was considered to be protected.

The Supreme Court narrowly affirmed the lower court's initial decision, 5–4, deciding that the aerial observation was not a violation of Ciraolo's Fourth Amendment rights. Citing *Katz*, the majority felt that the aerial observation was not a search because there should not be a reasonable expectation of privacy in a backyard that is clearly visible with the naked eye. "The Fourth Amendment protection of the home has never been extended to require law enforcement officers to shield their eyes when passing by a home on public thoroughfares. Nor does the mere fact that an individual has taken measures to restrict some views of his activities preclude an officer's observations from a public vantage point where he has a right to be and which renders the activities visible."[61] Using *Katz*, they considered that Ciraolo exposed his illegal activity to the public, because the fences were not so high as to completely shield his plants.

This decision was very similar to that of *Florida v. Riley* (1989), in which police used an aerial view from a helicopter to confirm that a man was growing marijuana in a greenhouse in his backyard. Even though the greenhouse partially obstructed the view of the plants, the Supreme Court still considered the aerial observation acceptable because a private citizen could fly in the same airspace and see the same thing the police officers saw. Also, the court felt it was an important distinction that the police helicopter had not interfered with the normal use of the property and that the helicopter flew at regulation distance from the ground.

The Supreme Court also used *Katz*'s reasonable expectation of privacy test to make a determination in *California v. Greenwood* (1988). In 1984, Laguna Beach police were using surveillance to see if Billy Greenwood was selling illegal drugs out of his home. Over time, the police began searching the trash Greenwood put outside his home every week. Eventually, the police discovered evidence that Greenwood was indeed selling drugs, and he was arrested. The state trial court overturned the conviction, reasoning that his Fourth Amendment rights were violated by the warrantless search of his trash. The appellate court affirmed this decision.

The Supreme Court reversed the lower court's ruling of a Fourth Amendment rights violation by applying the reasonable expectation of privacy test. The justices acknowledged that Greenwood had not anticipated that the Laguna Beach police would search his trash. Greenwood's attorneys claimed that he did show "an expectation of privacy with respect to the trash that was searched by the police: The trash, which was placed on the street for collection at a fixed time, was contained in opaque plastic bags, which the garbage collector was expected to pick up, mingle with the trash of others, and deposit at the garbage dump."[62] Still, the court did not find that there was definitely an actual (subjective) expectation of privacy by Greenwood. Additionally, the trash on the corner was exposed to the public, because other people and animals could get into it, and he left it there for someone else to pick up. This seemed to go against *Katz* ("But what he seeks to preserve as private, even if in an area accessible to the public, may be constitutionally protected"),[63] but the Supreme Court did not extend this to Greenwood's case. This case became an example of the abandoned property exception. "This exception to the warrant requirement holds that a person who voluntarily abandons his or her property forfeits any reasonable expectation of privacy in that property."[64]

In *O'Connor v. Ortega* (1987), the Supreme Court tackled the idea of reasonable expectation of privacy in regard to public employees' workspaces. In 1981, Dr. Magno Ortega, a doctor at a California State hospital, was accused of various incidents of misconduct. When the allegations first arose, he took two weeks of paid vacation, after which he was put on administrative leave. During that time, Dr. Ortega was barred from his office, but his supervisors entered his office and conducted a search. Evidence found during that entry was used as grounds to fire the doctor. Dr. Ortega sued, saying that his Fourth Amendment rights were violated, even though this was not a criminal search. The lawsuit made it to the United States Supreme Court in 1987.

The Supreme Court ruled, 5–4, that public employees do retain their Fourth Amendment rights. Justice Sandra Day O'Connor established the operating realities test for other courts to consider when reviewing similar cases. "An expectation of privacy in one's place of work is based upon societal expectations that have deep roots in the history of the Amendment. However, the operational realities of the workplace may make some public employees' expectations of privacy unreasonable when an intrusion is by a supervisor, rather than a law enforcement official."[65] However, the operating realities test did not become a binding precedent, because Justice Scalia felt it was not clear enough.

The court decided that a reasonable expectation of privacy for public workspaces would need to be decided on a case-by-case basis because some places are clearly public (like hallways), and some places carry a reasonable expectation of privacy (like a work desk). The court also decided that the probable cause standard was not applicable to a search conducted by a public employer. Instead, searches should be measured by a standard of reasonableness under the circumstances. The Ortega search was considered valid because there was state property in his office that might need to be protected. In the end, the court could not decide whether entry into Ortega's office was for search purposes, so it remanded the case to the district court. The case continued until the 1990s, when Dr. Ortega ultimately won a judgment and eventually received compensation. Clearly, the Supreme Court was changing its ideas about what could be considered private. This would seem to change again in the 21st century with cases such as *Bond v. United States* (2000).

United States v. Karo (1984)

The technology used to track suspects continued to be an issue for the United States Supreme Court. The issue of electronic beepers was analyzed in *United States v. Karo* (1984). Drug Enforcement Administration (DEA) agents were conducting an investigation on cocaine production and distribution. To obtain additional evidence, the DEA agents put an electronic beeper in a can of ether. This liquid can be used for extraction and production of cocaine. The owner of the ether was a government informant who gave his consent for the beeper's placement. The can was sold as part of a shipment of 50 gallons of ether to Karo and others. The beeper allowed DEA agents to track the can as it moved between different places, enabling investigators, in turn, to determine the location of the can and obtain arrest warrants. Karo and others were arrested for possession of cocaine with intent to distribute. Karo's attorneys wanted some of the evidence suppressed because it was obtained using the electronic beeper, which the attorney argued was an unlawful search method.

The decision in *United States v. Karo* (1984) was interesting in that the Supreme Court ruled that the use of the beeper without a warrant to conduct surveillance constituted an unlawful search and that the seizure was a violation of the Fourth Amendment. Still, the conviction of Karo and his accomplices was not overturned, because the justices felt that the affidavit that led to the issuance of the arrest warrant was valid. It contained other evidence not connected with the beeper to provide probable cause. The court also found that a beeper became an illegal search and seizure when the device was turned on and used to track information, not when it was placed on or in an item.

Presidential Wiretapping and Watergate

Law enforcement officials are not the only government officials to engage in warrantless surveillance. In *Inside the Oval Office*, William Doyle claimed that multiple U.S. presidents, from Franklin D. Roosevelt to Bill Clinton, have used wiretapping to record supposedly private conversations. "They drilled microphones into White House walls, drawers, and light fixtures, wired the microphones into recording devices in secret chambers near the Oval Office, rigged secret switches into tables, desks, and lamps, and tapped presidential phone lines with Dictaphone machines stashed away in closets."[66] Much of the wiretapping was used for legitimate purposes, such as recording conversations and decisions that could be used to write presidential memoirs. Roosevelt began electronic recording to protect himself in press conferences.

Some presidents used wiretapping infrequently, while others recorded thousands of hours of conversations. John Kennedy, Lyndon B. Johnson, and Richard Nixon were frequent recorders. Presidential wiretapping can lead to potential violations of the Fourth Amendment, of course.

Richard M. Nixon was the president most known for his wiretapping, probably because it contributed to his eventual resignation. Nixon was elected to the U.S. presidency in 1968. After his election, he found out that there was extensive wiretapping equipment in the White House, which had been installed during the Lyndon Johnson administration. Nixon had it removed; however, he had second thoughts about this during his first few years in office. Then in 1971, Nixon changed his mind and decided to build the most extensive White House taping system up to that time. Nixon did this for several reasons, including improving his relationship with Secretary of State Henry Kissinger. Nixon felt that Kissinger often changed his mind and was trying to pin down his opinions. The secret tapings of Richard Nixon showed a volatile, over-the-top chief executive. In meetings, Nixon would often get angry and order some drastic action be taken. Over time, his staff learned to ignore his orders; they knew he would often regret them later.[67]

The Nixon administration also ran into trouble with the law in regard to the Watergate break-in and cover-up. On June 17, 1972, there was a break-in at the Democratic National Committee headquarters at the Watergate complex in Washington, D.C. The purpose of the break-in was to photograph documents and wiretap the headquarters. The Nixon administration attempted to cover up its involvement. The subsequent investigation revealed Nixon's extensive tape-recording system in the White House. The administration did not want to release the recordings, because they revealed that the whole administration, up to the president himself, was involved in the cover-up. President Nixon was required to hand over the recordings. Nixon eventually resigned on August 9, 1974, and 43 other people were tried, convicted, and incarcerated for involvement in the Watergate scandal.

As a result of the Watergate scandal, the U.S. government took steps to ensure the public's privacy was not further compromised by the federal government. In 1974, the "U.S. Congress passed the *Privacy Act* to ensure public access and the right to correct information about oneself held in federal files."[68] The act required federal agencies to give the public notice, in the *Federal Register*, of their system of records. Also, it provides individuals with a means to seek access to their records, which they are allowed to amend. Additionally, the act prohibits the disclosure of information by these systems of records unless written consent is given by the subject individual, unless the information is considered appropriate due to one of the twelve statutory exceptions.[69]

Some changes have been made to the Privacy Protection Act over the years. In 1988, the Computer Matching and Privacy Protection Act amended the Privacy Act by adding extra protections "for the subjects of Privacy Act records whose records are used in automated matching programs."[70] In the 2000s, President George W. Bush provided an exemption for the Department of Homeland Security, the Arrival and Departure Information System (ADIS), and the Automated Targeting System. The Arrival and Departure Information System authorizes U.S. citizens for airplane travel after their personal information has been checked and cleared through a U.S. agency watch list.

As Nixon and his administration proved, violations of personal privacy by the government can happen by officials other than those in law enforcement. A way to prevent this is for Congress to pass laws to help protect privacy, but then they need to stop diluting the protections by adding exemptions to the laws. This has not happened, and the trend of privacy invasions has continued into the 21st century.

The Burger Court

The Burger Court (1969–1986) was supposed to be much more conservative than its predecessor, the Warren Court (1953–1969). Its chief justice, Warren Burger, was a Nixon appointee and a supposed strict constructionist, and some hoped that his court would repeal many of the decisions by the Warren Court. Although the Burger Court had conservative leanings, it did not reverse the work of the preceding court; in fact, it often went further with decisions on issues such as abortion and school desegregation.

The Burger Court also handed down a number of decisions that affected privacy and the interpretation of the Fourth Amendment. In *Bivens v. Six Unknown Federal Narcotics Agents* (1971), the Supreme Court weighed whether an individual could sue for civil damages if their Fourth Amendment rights were violated by federal agents. The case began when Webster Bivens's home was searched by Federal Bureau of Narcotics (FBN) agents without a warrant or probable cause. The search led to the discovery of narcotics, and Bivens was arrested. Although the charges were later dropped, Bivens claimed that his Fourth Amendment rights were violated. The FBN claimed that the violation only applied to a state law and that the Fourth Amendment did not provide any cause of action.

The court ruled 6–3 that an individual such as Bivens could sue for damages. The majority stated that even though the Fourth Amendment did not explicitly provide a way to enforce monetary damages, the federal courts could do what it took to right the wrong. Justice Blackmun wrote,

"The Fourth Amendment operates as a limitation upon the exercise of federal power regardless of whether the State in whose jurisdiction that power is exercised would prohibit or penalize the identical act if engaged in by a private citizen."[71] Later, Blackmun states that "historically, damages have been regarded as the ordinary remedy for an invasion of personal interests in liberty."[72]

A year after *Bivens*, in the *United States v. U.S. District Court* (1972), the Supreme Court ruled unanimously (8–0) to uphold the requirements of the Fourth Amendment in cases of domestic surveillance targeting a domestic threat. During the 1960s, John Sinclair, Larry Plamondon, and John Forrest were charged by the United States with conspiracy to destroy government property. Larry Plamondon was also charged with the dynamite bombing of an office of the CIA in Ann Arbor, Michigan. The defense requested in a pretrial motion for disclosure of all electronic surveillance information, including wiretaps. Attorney General John Mitchell said that he authorized the wiretaps due to Title III of the Omnibus Crime Control and Safe Streets Act of 1968 and that he was not required to disclose the information. The act allowed for warrantless wiretapping in order to prevent the overthrow of the government when a clear and present danger exists. The U.S. government claimed this came under the exception clause, because the defendants were members of a domestic organization attempting to subvert and destroy the government. Judge Damon Keith of the United States District Court for the Eastern District of Michigan did not agree with this assessment, and he ordered the government to disclose all the illegally intercepted conversations to the defendants.

The Supreme Court upheld Judge Keith's decision and ordered the electronic communications returned to the defendants. This decision established the precedent that a warrant needs to be obtained before beginning electronic surveillance, even if domestic security issues are involved. "Fourth Amendment protections become the more necessary when the targets of official surveillance may be those suspected of unorthodoxy in their political beliefs. The danger to political dissent is acute where the Government attempts to act under so vague a concept as the power to protect 'domestic security.'"[73]

Judge Keith's decision only applied to the domestic realm, though. The Foreign Intelligence Surveillance Act (FISA) of 1978 governs electronic surveillance of foreign intelligence information between or among foreign powers. It has been repeatedly modified since the attacks on September 11, 2001. This act provides rules and procedures for handling electronic surveillance, with and without a court order, and for physical searches of the "premises, information, material, or property used exclusively by" a foreign power. The act also created the Foreign Intelligence Surveillance Court (FISC). This court monitors and reviews the requests for surveillance warrants made by federal agencies against suspected foreign intelligence agents operating inside the United States. If a FISA application is denied, it can be appealed to the Foreign Intelligence Surveillance Court of Review, but this has only happed twice. Additionally, the FISC proceedings are confidential and have never been made available to the public.

Search and Seizures and the Press

The issue of searches and seizures in regards to the press was examined in the case *Zurcher v. Stanford Daily* (1978). In 1971, the police department of Palo Alto, California, searched the main office of *The Stanford Daily*, a student-run newspaper serving Stanford University. The police were seeking pictures of a violent clash between a group of protestors and the police, in hopes of identifying the assailants. *The Stanford Daily* had taken many photographs of the protest and had published them. Despite the search, no materials were taken from the office, because the only photos found had already been published in the newspaper. *The Stanford Daily* sued, arguing that their First and Fourth Amendment rights were violated. Their reasoning was that the warrant was not valid because the newspaper itself was not a suspect in any wrongdoing.

The Supreme Court ruled in favor of the police department, in a 5–3 decision. They cited that the police department did have a warrant and that the Constitution did not forbid warrants in regard to the press, even though it was a third-party search and the paper was not suspect. "We do hold, however, that the courts may not, in the name of Fourth Amendment reasonableness, prohibit the States from issuing warrants to search for evidence simply because the owner or possessor of the place to be searched is not then reasonably suspected of criminal involvement."[74] This ruling concerned many people because it opened up the opportunity for additional searches of the press.

The ruling led the U.S. Congress to pass the Privacy Protection Act of 1980. This act overturned *Zurcher*, as it protects journalists and newsrooms from government interference. Journalists will not be required to hand over to law enforcement officials any work product and documentary materials before they are made public. A subpoena must be ordered by the court to gain access to the information. This law especially protects journalists researching stories that might be controversial or about criminal activity that could be compromised by law enforcement. There are a few exceptions to the act (example: a search is permitted if the person holding the information is suspected of a crime), but overall, the act provides privacy protection to the press.

20th-Century Conclusion and Beyond

The 20th century began with questions over surveillance of telephones and searches due to illegal liquor distribution during Prohibition. The century ended with questions about searches of computers, public workspaces, and the press. U.S. presidents engaged in secret recordings and surveillance, and the abuse of this system contributed to the resignation of a president. Multiple important cases were heard and decided by the Supreme Court, as the pendulum of personal privacy swung back and forth with the court. The exclusionary rule was created and implemented, with some exceptions, and a reasonable expectation of privacy was defined and challenged. Still, the issue of privacy remained the same: What is a U.S. citizen allowed to keep private, and what should be detectable to law enforcement officers and other government officials?

The 21st century would have to deal with the same core question: How is the Fourth Amendment applicable to privacy in a society of rapidly changing technologies and new communication methods? The 20th century did not yet have Facebook and Twitter, but it did have the early days of the Internet and cell phones. During this time, the Supreme Court laid the legal foundation for what was considered acceptable for law enforcement in regard to surveillance.

In Jonathan Franzen's *The Corrections*, the character Alfred Lambert says, "Without privacy there was no point in being an individual."[75] This sentiment has been reflected throughout the course of U.S. history, from the American colonists' fight against the British and their writs of assistance, to the two attorneys writing a law review article on the right to privacy, to the Supreme Court justices whose rulings have asserted the principle of a reasonable expectation of privacy over the years. Communication technologies are continually evolving, but the American desire for personal privacy has remained constant for centuries.

NOTES

1. *Lopez v. United States*, 373 U.S. 427 (1963).
2. Christian Parenti, *The Soft Cage: Surveillance in America, From Slavery to the War on Terror* (New York: Basic Books, 2004), 91.
3. "Right of Privacy," *Merriam-Webster.com*, accessed July 5, 2013, http://www.merriam-webster.com/dictionary/right%20of%20privacy.
4. U.S. Const. amend. IX.
5. The Post Office Act 1710, 9 Ann c 11.
6. Frederick S. Lane, *American Privacy: The 400-Year History of Our Most Contested Right* (Boston: Beacon Press, 2008), 7.

7. Lane, *American Privacy: The 400-Year History of Our Most Contested Right*, 8.
8. Ibid., 10.
9. Ibid., 10.
10. Otis H. Stephens and Richard Glenn, *Unreasonable Searches and Seizures* (Santa Barbara, CA: ABC-CLIO, 2006), 36.
11. Richard B. Morris, *The Era of the American Revolution* (Gloucester, MA: Peter Smith, 1971), 75.
12. Andrew E. Taslitz, *Reconstructing the Fourth Amendment: A History of Search and Seizure, 1789–1868* (New York: New York University Press, 2006), 24.
13. *The Quartering Act of 1765*, accessed July 5 2013, http://www.ushistory.org/declaration/related/quartering.htm.
14. Stephens and Glenn, *Unreasonable Searches and Seizures*, 48.
15. U.S. Const. amend. IV.
16. Lane, *American Privacy: The 400-Year History of Our Most Contested Right*, 42.
17. John R. Vile and David L. Hudson Jr., eds., *Encyclopedia of the Fourth Amendment* (Thousand Oaks, CA: CQ Press, 2012), 244.
18. Ex Parte v. Burford, 7 U.S. 448 (1806).
19. Taslitz, *Reconstructing the Fourth Amendment*, 227.
20. Lane, *American Privacy*, 26.
21. Ibid., 26.
22. Ibid., 26.
23. U.S. Const. amend. XV, § 1.
24. U.S. Const. art. IV, § 2 cl. 1.
25. Taslitz, *Reconstructing the Fourth Amendment: A History of Search and Seizure, 1789–1868*, 248.
26. U.S. Const. amend. XIV, § 1.
27. Lane, *American Privacy: The 400-Year History of Our Most Contested Right*, 34–35.
28. Ibid., 33.
29. Stephens and Glenn, *Unreasonable Searches and Seizures*, 250.
30. Ibid., 53.
31. Boyd v. United States, 116 U.S. 616 (1886).
32. Amy Gajda, "What If Samuel D. Warren Hadn't Married a Senator's Daughter?: Uncovering the Press Coverage that Led to 'The Right to Privacy,'" *Michigan State Law Review*, v. 35 (2008): 37.
33. Louis D. Brandeis and Samuel D. Warren, "The Right to Privacy," *Harvard Law Review* v. 4, no. 5 (1890).
34. Brandeis and Warren, "The Right to Privacy."
35. Gajda, "What If Samuel D. Warren Hadn't Married a Senator's Daughter?".
36. Jill Lepore, "The Prism: Privacy in an Age of Publicity," *The New Yorker*, June 24, 2013, http://www.newyorker.com/reporting/2013/06/24/130624fa_fact_lepore.
37. Stephens and Glenn, *Unreasonable Searches and Seizures*, 68.
38. Ibid., 69.
39. Ibid., 69.

40. Ibid., 70.
41. Ibid., 72.
42. *Weeks v. United States*, 232 U.S. 383 (1914) at 393.
43. Stephens and Glenn, *Unreasonable Searches and Seizures*, 161.
44. *Wolf v. Colorado*, 338 U.S. 25 (1949).
45. *Mapp v. Ohio*, 367 U.S. 643 (1961).
46. *Carroll v. United States*, 267 U.S. 132 (1925).
47. Claude S. Fischer, *America Calling: A Social History of the Telephone to 1940* (Berkeley, CA: University of California Press, 1992), 26.
48. J.K. Peterson, *Understanding Surveillance Technologies: Spy Devices, Privacy, History & Applications*. 2nd ed. (Boca Raton, FL: Auerbach Publications, 2007), 50.
49. "Tracking Katz: Beepers, Privacy, and the Fourth Amendment," *Yale Law Journal* 86, no.7 (June 1977): 1461.
50. Peterson, *Understanding Surveillance Technologies*, 170.
51. Stephens and Glenn, *Unreasonable Searches and Seizures*, 146.
52. *Olmstead v. United States*, 277 U.S. 438 (1928) at 472.
53. *Olmstead*, 277 U.S. at 487.
54. *Katz v. United States*, 389 U.S. 347 (1967).
55. Stephens and Glenn, *Unreasonable Searches and Seizures*, 335.
56. *Katz*, 389 U.S. at 351.
57. Stephens and Glenn, *Unreasonable Searches and Seizures*, 7–8.
58. *Katz*, 389 U.S. 347.
59. *Terry v. Ohio*, 392 U.S. 1 (1968) at 27.
60. *California v. Ciraolo*, 476 U.S. 207 (1986).
61. *California*, 476 U.S. at 213.
62. *California v. Greenwood*, 486 U.S. 35 (1988) at 39.
63. *Katz*, 389 U.S. at 351.
64. Stephens and Glenn, *Unreasonable Searches and Seizures*, 245.
65. *O'Connor v. Ortega*, 480 U.S. 709 (1987).
66. William Doyle, *Inside the Oval Office: The White House Tapes from FDR to Clinton* (New York: Kodansha International, 1999), x.
67. Doyle, *Inside the Oval Office*, 170.
68. Peterson, *Understanding Surveillance Technologies*, 42.
69. The Privacy Act of 1974, 5 U.S.C. § 552a.
70. "Internal Revenue Manual—11.3.39 Computer Matching and Privacy Protection Act," *Internal Revenue Service*, accessed July 17,2013, http://www.irs.gov/irm/part11/irm_11-003-039.html.
71. Bivens v. Six Unknown Federal Narcotics Agents, 403 U.S. 388 (1971).
72. *Bivens*, 403 U.S. 388 (1971).
73. *United States v. United States District Court*, 407 U.S. 297 (1972).
74. *Zurcher v. Stanford Daily*, 436 U.S. 547 (1978) at 560.
75. Jonathan Franzen, *The Corrections* (New York: Picador, 2001), 463.

2

Wiretaps, Electronic Surveillance, and the Fourth Amendment

Jason Hochstatter

Is it better to be safe, or is it better to have privacy? This is an ongoing debate in American politics that is inherently tied to the Fourth Amendment, as the Fourth Amendment guarantees protection from unwarranted searches. With advances in technology, law enforcement has struggled, at times, to keep up with its ability to investigate. At other times, law enforcement has overreached, causing controversy and distrust, particularly in the realm of surveillance.

The FBI came under public scrutiny for surveillance abuses during the reign of J. Edgar Hoover. Hoover was known for expanding scientific study of law enforcement, but he was also known for infringing on the Fourth Amendment by conducting a wide-reaching surveillance program and for keeping records on numerous American citizens, called COINTELPRO. Hoover famously had civil rights leader Dr. Martin Luther King Jr., actress Marilyn Monroe, and even President John F. Kennedy under surveillance. The FBI stated, in their defense, that it was done to protect the nation from subversive groups that threaten national security. When discovered, these activities were widely criticized, and Congress formed an investigative committee, the Church Commission, which ultimately harshly criticized the FBI's actions.[1] Today, the FBI admits that its actions under these programs were abusive, and they say that it was right for Congress to criticize their behavior.[2]

Law enforcement has been concerned that increasing digitalization threatens their ability to conduct investigations. In the early 1990s, the FBI and others pushed for legislation that would prevent them from losing their ability to tap phone lines, even in the face of emerging technology that radically changed the tools they relied on to listen in on phone conversations.

In response to this, Congress passed a law referred to as CALEA, which mandates that all telephones be designed so that the FBI and other law enforcement agencies can easily put a wiretap in place.[3] This has resulted in criticism from civil liberties organizations, which say that CALEA goes too far.

The NSA has come under a great deal of criticism lately, due to revelations that it keeps massive databases of information about Internet traffic and phone call records. According to documents obtained by newspapers from a whistleblower, the NSA has direct access to all information that passes through the servers of seven major Internet companies, including Google, Apple, Yahoo, and Facebook.[4]

This chapter will investigate the past and ongoing uses of electronic surveillance and wiretaps. It will examine what different forms of surveillance do, under what circumstances each is permissible, and how each has been used and is being used.

TYPES OF ELECTRONIC SURVEILLANCE

Wiretaps

A wiretap is a method of electronic surveillance that allows one person on the phone, or even a third party, to listen in on telephone conversations. Wiretaps have been in existence very nearly as long as telephones have been used, and as people generally do not think of a conversation over the phone as being recorded, wiretaps are a touchy subject.

These devices are called wiretaps because they were originally separate phone lines that directly tapped into another phone line, picking up any conversations that were being had on that line.[5] The requirements for police who would like to wiretap someone have been back and forth on the issue. Originally, police were not required to obtain a warrant because phone lines crossed over public property.[6] In the 1960s, however, a court case ruled that the police do require a warrant, because those on the phone have some reasonable expectation of privacy.[7] Some types of surveillance, such as a pen register, are able to bypass the warrant requirement because they only collect information that an individual is sending to a phone company; therefore, the individual does not have a reasonable expectation of privacy.[8]

Government agents do not necessarily have to notify individuals on a phone call that the call is being monitored, but in most cases, they must acquire a warrant from a judge before they can record phone calls.[9] Standard law enforcement agencies, like the police, the DEA, or the FBI, may go to a normal judge in order to get a search warrant. But the NSA, the federal intelligence organization that specializes in cryptography and intercepting communications, may have a need to conduct domestic surveillance, and in most cases, they want to avoid letting the target know that he or she will be under surveillance. The NSA has a special court, known as the FISA Court, named after the legislation that created it, that can provide the NSA with a discreet warrant.[10] The NSA has recently been in trouble for allegedly breaking these rules and collecting information that it should not have, under PRISM and similar programs.[11] The NSA, FISA, and PRISM will be discussed at greater length later in this work.

As a wiretap is an actual recording of any phone call taking place, it is a very valuable tool for law enforcement. Originally, wiretaps were put in place by adding additional wires that would transmit the contents of a conversation to a voice recorder or someone wearing headphones.[12] In the 1990s, law enforcement became more and more concerned that if digital technology advanced further, they would not be able to continue to easily wiretap those they wished to listen in on as they could have in the past, and law enforcement would become significantly disadvantaged.[13] At that time, telecommunications companies were beginning to make the switch from analog telephone signals to digital ones. While this worked well for the telephone companies, the police had previously depended on the fact that telephone signals were not digital to put their taps in place. The police had set up an easy system for wiretapping any phone that they deemed necessary at the actual phone company. They were not equipped to do the same for digital connections.[14] The police were also concerned that individuals making illegal business deals, like drug dealers, could simply buy a cell phone, use it once for a transaction, and then throw it away, making it impossible for the police to obtain a tap on that phone before the target had stopped using it.[15]

Law enforcement began pushing for legislation that that would require that all new communications devices, including all new digital landlines and cell phones, would be built in such a way that they could be easily tapped by the government. In response, the Communications Assistance for Law Enforcement Act (CALEA) was passed to do just this. CALEA was a controversial act that was passed to give the police even footing with advancing technology.[16] Over time, CALEA has advanced to include digital phone calls, instant messaging, and several other types of electronic communication. Recently, the FBI has been pushing for CALEA to be expanded to include all types of online communications, including Facebook, Twitter, Reddit, and other online forums and social networks.[17] Opponents argue that law enforcement already has ways to deal with this, that these expansions to CALEA are giving law enforcement more than even footing, and that these new regulations are stifling new businesses from introducing new products.[18] CALEA and its related issues will be discussed later on, at length.

Pen Registers

Pen registers are a type of surveillance mechanism that is not as highly monitored as wiretaps are. Pen registers log all of the outgoing phone calls made from a phone.[19] Installing a pen register requires a court order, which is a lesser requirement than a warrant.[20] Pen registers are not considered a search, and as such, are not subject to the exclusionary rule. The reason for this is that nothing is being overheard. A caller needs to tell the phone company what number they want to call in order to call it, and as such, the caller does not really have an expectation that the phone company is unaware of the phone number that he or she called. The same applies to the length of the call and the time the call was made. The phone company tracks when calls are made for billing purposes, so the caller does not have an expectation that this information will be kept a secret.

With the passing of the USA PATRIOT Act (truncated to the PATRIOT Act hereon out), the requirements for placing pen registers and similar devices have changed. The federal government secretly acquired court orders that would allow it to obtain unlimited access to all of the information that a pen register would normally provide. These court orders were secret orders that required telephone companies to willingly turn over this information.[21] As was hinted at in previous court cases in the late 2000s and later confirmed by information leaked by Edward Snowden, the NSA has rooms in major phone companies, including Verizon, that collect all of this information and store it in a database.[22] Prior to this act, the understanding was that the police had to get permission to place each individual pen register, but now they have a database that contains all the information that they need, so they would not have to acquire an additional warrant.

The amount of information that is collected by a pen register is very limited, which is part of the reason that it only takes a court order to acquire one. A pen register makes a record of all phone numbers that an individual calls, what time the phone call is made, and whether or not the call went to voice mail.[23]

Imagine that someone named George is a known terrorist, and the police have put a pen register on his phone. George is planning to bomb a building with the help of Steven, a terrorist whom the police are unaware of. George calls Steven a total of seven times. The police see that George called Steven three times on Monday morning and four times on Tuesday. They can see the exact times that he called Steven and whether or not the call was received. The police would not be able to hear the content of any of the phone calls that George and Steven had. Based on this information, the police may decide that Steven is a potential terrorist, who could be planning something with George, and he is someone who should be watched.

After the PATRIOT Act was passed, this use of pen registers has become less important for the NSA in particular and for other law enforcement agencies too, if the NSA is providing them with access to their call database. Through this database, the NSA can easily locate every single call that a person has been involved with since the program began. A computer program would analyze all of George's phone calls, and it would easily tell the NSA that Steven has a relationship with George and that the agency may have a reason to listen to these phone calls.

In another example, the police have placed a pen register on Michael's phone. The police know that Michael is a drug dealer. The police begin keeping track of who is calling Michael on a regular basis. James buys drugs from Michael regularly, and he calls Michael to set up meetings. The police, being able to see that James calls Michael on a regular basis, may soon suspect that he is either buying drugs from Michael or supplying drugs to him.

The NSA would not need to acquire a pen register for both men, as they already have all of the data that they need to analyze Michael's phone and all of his regular contacts. The only issue would be that the NSA may not be sharing their data with the DEA or local law enforcement to an extent that the call database can be used in cases like these. The information, however, would still be collected.

Michael, from the previous example, is friends with Phil. Phil does not know about Michael's business selling drugs, but he does go to the movies with him each week and calls him each Saturday to decide what movie they would like to see. The police may begin to suspect that James may be buying drugs from Michael due to their frequent contact and may begin to watch James even though he has done nothing wrong.

Currently, the NSA would already know that Michael regularly contacts James. They could determine the frequency with which they contact each other and would not know whether James is a threat. The NSA, however, may not use its databases for drug-related offenses. At the time of this writing, exactly what the NSA uses its database for is unclear.

Online Surveillance

The PATRIOT Act has extended the use of pen registers to online forms of communication like e-mail. The police can see all the header information except for the subject line. They also see the time each e-mail was sent, the size of each e-mail, and the IP address where the e-mail was sent from and received.[24] This information allows the police to guess the physical location of both parties, to determine how often they contact each other, and to have a good idea of whether or not they are sending attachments.

Recently leaked documents have revealed that online spying by the government, and the NSA in particular, has been much more widespread than the simple pen register style of taps that the police were known to be placing on specific individuals. As part of a program called PRISM, the NSA has been collecting a great deal of data on Internet traffic directly from the servers of major Internet companies, such as Google, Facebook, Microsoft, and Yahoo.[25] The government claims that this surveillance was put in place for security purposes, but their actions have been criticized by the ACLU and other civil liberties groups as being illegal and undermining the Fourth Amendment. Specifics on the history of this program will be discussed later in this chapter. The following examples will first discuss what was thought by the general public to be legal under the existing laws, followed by an example describing the way the government has actually been monitoring Internet traffic.

George the terrorist is still planning a bombing with Steven, but he has decided that calling him on the phone is too risky, so he is contacting Steven by e-mail. The police have predicted that George might be using e-mail in the future, so they have installed software to track his e-mail account. They see that George is frequently sending e-mails to steventheterrorist@email.com, and they see that Steven's e-mail address is in the same general area as George from his IP address. The police are not able to see what sort of information is being sent between the men, but they may suspect that whoever is sending mail from steventheterrorist@email.com is a terrorist.

In this case, the reality of the matter was that the NSA would have an easier time determining whom George was contacting. The NSA has installed software directly into the computer systems of both George and Steven's e-mail providers and thus does not have to install software or make requests to gain information on specific individuals. The NSA has a database, and upon a person becoming interesting to the NSA, all the agency has to do is look in the database for whomever that person has been contacting.

Phil is buying child pornography online, and the police have begun to suspect this. The police install a pen register to track his online activities. They see that he is receiving a lot of very large e-mails from Victor. The police are not able to tell what the large e-mails contain, but if the size is consistent with picture or video files, the police may suspect that Victor is involved in trading child pornography and may begin keeping an eye on him as a result.

In this case, the police, with cooperation from the an Internet database, do not need to make a specific request for this information because they have already obtained all the data that they need to make these types of determinations from e-mail providers. The police can easily track the individuals that Phil and Victor contact and how often, and they may be able to discover an entire ring of child pornography dealers in this way.

Victor loves to look at funny pictures and videos of cats on the Internet, and he loves sending picture and video attachments to his friends. Victor is friends with Phil and frequently sends him funny pictures and videos of cats. The police do not know Victor's fondness for cats, and they see that he is frequently sending files that are consistent with the size of pictures or videos to Phil, who they suspect of trading in child pornography. The police may suspect that Victor is trading child pornography, even though he is only sending Phil pictures of cats.

These types of false connections, while conjecture, serve as an illustration of what might occur under constant surveillance. Is the additional security that such a surveillance system could provide worth the risk of being falsely suspected as a criminal?

Telegraph Surveillance

The telegraph was a technology that was widely used for long-distance communications prior to the advent of the Internet. A telegraph sends messages in the form of electronic signals over long distances.[26] The telegraph originally used wires to send messages by Morse code, but it evolved when it was discovered that messages could also be sent wirelessly by radio waves. These radio waves, although encrypted, could be easily intercepted by intelligence agencies and then read. From the 1950s to the 1970s, law enforcement frequently intercepted telegrams without warrants, prompting Congress to conduct an investigation and implement new laws that made it more difficult for the agency to legally intercept that information.[27]

ORGANIZATIONS

The United States has several different organizations that may conduct surveillance, and these organizations all fall within the realms of foreign intelligence gathering for the purposes of national security and domestic law enforcement. This section will examine each of the organizations that may engage in domestic surveillance that may fall under the scope of the Fourth Amendment and the circumstances under which these organizations may apply for surveillance, and it will also provide examples of cases in which these organizations may be able to apply for surveillance.

The FBI

The Federal Bureau of Investigation (FBI) is a federal law enforcement agency that extensively uses wiretaps. It began wiretapping extensively during the prohibition era to arrest bootleggers.[28] The FBI later began extensive domestic wiretapping of individuals linked to groups that the bureau deemed to be a threat to national security under a controversial program called COINTELPRO, or Counter Intelligence Program.[29] The FBI underwent a technological upgrade in 1998 so that it could conduct surveillance on electronic communications, and it continues to extensively wiretap phones and conduct surveillance of the Internet today.[30]

George is a terrorist, and he is planning to bomb an airport. The FBI has learned of the plot and has narrowed down their search to someone living on George's street, but they have not yet discovered that George is the terrorist. The Fourth Amendment affords everyone living on George's street some degree of privacy, and because of this, the FBI cannot simply search every house on the street to discover that George is the terrorist. Wiretaps work the same way, in that the police need to have probable cause to place a wiretap on someone's phone.

The FBI has the capability to wiretap anyone's phone with the push of a button, and everyone's phone is connected to a larger system that is required by law to be built with back doors that allow the police to put a wiretap in place. The police could not simply suspect that someone somewhere is going to commit a crime and monitor every single phone call that takes place. Even if police had the necessary resources to do this, police must have a reason to tap a phone, just as they would need a defendable reason to search someone's home.

The NSA

The National Security Agency (NSA) is a federal cryptology and intelligence agency. Their intelligence gathering is generally required to be limited to foreign communications, although occasionally they are allowed to conduct domestic surveillance if they meet the guidelines outlined in the Foreign Intelligence Surveillance Act of 1978 (FISA).[31] The NSA is considered to be similar to the CIA, in that its primary activities are meant to be the monitoring of foreign communications. There have been several cases in which the NSA has conducted domestic surveillance even though it was questionable as to whether or not it is supposed to.

There have been several cases in which the NSA has conducted domestic surveillance even though it wasn't supposed to in the 1960s and 1970s, more recently during the warrantless wiretap controversy in the 2000s, and as of this writing, the NSA has been conducting wide-reaching surveillance that most Americans were completely unaware of.[32]

The DEA

The Drug Enforcement Administration (DEA) is a government organization that is tasked with enforcing drug laws in the United States. The DEA was formed in response to growing fears about the abuse of drugs, and its creation was one of the first acts in the war on drugs in the United States. The DEA frequently uses all manner of wiretaps in the ongoing war on drugs.[33]

LAWS REGARDING SURVEILLANCE

The Right to Privacy

Katz v. United States established that the Fourth Amendment provides the right to privacy. When Katz entered a phone booth that the FBI was tapping and closed the door, he had reason to believe that his conversation was private. The right to privacy protects individuals' conversations and personal effects from searches if they have a reasonable expectation of privacy.[34]

This requirement clearly applies while taking phone calls within one's home or within a phone booth, but with the advent of widespread cell phone usage, payphones have mostly been replaced with cell phones. Cell phones do not require closing out the rest of the world with a glass door and invariably broadcast speech over publicly owned airwaves. Cell phone conversations may take place in public spaces, and neither individual can know for certain that the other party is not in a public space. Should individuals be required to ask if the other person is in a public place or on speakerphone in order to have the benefit of privacy?

The Exclusionary Rule

The exclusionary rule demands that evidence be gained in a legal manner. Any evidence gained in an illegal manner will be deemed inadmissible and cannot be used against a defendant.[35] This rule makes the distinction between a warranted wiretap and a warrantless wiretap essential to law enforcement, because if a wiretap is obtained illegally, any evidence obtained from it cannot be used against the person that the authorities are trying to convict.

Wiretaps were not covered under this exclusionary rule for a long time as a result of the 1928 court case of *Olmstead v. United States*. In this case, decided during the prohibition of alcohol, Roy Olmstead was convicted of conspiracy to sell and transport alcohol. Law enforcement officers had placed a wiretap and listened in on conversations between Olmstead and his associates for several months. The wiretap was placed outside of both individuals' homes, on the telephone wires between houses. Olmstead argued that his Fourth and Fifth Amendment rights were being violated, but the courts did not agree.[36] The Supreme Court argued that because the conversation travelled on wires that were outside of both residences, no search had taken place, and there was no expectation of privacy.[37]

The case of *Olmstead v. United States* was overruled in 1967 with the decision in *Katz v. United States*.[38] Charles Katz was using a phone booth to transmit illegal gambling wagers across state lines. Katz did not realize that the phone booth he was using was being listened to by the FBI, and was arrested. Katz argued that the recordings were made in violation of his rights under the Fourth Amendment, and the Supreme Court agreed with him. This time, the Supreme Court ruled that even though phone lines pass through public space, by closing the door to the phone booth, Katz had no expectation that his conversation would be broadcast to a larger audience.[39] *Katz v. United States* extended the Fourth Amendment's exclusionary rule to all places where a person could have a reasonable expectation of privacy.

Thomas is a drug dealer, and the DEA suspects that he is. The DEA puts a wiretap on his phone, but they neglect to get a warrant first. Over the course of the wiretap, the DEA obtains constant confessions from Thomas that he is a drug dealer. The DEA, however, obtained all of these confessions illegally, as they did not have a warrant to wiretap his phone. The exclusionary rule states that the DEA will not be able to use the evidence they obtained from these wiretaps in court, because they were obtained from an illegal search.

Federal Stored Communications Act

The Stored Communications Act is a safeguard against law enforcement working together with communications companies to get access to information without a warrant. This act, established in 1986, prevents Internet service providers (ISPs) from willingly giving information to the federal government without a warrant.[40] AT&T, Comcast, Time Warner, and other ISPs cannot volunteer information that they have gained about your online activities, even though they are required to store this information for one year.

For example, if George's Internet service provider discovers evidence that George is illegally downloading a television program, movie, video game, or other copyrighted material, George's ISP would not be able to hand evidence of George's activities over to law enforcement, despite the fact that George is breaking copyright law.

Concerns about Internet piracy have sparked media companies and organizations to find workarounds to this rule. Recently, many Internet service providers have signed an agreement stating that they will be on the lookout for illegal piracy and police it themselves, instituting a six-strike program in which they may reduce Internet speeds for repeated piracy, eventually providing the information that they have found to media companies who can then sue the pirates.[41]

If an Internet service provider finds evidence that links a user to child pornography, they are not able to give this information to law enforcement. This law has been controversial because it appears to provide shelter to those committing the heinous act of trading and making child pornography. In 2011, the Protecting Children from Internet Pornographers Act was passed, which, in addition to putting harsher penalties in place for trading child pornography, created the requirement that all ISPs keep all of their stored data of who visits what sites for at least one year, so that law enforcement may access it if it becomes necessary.[42]

Should an Internet service provider discover suspicious traffic indicating possible criminal or terrorist activity, they would not be able to provide that to the government. This law protects Internet traffic from being willingly handed over to the government, regardless of criminal activity being discussed. This is to protect the privacy of individuals. If George were to anonymously post on an Internet forum or community, such as Reddit or Facebook, that he intends to bomb a building or commit a murder, the Internet service provider may have information to link the post to George, but they would not be able to willingly tell the government who George is. Law enforcement would need to ask a judge for a warrant before any information could be provided.

The Stored Communications Act makes communications companies unable to willingly comply with federal demands unless they are presented with a court order, but recent allegations say that the federal government may make the compliance issue a moot point. If these allegations are correct, the courts can simply issue a court order for ongoing surveillance of all information regarding all phone calls passing through a telecommunications company.[43] This bypasses the compliance issue by forcing the telecommunications companies to always provide the information that the government wants.

In the previous example, George's telecom company was unable to willingly provide the government with any information about his phone or Internet activities. The government has recently been interpreting the PATRIOT

Act to mean that it can issue a court order that forces the telecom company to allow ongoing access to all of its incoming and outgoing records.[44] In the previous example, the government would not need to provide additional documents to access George's records, because it already has complete access to them.

FISA

The Foreign Intelligence Surveillance Act of 1978 (FISA) is a law that clarifies when the National Security Agency (NSA) is able to conduct domestic surveillance. It was put into place in 1975 as a clarification of how wiretap laws apply to foreign agents. FISA was updated by the PATRIOT Act to include terrorists as foreign agents.[45]

FISA gives the NSA the ability to conduct surveillance, including wiretaps and data taps, on foreign agents domestically. Ordinarily, the NSA is not allowed to conduct domestic surveillance, but the fact that international concerns may occasionally present themselves domestically, especially in the case of foreign agents, prompted the passing of FISA. The NSA may electronically conduct surveillance for a period of up to one year without a court order, and if they acquire a court order, the surveillance may be ongoing.[46]

Lauren is a foreign agent representing the country Corneria, and the NSA would like to keep an eye on her activities. They tap her phones for a period of five months and decide that her conversations are not interesting enough to spend resources on, so they decide to remove the wiretaps. In this case, the NSA does not need to acquire a court order, because their surveillance lasted for less than a year.

The PATRIOT Act expanded the scope of FISA to apply to terrorists. Prior to this law, FISA only applied to foreign agents, and because terrorists were not official agents of another country, the NSA was unable to conduct surveillance on them even though their activities were a foreign security concern.[47] The Patriot Act also put into place a lone wolf provision, meaning that a terrorist does not have to be associated with a terrorist organization in order to be considered a terrorist for purposes of FISA.[48]

Ted Kaczynski conducted a bombing campaign from 1978 until the time he was eventually caught in 1995. He was a child prodigy, who was accepted into Harvard at the age of 16 and resigned from the position of assistant professor after two years to live in a cabin in the wilderness of Montana, free of electricity and running water. As Kaczynski saw developers destroying the wilderness around him, he became frustrated and decided that drastic action was required. During the years that he was active, he mailed 16 bombs to universities and airlines, killing 3 people and injuring over 20. Kaczynski was considered to be a terrorist and was highly sought after by the FBI.[49] Had the PATRIOT Act been in effect and had Kaczynski had a phone, the NSA would have been able to tap his communications under the lone wolf provision, as he was not associated with an organization but was clearly a terrorist. This example also highlights an issue with electronic surveillance, in that it does not help against threats that choose not to use electronic communications.

The Protect America Act legalized the warrantless wiretapping of international phone calls that had sparked the NSA warrantless wiretap controversy. Under this law, the NSA can legally wiretap international calls without a warrant, even if one party is on American soil, if one of the parties on the phone call is under suspicion of being a terrorist.[50] It does not allow the warrantless wiretapping of calls where both parties are on American soil. This act has been allowed to expire; however, law enforcement is working to get it back.[51]

CALEA

The Communications Assistance for Law Enforcement Act (CALEA) was passed in 1994 in an attempt to help law enforcement officials track communications in the digital age. Law enforcement became concerned that as phone companies switched to new digital phone lines, they may not be able to access conversations as easily as they once were.[52] CALEA requires that communications agencies equip all types of telecommunications devices with some sort of means to be easily tracked and listened in on, and it was later expanded to include broadband network connections and voice over Internet protocol services.

Law enforcement felt that CALEA was necessary because the 1990s were a time of change for communication, and phones, in particular, were changing. While cell phones had existed for years, the early '90s saw cell phones beginning to shift from using analog radio waves to digital radio waves. This shift also saw the introduction of text messages and other data-based services. Law enforcement officials worried that if they needed a wiretap, it would become much more difficult, if not impossible, to quickly put one into place. A law like CALEA, which requires phone companies to help authorities put a system in place before it is required, means that law enforcement will never have this problem.

The original law was controversial, and it met with protest from telecommunications companies, who lobbied against it. Telecommunications companies argued that it would take great expense to make their products compatible with the new law. This would in turn drive up the cost of new technology, making people less likely to adapt, and providing the telecom companies with less incentive and ability to innovate in their respective fields.[53] In addition, because these restrictions would only apply to providers in the United States, it would put them at a disadvantage to foreign providers.

CALEA was implemented by a joint effort of the Federal Communications Commission and the FBI. The federal government provided funding to implement the changes and offered time frames for doing so, complete with extensions.[54]

CALEA was expanded in 2004 to include broadband networks and VoIP communications, including instant messages. This was an attempt to keep CALEA up to date with emerging technology. This was a controversial move, and many Internet activist organizations are still working against it and its expansion.[55]

This expansion required many broadband providers, including those that offered VoIP services and instant messaging, to rebuild their networks. While money was provided to the telecommunications companies to update their networks when CALEA was first passed, no similar assistance was provided to those required to adapt after 2004.[56] If the company felt that it would be too expensive to put such measures into place themselves, the FCC and FBI allowed them to hire companies to provide the equipment and carry out monitoring, should the FBI make the request.[57]

The FBI argued that more groups were communicating online, and they needed to make sure that advancing technology did not prevent them from listening in on conversations. They would still be required to obtain a warrant, but with this expansion, the FBI was able to force broadband providers to make it easy for the bureau to tap into connections by pushing a button.[58] This expansion faced a lot of criticism from groups like the American Civil Liberties Union (ACLU) and the Electronic Frontier Foundation (EFF).

Opposition argued that this expansion incorrectly interprets the spirit of the original CALEA legislation. The original legislation was meant to keep the status quo, to prevent digital technologies from keeping law enforcement from obtaining needed wiretaps. They argued that the FBI could already easily obtain and put into place surveillance of data transmissions without having to disrupt the industry and stifle innovation by forcing them to put into place new restrictions.[59] They also argued that the expansion went against the letter of the law. The original CALEA legislation exempted the transfer of digital information from being covered under CALEA.

The opposition to this expansion of CALEA has argued that the expansion is unwarranted. The EFF and ACLU argued that just because a wiretap could be put into place does not mean that it should be. They argued that by the FBI's logic, houses should come with microphones and cameras preinstalled just in case law enforcement ever needs to use them.[60]

The PATRIOT Act

The PATRIOT Act was a controversial law passed shortly after the terrorist attacks on 9/11. This act provided law enforcement with sweeping new powers in an attempt to prevent a similar attack from happening again. The PATRIOT Act made several changes to wiretapping, including making adjustments to the Foreign Intelligence Surveillance Act, redefining who can give permission to search a computer, and modifying roving wiretaps.[61]

The Foreign Intelligence Surveillance Act (FISA), which allows the NSA to conduct surveillance on foreign intelligence operatives operating within the borders of the United States, was changed to include terrorist organizations. Normally, the NSA is not allowed to conduct surveillance, like wiretaps, domestically, but because Stephanie is a spy from the country of Latveria, the NSA could tap her phone if they acquire a court order. A member of al-Qaeda or another terrorist organization may be a citizen of a country, but they are not officially operating on their country's behalf and so are not covered by FISA rules. George is not a spy working for his home country of Blueland; he is a terrorist working for al-Qaeda. Because al-Qaeda is an organization and not a country, the NSA would not have been able to legally wiretap George. Terrorist organizations were added to FISA under the PATRIOT Act, so now George can be legally wiretapped with a court order.

Roving wiretaps are a tool put in place by the PATRIOT Act that law enforcement uses to make sure wiretaps on cell phones stay in place. Before the PATRIOT Act, the police were required to apply for a wiretap for each and every phone that they intended to tap. This could lead to confusion and loss of evidence, as it created a lot of extra work if the police wanted to monitor multiple phones.[62] Civil rights groups opposed this provision, as they argued it was an unconstitutional overstep in the rights of the individual. They argued that the PATRIOT Act did not provide oversight or enough clarity about what phones could be tapped and what phones could not be tapped under the legislation.[63]

George is a terrorist, and the police are well aware of this. The police have placed a wiretap on his phone to keep track of what he is planning. George decides to get a cell phone, and the police find out. Before the PATRIOT Act, the police would have needed a second warrant to put the second tap in place. However, because the PATRIOT Act is already in effect, the police can tap his cell phone as soon as they find out about it.

Michael, the drug dealer, arranges business meetings on a cell phone. Michael is concerned that the police might wiretap him and arrest him or show up at one of his business meetings, based on the information they hear from his phone. In order to combat this, Michael buys cheap disposable cell phones and throws them away after using them to arrange a drug deal. Before the PATRIOT Act was passed, this may have been a good strategy for avoiding wiretaps. Today, the police can easily put a wiretap on any phone that Michael has, meaning that his strategy is not nearly as effective anymore.

Tim works in the same office as Amanda, who is suspected of insider trading and has her phones tapped by law enforcement agents. The police want to be sure that they record all of her phone calls. Before the PATRIOT Act, the police would have needed to be very selective about who they tapped, because they would need a warrant for each wiretap they wanted to get. Under the PATRIOT Act, if the police interpret it very broadly, the police might tap all the phones in the office, assuming that Amanda might use any of them. Because of Amanda's wiretapping, all of Tim's phone calls at work are tapped by the police.

SURVEILLANCE PROGRAMS AND CONTROVERSIES

Project Shamrock and Minaret

Project Shamrock was a spying program begun by the National Security Agency in 1945. Under this program, copies of all telegraphs that left or entered the United States were collected and read by the NSA.[64] If the NSA felt that another agency, like the FBI, would be interested in the contents of a telegram, they would pass it on to that agency.[65] Project Minaret was a similar program that collected all the telegrams sent to and from a list of persons of interest. The NSA did not acquire any warrants for reading these telegrams, and it became controversial once the program was revealed. These programs ended in 1975, and their existence helped spark the passing of FISA laws.

COINTELPRO

COINTELPRO, short for Counter Intelligence Program, was a program to expose, disrupt, misdirect, discredit, or otherwise neutralize organizations and individuals that the FBI considered to be enemies of the United States.[66] This program included covert spying on suspected individuals and groups, including illegal wiretapping. The program was started by J. Edgar Hoover in August of 1956 and headed by William C. Sullivan. The vast majority of the program's resources were targeted at left-wing groups, including socialist organizations and civil rights groups, while the remaining funding was directed at white hate groups.[67] The program was exposed in 1971, and the Church Committee was launched to investigate its misconducts. The FBI claims it no longer engages in activities like this.

The NSA Call Database

The NSA has been building an enormous repository of phone records, called the NSA call database. This database was started approximately seven months before the terrorist attacks of 9/11 and has gathered information on over 1.9 trillion phone calls. The database of information was built with the help of telecommunications companies, including AT&T, SBC, Bell South, and Verizon, although Bell South and Verizon have denied this.[68] Some groups view this database as an unconstitutional warrantless search and a violation of FISA's pen register provisions. Qwest was asked to participate in this program but declined, asking to see the warrant first. The NSA told Qwest that the warrant was unnecessary, and Qwest declined to participate in the program, doubting that the warrant was unnecessary.[69] It seems likely that this program is connected with the allegations of the NSA overreaching its boundaries in 2013, with programs like Prism.[70]

Hemisphere

It has recently been discovered that the DEA has an ongoing agreement with the phone company AT&T to access a gigantic database of call information. This program, called Hemisphere, was first put into place in 2007, and is very similar to the NSA's call database, but it contains all calls made through AT&T for a longer period of time. All calls that pass through an AT&T system are recorded into this database; reportedly, it contains phone calls that date back 26 years, which is a much wider scope than the NSA program. This program also differs from the NSA call database in that the DEA does not store the phone call's information; AT&T does it for them. AT&T manages a massive database of its own calls, and then it provides information about these calls to the DEA. Under this program, the DEA is not required to obtain a court order of any sort to obtain information from Hemisphere; they only need to acquire an administrative order from within the DEA. The DEA argues that this is an essential part of their investigative powers. The administrative subpoenas they use to obtain information from Hemisphere streamline the process of obtaining phone call information, and this makes it easier to trace callers who routinely use disposable phones. In a sense, then, the argument used is like that of CALEA; the government is just trying to catch up with criminals. The Hemisphere program has been used successfully to seize a drug ring running cocaine in 2011, a woman who made bomb threats in 2013, and a man who impersonated a general in the same year.[71]

THE PRESIDENT'S SURVEILLANCE PROGRAM

The President's Surveillance Program is an ongoing, controversial call-monitoring program put into place by President George W. Bush after the terrorist attacks on 9/11. This program authorized the warrantless wiretapping of any international call where one party is believed to be affiliated with a terrorist organization, in a subsection referred to as the Terrorist Surveillance Program.[72] The full extent of the President's Surveillance Program is unknown; the only publicly revealed section is the Terrorist Surveillance Program.

This Terrorist Surveillance Program was the source of much scandal when it was discovered, and it sparked the NSA warrantless wiretap controversy. It provided a framework with which the NSA could wiretap international calls if one of the parties was believed to be a terrorist.[73] Purely domestic calls cannot be tapped using this program, although the NSA has admitted that some were accidentally tapped. The exact nature of the taps and how much information is gathered is unknown. Some experts have speculated that it is only a pen register.[74]

George is suspected of having ties with a terrorist organization. George places a call to a friend of his, Steven, who lives in another country, Orangeland. Under the President's Surveillance Program, the NSA would be able to record the conversation and listen for any information that they see fit. The NSA would not need a warrant for this type of activity and would not need permission of any kind from the FISA court.

George is back in his home country of Orangeland to meet with some other terrorists. While he's away, George hires a cleaning service. Stephanie, an American citizen, has a question about the notes that George left her, so she calls him at his number in Orangeland. The NSA could wiretap Stephanie's call to George and would not need any sort of court-granted permission to do so.

Unlimited Access to Phone Records

In June of 2013, it was verified that the NSA's domestic surveillance reaches farther than many expected. The NSA has submitted a court order to

Verizon for complete unlimited access to all records of all calls that are made on Verizon's network.[75] This includes the date and time each call was made, the location the calls were made from, the identities of all callers involved, but not the content of the calls themselves. The NSA has interpreted certain sections of the PATRIOT Act to mean that they can do this, although there are some who consider this to be an overreach on the part of the NSA. According to the NSA, they do not need a separate warrant to search each individual, because like a pen register, this information is provided to the phone carrier. Additionally, many suspect that the NSA has filed similar orders to other wireless carriers.[76]

The information on every single phone call that is made through each of these carriers is transmitted directly to the NSA. If George calls Steven, the NSA now has a record of it that they can access at any time without a warrant. This raises a great deal of privacy concerns, as many people do not want the NSA to have access to their phone records without permission. Some may question whether the NSA has the ability to request a court order that operates on such a grand scale. In any case, each phone call that is made on one of the tracked phone carriers is entered into the NSA's database of phone calls.

PRISM

The same documents that verified the existence of limitless pen register capabilities also verified the existence of PRISM, a program under which the government obtains all information about the online activities that pass through a number of popular Internet services, including those owned by Microsoft, Google, Apple, Facebook, and others.[77] Some reports suggest that the NSA has direct access to the servers of these companies, though most of these companies have denied it.[78]

The implication of this program is that nothing done online is anonymous, and everything that you do on the Internet is easily viewable by the NSA. While someone may feel comfortable in that all they do on the Internet is look at cat pictures or read the news, some are uncomfortable with the NSA being able to look at their private e-mails, pictures that were supposed to be private, or any interests that they may be embarrassed about.

Even if privacy from the government is not an issue, misuse by the employees of the NSA could be. With great power comes great responsibility, and the temptation to misuse that power becomes greater. There have already been cases in which NSA employees have misused the surveillance system to spy on former lovers and an ex-spouse.[79] Would you feel comfortable if an ex-boyfriend or ex-girlfriend could track your entire web history and could trace your phone calls?

Successful Use of Wiretaps

Wiretaps have proven to be an exceptional tool for law enforcement. Law enforcement can use wiretaps to quickly get information about a suspected criminal's plans without that person's knowledge, and later on, law enforcement can use the suspect's own words against him or her in court. There have been many famous cases in which wiretaps, voice recordings, and other forms of surveillance have been very useful to law enforcement, including the impeachment of President Richard Nixon and the arrest of Illinois governor Rod Blagojevich. The following portion of this chapter will examine several real cases of when wiretaps have been put to exceptional use.

President Richard Nixon

President Richard Nixon tripped himself up. In 1972, there was a break-in at the Democratic National Headquarters in the Watergate hotel and office complex. A group of five men were arrested for the burglary, and some of the connections that these five men had were connected to the committee to reelect the president. The investigation, granted a high degree of public exposure by the extensive media coverage, particularly the *Washington Post*, eventually discovered that members of the Nixon campaign were, in fact, involved in the burglary and that the president himself was involved in an extensive cover-up operation.[80] As the extent of the cover-up came out and as more people became critical of the president's behavior during the cover-up, it became clear that he would be impeached. Rather than face impeachment, Nixon resigned the presidency, becoming the first American president to do so. Nixon was later pardoned of any wrongdoing by his successor, President Gerald Ford.

Some of the strongest evidence that the prosecutor had to use against Nixon included taped conversations that Nixon made, which implicated him in the cover-up. Nixon had placed an extensive number of listening devices and wiretaps in his office, taping most business and nonbusiness conversations that he had while president. His reasoning for doing so was to create a lasting record of his presidency.[81] When the court was made aware of these tapes, they were subpoenaed, and the transcripts became evidence.

These tape recordings were crucial evidence against him. The transcripts assured the court that Nixon knew that a cover-up was going on and that it occurred with his consent, making him look complicit in the act itself. Under the laws that currently exist, law enforcement could not have acquired these recordings had Nixon not made them himself. If recordings of all phone calls were made, using an enhanced version of the vast pen register system that currently exists, law enforcement would be much better equipped to prosecute criminals.

Governor Rod Blagojevich

Illinois governor Rod Blagojevich found himself in serious legal trouble after it turned out that he was being wiretapped by the FBI. Blagojevich was elected governor in 2002, and he was removed from office in 2009, due to charges of corruption.[82] Blagojevich's predecessor had also been convicted of corruption, and Blagojevich had passed legislation that he said would curb corruption, but he found himself in jail related to the selling of President Barack Obama's former seat in the Senate after his presidential election, in part due to evidence from recorded conversations.

Patrick Fitzgerald, a United States attorney, began a fraud investigation called Operation Board Games related to the Illinois teacher's pension fund in 2003.[83] Blagojevich himself was not under investigation until 2006, and as Blagojevich came under investigation, federal investigators requested and received permission to obtain wiretaps on Blagojevich's personal and office phone lines.

Blagojevich suspected that his conversations might be recorded, telling an associate to speak as though the whole world was listening, but he was not able to keep to his own advice.[84] He attempted to keep the transcripts of his phone conversations out of court, stating that they were unreliable and taken out of context. There were gaps in the recordings because the investigators were not allowed to record conversations that were not relevant to the investigation, but ultimately, the wiretap recordings were admitted into court.[85] The recordings held statements made by Blagojevich that were extremely damaging to his case, including the following statement in reference to the Senate appointment: "I've got this thing and it's (expletive) golden, and I'm not just giving it up for (expletive) nothing."

Political corruption and conspiracy can often be extremely difficult to prove. Often, those involved are making a great deal of profit from it or are unwilling to expose themselves to backlash by coming forward as a witness. Without proof, wrongdoing and intention of wrongdoing can be difficult to demonstrate beyond a reasonable doubt. Use of a wiretap can make those involved in wrongdoing unknowingly testify against themselves. In this case, Blagojevich got careless and made his intention of wrongdoing widely known. Blagojevich was convicted of 17 charges and will serve at least 12 years in prison.[86]

John Gotti

John Gotti, a famous leader in a New York crime family, was arrested and convicted due to wiretaps. Gotti became the leader of the Gambino mafia family in 1986. Gotti became famous as "the Teflon Don," as it seemed that no charges would stick, despite accusations of racketeering, jury tampering, loan sharking, and a litany of other charges associated with organized crime.[87]

John Gotti made for a very high-profile target for law enforcement, and as they had very good reason to suspect that he was involved in organized crime, they were able to gain access to a variety of wiretaps to use against him.[88] In 1990, the FBI arrested Gotti, Salvatore Gravano, and Frank Locasio for five murders and a variety of other charges. The FBI played several tape recordings for Gravano that they had obtained through wiretaps of Gotti. These included statements by Gotti, disparaging Gravano and suggesting that Gotti was considering framing Gravano for the murders. Gotti attempted to make up with Gravano, but Gravano eventually agreed to testify against Gotti. The trial included a fully anonymous and sequestered jury, due to Gotti's reputation for jury tampering, and he was convicted and sentenced to life in prison without parole, where he died of cancer in 2002.[89]

This is a case where a criminal who was very difficult to pin down was finally convicted. It is very possible that without the wiretap evidence that was lawfully obtained and used to convince Gravano to testify against him, Gotti would not have been convicted for the murder charges.

CONCLUSION

The proper use of surveillance is a complex issue. People do not like other people listening in on their conversations, but conversations may need to be listened in on in order to promote security. The balance between privacy and security is a delicate one in which both sides continually attempt to pull the balance further in their favor.

NOTES

1. Tim Weiner, *Enemies: A History of the FBI* (New York: Random House, 2012)

2. Federal Bureau of Investigation, "COINTELPRO," accessed August 6, 2013, http://vault.fbi.gov/cointel-pro.

3. Federal Communications Commission, "Communications Assistance for Law Enforcement Act," *Federcal Communications Commission.* accessed June 17, 2013, http://www.fcc.gov/encyclopedia/communications-assistance-law-enforcement-act.

4. Timothy B. Lee, "Here's everything we know about PRISM to date." *Washington Post*, June 12, 2013.

5. Howard J. Kaplan, Joseph A. Mateo, and Richard Sillett, "The History of Law and Wiretapping," *American Bar Association*, April 2012, http://www.americanbar.org/content/dam/aba/administrative/litigation/materials/sac_2012/29-1_history_and_law_of_wiretapping.authcheckdam.pdf.

6. *Olmstead v. United States*, 277 U.S. 438 (1928).

7. *Katz v. United States*, 389 U.S. 347 (1967).

8. Robert B. Parrish, "Circumventing Title III: The Use of Pen Register Surveillance in Law Enforcement," *Duke Law Journal* (1977).

9. U.S. Courts. "2010 Wiretap Report," accessed August 6, 2013, http://www.uscourts.gov/uscourts/Statistics/WiretapReports/2010/2010WireTapReport.pdf.

10. Federal Judicial Center. "Foreign Intelligence Surveillance Court." accessed September 10, 2013, http://www.fjc.gov/history/home.nsf/page/courts_special_fisc.html.

11. American Civil Liberties Union. "Fix FISA—End Warrantless Wiretapping," accessed August 6, 2013, http://www.aclu.org/national-security/fix-fisa-end-warrantless-wiretapping.

12. Kaplan, Mateo and Sillett, "The History of Law and Wiretapping."

13. Federal Communications Commission, "Communications Assistance for Law Enforcement Act," accessed June 17, 2013, http://www.fcc.gov/encyclopedia/communications-assistance-law-enforcement-act.

14. Ibid.

15. Ibid.

16. Ibid.

17. Electronic Frontier Foundation, "CALEA," accessed August 6, 2013, https://www.eff.org/issues/calea.

18. American Civil Liberties Union, "CALEA," September 12, 2005, http://www.aclu.org/technology-and-liberty/calea-feature-page.

19. Electronic Frontier Foundation, "Pen Registers and Trap and Trace Devices," *Surveillance Self-Defense*, accessed August 6, 2013, https://ssd.eff.org/wire/govt/pen-registers.

20. 18 U.S.C. § 3122

21. Lee, "Here's everything we know about PRISM to date."

22. Ibid.

23. Electronic Frontier Foundation, "Pen Registers and Trap and Trace Devices."

24. Ibid.

25. Lee, "Here's everything we know about PRISM to date."

26. Kaplan, Mateo and Sillett, "The History of Law and Wiretapping."

27. Ibid.

28. Weiner, *Enemies: A History of the FBI*.

29. Ibid.

30. Ibid.

31. Federal Judicial Center, "Foreign Intelligence Surveillance Court."

32. Jame Risen and Eric Lichtblau, "Bush Lets U.S. Spy on Callers Without Courts," December 16, 2005, http://www.nytimes.com/2005/12/16/politics/16program.html?pagewanted=all&_r=0; Lee, "Here's everything we know about PRISM to date"; Britt Snider, "Recollections from the Church Committee's Investigation of the NSA: Unlucky Shamrock," accessed September 10, 2013, https://www.cia.gov/library/center-for-the-study-of-intelligence/csi-publications/csi-studies/studies/winter99-00/art4.html.

33. Drug Enforcement Administration, "DEA History," accessed September 10, 2013, http://www.justice.gov/dea/about/history.shtml; John Shiffman, "How DEA Program Differs from Recent NSA Allegations," August 5, 2013, http://www.reuters.com/article/2013/08/05/us-dea-sod-nsa-idUSBRE9740AI20130805.

34. *Katz*, 398 U.S. 247.

35. 18 U.S.C. § 121

36. *Olmstead*, 277 U.S. 438.

37. Kaplan, Mateo and Sillett, "The History of Law and Wiretapping."

38. Ibid.

39. *Katz*, 398 U.S. 247.

40. 18 U.S.C. § 121

41. "US Internet 'Six Strikes' Anti-Piracy Campaign Begins," February 26, 2013, http://www.bbc.co.uk/news/technology-21591696.

42. Martin Kaste, "Child Pornography Bill Makes Privacy Experts Skittish," August 24, 2011, http://m.npr.org/story/139875599?url=/2011/08/24/139875599/child-pornography-bill-makes-privacy-experts-skittish., *Katz v. United States*, 389 U.S. 347 (1967).

43. Glen Greenwald, "NSA collecting phone records of millions of Verizon customers daily," June 5, 2013, http://www.theguardian.com/world/2013/jun/06/nsa-phone-records-verizon-court-order.

44. Lee, "Here's everything we know about PRISM to date."

45. James G. McAdams III, "FISA: An Overview," accessed September 10, 2013, http://www.fletc.gov/training/programs/legal-division/downloads-articles-and-faqs/research-by-subject/miscellaneous/ForeignIntelligenceSurveillanceAct.pdf/view.

46. Ibid.

47. Ibid.

48. Ibid.

49. CNN, "Year in Review 1906: Unabomber Suspect Is Caught, Ending 18 Year Manhunt," accessed September 10, 2013, http://www.cnn.com/EVENTS/1996/year.in.review/topten/unabomb/unabomb.index.html.

50. United States Department of Justic, "What is the Protect America Act?" accessed September 10, 2013, http://www.justice.gov/archive/ll/.

51. Ibid.

52. Federal Communications Commission, "Communicatons Assistance for Law Enforcement Act."

53. Electronic Frontier Foundation, "CALEA."

54. Federal Communications Commission, "Communications Assistance for Law Enforcement Act."

55. Electronic Frontier Foundation, "CALEA."
56. Ibid.
57. Federal Communications Commission, "Communications Assistance for Law Enforcement Act."
58. American Civil Liberties Union, "CALEA."
59. Electronic Frontier Foundation, "CALEA."
60. American Civil Liberties Union, "CALEA."
61. United States Department of Justice, "What Is the U.S. Patriot Act," accessed September 10, 2013, http://www.justice.gov/archive/ll/highlights.htm.
62. Ibid.
63. American Civil Liberties Union, "Reform the Patriot Act," accessed September 10, 2013, https://www.aclu.org/reform-patriot-act.
64. Snider, "Recollections Recollections from the Church Committee's Investigation of the NSA: Unlucky Shamrock."
65. Ibid.
66. Weiner, *Enemies: A History of the FBI*.
67. Ibid.
68. Leslie Cauley, "NSA Has Massive Database of Americans' Phone Calls," *USA Today*, May 11, 2006.
69. Ibid.
70. Lee, "Here's everything we know about PRISM to date."
71. Scott Shane and Colin Moynihan. "Drug Agents Use Vast Phone Trove, Eclipsing NSA's." *New York Times*. September 1, 2013.
72. Risen and Lichtblau, "Bush Lets U.S. Spy on Callers Without Courts."
73. Ibid.
74. Electronic Frontier Foundation, "Pen Register and Trap and Trace Devices."
75. Greenwald, "NSA collecting phone records of millions of Verizon customers daily."
76. Ibid.
77. Lee, "Here's everything we know about PRISM to date."
78. Ibid.
79. Adam Gabbat, "NSA analysts 'wilfully violated' surveillance systems, agency admits," August 24, 2013, http://www.theguardian.com/world/2013/aug/24/nsa-analysts-abused-surveillance-systems.
80. "Select Committee on Presidential Campaign Activities," accessed September 10, 2013, http://www.senate.gov/artandhistory/history/common/investigations/Watergate.htm.
81. Ibid.
82. Monica Davey and Emma Fitzsimmons, "Jury Finds Blagojevich Guilty of Corruption," June 27, 2011.
83. "Operation Board Games Timeline," accessed September 10, 2013, http://illinoisissues.uis.edu/archives/2009/01/boardgames.html.
84. Ibid.
85. Ibid.
86. Davey and Fitzsimmons, "Jury Finds Blagojevich Guilty of Corruption."

87. Selwyn Raab, "John Gotti Running the Mob," April 9, 1989, http://www.nytimes.com/1989/04/02/magazine/john-gotti-running-the-mob.html?pagewanted=all&src=pm; Federal Bureau of Investigation, "A Byte out of History: John Gotti: How We Made the Charges Stick," accessed September 10, 2013, http://www.fbi.gov/news/stories/2007/april/gotti040207.

88. Raab, "John Gotti Running the Mob."

89. Ibid.; Federal Bureau of Investigation, "A Byte out of History: John Gotti: How We Made the Charges Stick."

3

Forensic DNA Analysis, the Fourth Amendment, and Personal Privacy

Wendy Watson

The Fourth Amendment protects U.S. citizens against government intrusion by prohibiting unreasonable searches and seizures. From the mid-20th century to today, the courts have struggled to define the notions of search, seizure, and unreasonableness. During those decades, a general framework for the analysis of Fourth Amendment issues has arisen. A search occurs when the police invade a sphere in which individuals have a reasonable expectation of privacy,[1] and the seizure of a person (placing a person in custody) occurs when an individual would objectively believe that he or she is not free to leave police presence.[2] "Fourth Amendment reasonableness is generally defined as the conjunction of probable cause to search plus either a warrant or an exigency that waives the warrant requirement."[3] While courts are consistently testing the boundaries and limitations of that framework, it has been adequate to address a wide variety of government action and has even taken into consideration changes in social norms and technology.

DNA analysis, however, may have exposed the chinks in the current Fourth Amendment jurisprudence. Advances in the area of DNA analysis have revolutionized the investigation and prosecution of violent criminal offenders.

> Modern DNA testing can provide powerful new evidence unlike anything known before. Since its first use in criminal investigations . . . there have been several major advances in DNA technology . . . It is now often possible to determine whether a biological tissue matches a suspect with near certainty. While of course many criminal trials proceed without any forensic and scientific testing at all, there is no technology comparable to DNA testing for matching tissues when such evidence is at issue.[4]

As this technology and its uses have expanded, the law has struggled to keep up with the new reality of law enforcement. Courts and scholars have struggled to compare DNA analysis to other forms of evidence and investigatory tools, those comparisons often seeming, at best, flat-footed. If DNA analysis cannot be analogized to some other existing form of evidence, the courts have their work cut out for them in creating a new framework for constitutionally regulating this "super evidence" and the many ways in which it can be abused.

This chapter will discuss the ways in which DNA analysis is used by law enforcement. After a general overview, the chapter will examine the Fourth Amendment and privacy implications of the various uses of DNA analysis, using existing case law as a guide. Finally, it will conclude with thoughts about the ability of existing jurisprudential analysis to address this rapidly changing area of the law.

THE USES OF FORENSIC DNA ANALYSIS

DNA is the genetic blueprint for an individual. Every facet of our physical selves—and potentially facets of our personality—is written in the intertwined strands of deoxyribonucleic acid found in every cell in our bodies. Other than identical twins, no two people have exactly the same sequence of DNA. It is a unique identifier of a single individual.

Given the current state-of-the-art technology, we cannot compare the entire genomes of two samples of DNA. Instead, scientists have identified a number of fragments—loci—that show broad variation among individuals. Forensic DNA analysis involves comparing the loci of two different DNA samples, one from a suspect (suspect DNA) and one left by the perpetrator at a crime scene (scene DNA), to determine whether the chemical structures at the loci are identical. Specifically, the analysis looks at the size and frequency of "short tandem repeats," or STRs: "'repeated DNA sequences scattered throughout the human genome[.]'"[5] Because of the degree of variation in these loci, analysis can yield results with surprising power. The odds that an innocent individual's DNA will match the DNA left at a crime scene run from one in millions to one in billions.[6]

DNA is a remarkable tool in law enforcement, one that can be used both to the advantage and to the disadvantage of suspects, arrestees, and convicts. It can serve as compelling evidence against a guilty individual, yet it can also be used to exonerate the innocent.[7] Moreover, because DNA is found throughout the human body, a sample from a potential suspect can be matched against DNA extracted from a variety of materials found at a crime scene: skin cells, semen, blood, saliva, and even sweat.

According to *Frye v. United States*,[8] technical and scientific evidence cannot be admitted in court unless it is both probative and generally accepted as reliable by the scientific community. While *Frye* has been implicitly overturned by *Daubert v. Merrell Dow Pharmaceuticals*,[9] it was the standard at the time DNA analysis was beginning to make its way into the world of law enforcement. DNA analysis was first subjected to a rigorous Frye analysis in *People v. Castro*.[10] The judge in *Castro* determined that forensic DNA analysis, in general, had been sufficiently proven reliable to allow its admissibility as evidence, even though the particular procedures used in *Castro* were not rigorous enough to justify the use of DNA evidence in that particular case. Since then, forensic DNA analysis has been generally accepted as a legitimate tool of law enforcement; however, there have been occasional controversies over techniques and procedures employed by specific laboratories or in specific cases. For example, the Houston Police Department's DNA lab closed in 2002, following a string of wrongful convictions that were traced to poor management, haphazard testing practices, and contamination of samples due to a leaky roof.[11]

It should be noted here that DNA analysis may not be the holy grail that law enforcement officials portray it to be. Since the early days of its use as evidence in criminal trials, the technology for the analysis has evolved considerably. Results that were considered rock-solid in 1990 are now coming under scrutiny as newer tests suggest some of those results might have been flawed.[12] Today, we place tremendous faith in the accuracy of our DNA analysis, but so did scientists and law enforcement officials in 1990. It is hard to predict whether future refinements of DNA technology might someday cast a shadow of uncertainty over the results we are relying upon today.

Forensic DNA analysis can be used in two primary ways: confirmation and "trawling."[13] Confirmation involves the comparison of a known scene sample to a known suspect sample, with the match used to confirm the suspicion of law enforcement that the particular suspect is guilty of the particular crime. Trawling, on the other hand, involves using DNA analysis to find a suspect for a case. In other words, confirmation takes place when there is already evidence linking Joe to crime A, and the DNA is a way of confirming that link, while a trawl case is when the DNA match is the basis for connecting Joe to crime A.

For example, imagine that law enforcement is trying to solve a series of rapes. If the police believe that Joe may have committed the crime, they may compare a sample of Joe's DNA against samples retrieved from the rape scenes. This is a confirmation case.

In the alternative, law enforcement may compare DNA retrieved from the rape scenes to a database of DNA samples from former convicts to identify one of those convicts as the perpetrator of this series of rapes. Conversely, law enforcement may compare DNA from the suspect in a current case against a database of DNA samples taken from other crimes to see if this suspect may be guilty of other offenses. These are examples of trawling.

The compilation of DNA indices—both scene samples and suspect samples—makes trawling possible. The DNA Identification Act of 1994 provided financial incentives for states to create their own DNA sample databases.[14] The federal government then created CODIS, the FBI's Combined DNA Index System, to integrate all of the state indexing systems as well as the federal government's own index (NDIA, the National DNA Index System). Among other things, CODIS has standardized the manner in which DNA is analyzed, specifying the 13 loci at which the STRs should be examined. This integrated state and federal online network of DNA indices allows law enforcement offices across the nation to engage in trawling to solve their most serious crimes.

As early as 1989, when DNA analysis was first being used in a forensic context and the state and federal indices of genetic profiles were still in the early planning stages, scholars and public officials recognized the potential privacy concerns associated with the technology.[15] Those privacy concerns included the risk of disclosure of health information, including the predisposition toward certain diseases; discovery and disclosure of family relationships; and the possibility of community-wide genetic testing in the effort to solve a specific crime.[16]

Indeed, DNA analysis was used for forensic purposes in England before it was adopted in the United States. In 1987, police near Leicester, England, were attempting to solve the rape and murder of two teenagers. DNA testing—used to exonerate a mentally disabled man who had confessed to one of the crimes—had proven that the crimes were committed by the same person. Police requested all men between the ages of 17 and 34 to submit to a voluntary DNA test. More than 98 percent of the men asked agreed to the testing.

Scholars in the United States pointed to the incident in England as an example of how forensic DNA analysis could be abused by law enforcement. The high turnout rate, they argued, was evidence that the genetic dragnet was inherently coercive. "Significant social pressure to submit to testing apparently existed, and over 5000 males in the small community's target age group provided blood samples. Only two refused, one of them the man eventually convicted of the crimes."[17]

Both confirmation and trawling uses implicate the Fourth Amendment and concomitant privacy concerns. In the next section, I will explore the issues associated with confirmation cases, where the collection and use of DNA fits more squarely within the framework of Fourth Amendment jurisprudence.

I will then turn the spotlight on trawling policies, where the boundaries of the Fourth Amendment are stretched to their limits, with special attention paid to the recent Supreme Court decision in *Maryland v. King*.[18]

DNA CONFIRMATION AND THE FOURTH AMENDMENT

When DNA analysis is used for confirmation as part of the investigatory process, perhaps the most significant issue that arises relates to the actual collection of the evidence. That evidence may be collected using a buccal swab or a blood test, in which case police will first obtain a warrant based on probable cause. Because DNA is found throughout the human body and because a laboratory can turn even trace amounts of DNA into a usable sample, law enforcement can collect incriminating DNA from the detritus of our daily lives.[19]

> Imagine this fanciful (but plausible) scenario:
> Last week, you killed someone. You meticulously removed any evidence which might link you to the crime. You are still the prime suspect, but you're not worried. The police have nothing, not even probable cause to search your apartment. You are literally getting away with murder. But you have a nagging cold with an awful cough. On your way out the door one morning, you spit some phlegm on the sidewalk. A few days later, you find yourself under arrest and wondering what went wrong.[20]

Under traditional Fourth Amendment jurisprudence, the police need no warrant or even probable cause to collect evidence that a suspect is deemed to have abandoned.[21] For example, if police suspect me of a robbery, and I throw an incriminating receipt into the trash and then abandon that trash at the curb for removal by solid waste collectors, the police may retrieve the receipt without probable cause. As a matter of course, we abandon our genetic material everywhere: when we discard a paper coffee cup, a used tissue, a cigarette butt, or even spittle on the sidewalk (to name just a few common examples). Those discarded items, when collected and analyzed by law enforcement, can form the basis for probable cause to support an arrest warrant or, at a minimum, to support a warrant for a more controlled DNA collection. Indeed, this police tactic has become commonplace in police procedural novels and television shows.

Police use of abandoned DNA is not just a plot device. In 2003, the Los Angeles Police Department used DNA recovered from a coffee cup discarded by Adolph Laudenberg to solve a series of cold murder cases.[22] Indeed, the police don't need to wait for a suspect to leave behind his or her DNA; they can use deceit to trick a suspect into handing over genetic material:

> Seattle police devised a clever ruse to obtain a DNA sample from John Athan, whom they long suspected in the 1982 murder of a thirteen-year-old girl. Writing on the stationery of a fictitious law firm, the police sent a letter in 2003 to Athan, then living in New Jersey, asking him to join a class action lawsuit to recover overcharged traffic fines. Athan complied, and by licking the return envelope, he provided the detectives with the DNA sample they needed. Athan's DNA matched that found at the crime scene, and in 2004 he was convicted of second-degree murder.[23]

As this case illustrates, the abandonment rule allows law enforcement to take extraordinary steps to obtain DNA from criminal suspects without ever triggering a Fourth Amendment inquiry.

The abandonment rule is based on the fundamental Fourth Amendment analysis articulated in *Katz v. United States*.[24] In *Katz*, the Supreme Court held that the Fourth Amendment "kicks in" when the police intrude upon an area or activity in which a citizen has a reasonable expectation of privacy, where "reasonableness" depends on both subjective and objective criteria. If either the subjective or objective expectation of privacy is absent, then there is no search, and there are no Fourth Amendment constraints on police behavior.

When someone discards their trash and places it on the curb for collection, that person has no reasonable expectation of privacy. Trash bags might be torn apart by animals and the trash strewn across the street for anyone to see. The trash collector might look into the bags or cans at the time of delivery. Once the trash is delivered to a landfill, any number of people may have access to the contents. Similarly, in *U.S. v. Miller*,[25] the Court "held that a subpoena for information held by Miller's bank was not a search because Miller could not reasonably expect his bank information to remain private."[26]

What about collection of DNA evidence? The comprehensive workability of the *Katz* decision has been called into question, especially as it pertains to searches and seizures involving highly technical police procedures.

> [T]he constitutional law of search and seizure is largely indeterminate. This is especially true with respect to governmental observations aided by technological surveillance devices. The judicial assertion of authority in this area has benefited society by creating a legal monitor of some of the ill effects of scientific advancement, but it has done so at the expense of clarity, guidance, and coherence.[27]

The *Katz* paradigm was founded on the need for Fourth Amendment jurisprudence to tackle a new, technological law enforcement tool—trap and trace technology—yet it has encouraged analogies between old technology and new technology. Those analogies often fail to consider the true scope and possibility for abuse inherent in these new technologies. For good or ill, however, *Katz* remains the law of the land.

There is no question that our daily activities put our DNA in the public, exposing it to potentially prying eyes, but that doesn't mean DNA collection from discarded items is immune from Fourth Amendment protections. As we go about our daily lives, we surely have a subjective expectation of privacy in our DNA. Most of us would never consider the possibility that the government might be collecting our refuse in an effort to gain the most fundamental information about ourselves: our genetic profile.

> The volition that is implied in abandonment is simply unrealistic here. Courts may readily find that criminals have clearly intended to renounce all privacy claims to bags containing illegal firearms or to packages of drug paraphernalia when fleeing the police, but we hardly have a realistic choice in shedding DNA. One can shred private papers or burn garbage so that no one may ever delve into them, but leaving DNA in public places cannot be avoided.[28]

Is that subjective expectation of privacy reasonable?

The answer may hinge on the way in which that collected material is converted into meaningful information. The receipt I discard in my trash is meaningful on its face; police look at the receipt, and they know where I shopped, what I bought, when I bought it, and what I paid. The financial records obtained from Miller's bank may have been slightly more opaque, but the meaning behind the records could still be read by a trained eye. In the case of the DNA found on my discarded coffee cup, though, the trash has no meaning until it is manipulated and analyzed using a highly technical and specialized chemical process.

In *Kyllo v. United States*,[29] the Supreme Court considered the constitutionality of a thermal scan of Kyllo's house taken without a warrant. The state claimed that no warrant was necessary because they had taken the scan from the street outside of Kyllo's house. As a result, they argued, Kyllo could not have had a reasonable expectation of privacy, because anyone in the public could have obtained the same information obtained by law enforcement. The Court disagreed. Specifically, they noted that the thermal scan constituted a search within the meaning of the Fourth Amendment because it required the use of high-tech, specialized equipment that was not available to the general public. While *Kyllo* dealt with the search of a home, which has been given special protection in Fourth Amendment jurisprudence, surely the search of our most fundamental selves, our genetic blueprint, should be similarly protected. Although the saliva, blood, or skin sample that was discarded may itself be abandoned and thus in the public sphere, failing to implicate the Fourth Amendment, the genetic material *within* the sample should still be considered private and require a warrant for analysis because the genetic material cannot be extracted from the abandoned biological material without the use of highly technical procedures and equipment.

Similarly, in *Skinner v. Railway Labor Executives' Association*,[30] the Supreme Court considered the constitutionality of a federal rule that authorized railroads to require employees who violated safety regulations to submit to alcohol breath tests and urinalysis. While the Supreme Court ultimately upheld the regulation on the basis of the special needs doctrine—which recognizes an exception to the Fourth Amendment where the invasion into private space or personal integrity serves a compelling interest beyond the scope of regular law enforcement, such as promoting the safety of our railroads—it did hold that the tests themselves were searches within the definition of the Fourth Amendment. In doing so, they noted that "[t]he ensuing chemical analysis of the sample to obtain physiological data is a[n] . . . invasion of the tested employee's privacy concerns."[31] In other words, the collection of the biological material constituted one invasion, while its testing—extracting the meaning from the material—constituted a separate, additional Fourth Amendment event.

Interestingly, at least one scholar has suggested distinguishing between the sample (the biological material obtained for testing) and the profile (the blueprint of alleles at the 13 CODIS loci that are ultimately used to match scene samples to subject samples), but with an eye toward protecting the biological material rather than the profile.[32] Leigh M. Harlan suggests that the law distinguish between the biological sample—in which the subject should retain property rights and that should thus be destroyed once it is no longer needed for legitimate law enforcement purposes—and the profile, which is used solely for law enforcement purposes and could be kept indefinitely. Harlan's concern is with the potential for the biological sample to fall into the wrong hands and be used by third parties, such as employers and insurance companies for their own analysis, which might result in discrimination. Still, argument about the distinction between the biological material, which is surely abandoned, and the profile, which arguably has not been abandoned, casts an interesting light on this discussion.

To date, no court has applied either the technological enhancement argument from *Kyllo* or the technical manipulation argument from *Skinner* to the context of analyzing abandoned DNA. Rather, the courts have treated abandoned DNA in the same manner as abandoned receipts and discarded gloves. The special nature of DNA—both the content of DNA and the way in which it is used by police—suggests that a more rigorous analysis should be employed, and the *Kyllo* and *Skinner* cases provide a framework for that analysis.

DNA TRAWLING AND THE FOURTH AMENDMENT

As briefly outlined above, DNA trawling involves the use of DNA indices to match individuals to crimes without any specific, prior expectation that the individuals are connected to those crimes. It is a genetic fishing expedition. It is made possible by the maintenance of state and federal genetic indices, all of which have been consolidated into the federal CODIS system.

It should be noted that CODIS is hardly the only government project aimed at data mining. Perhaps the most comprehensive effort is a program called ADVISE (Analysis, Dissemination, Visualization, Insight, and Semantic Enhancement), a system designed to trawl a huge array of information sources for "chatter" and other suspicious behavior indicative of terrorist threats.[33] Indeed, "[a]ccording to a GAO report issued in 2004 . . . 52 federal agencies were using or were planning to use data mining, for a total of 199 data mining efforts, 68 planned and 131 operational. Of these programs, at least 122 are designed to access '"personal' data."[34] CODIS, however, is unique in its collection of biological material.

From the perspective of the state, these genetic indices are a windfall, an enormous leap forward in the ability of law enforcement to protect society. First, they allow the state to close cold cases, bringing resolution to victims and their families. Second, they allow the state to identify serial offenders, those who pose the greatest risk to public safety. Finally, they allow the state to better identify an individual's criminal history prior to trial and sentencing on the crime for which that person is currently being investigated, thus allowing for more appropriate sentencing.

While some may cringe at the expense associated with handling and analyzing a huge number of DNA samples, which require careful storage and manipulation, there is some evidence to suggest that use of DNA indices can actually save states money.[35] Specifically, the argument goes that the increased convictions and sentences that flow from the use of genetic indices result in savings by avoiding the costs (officer response, investigation, and prosecution) of future crimes—those that never take place because the offender is in prison. Increasing the number of known samples in the databases by extending collection from convicts to arrestees increases the cost savings the DNA indices provide by increasing the likelihood of matching a scene sample to an existing suspect sample. While this analysis is not universally accepted,[36] it seems to have gained some traction in public policy circles.

Whether they make good policy sense or not, the creation of these indices raises significant Fourth Amendment concerns. As Pearsall notes, when forensic DNA analysis was first being used in the United States and most criminal DNA indices were still in their planning states, scholars and political figures were already expressing concerns for the ways in which the indices might be used. For example, even as the state of California was collecting DNA from sex offenders with an eye toward launching their statewide database, the California attorney general, John Van de Kamp, was expressing his concern over its potential abuses:

> It is one thing to have fingerprints and criminal histories easily accessible to tens of thousands of peace officers. It is quite another to have information on-line that can mark you as a carrier of A.I.D.S.; or prove that you are not genetically related to either of your parents. Which of us would like to know that we are genetically predisposed to [a disease]? And which of us would be willing to have such information easily available to others?
>
> [Although] we envision no such intrusive databank . . . in this era of interlocking databases . . . [troubling issues will arise].
>
> Some of the most predictable [issues] involve the temptation to engage in genetic fishing expeditions. For example, researchers now postulate that certain types of chromosomal deficiencies may incline people toward violent crimes. If D.N.A. analysis becomes commonplace, there will surely come a date when a desperate detective tries to run a search for every person with that deficiency in the vicinity of a series of unsolved murders.[37]

While the specific examples raised by the attorney general have not come to pass—for example, DNA analysis cannot identify people who are carrying a virus, and police do not use the indices to identify people with a predisposition toward crime—his fundamental concern has borne fruit with law enforcement trawling CODIS to solve cold cases or accumulate additional charges against a suspect in custody.

What type of genetic information finds its way into CODIS? It is unclear whether abandoned DNA collected during law enforcement confirmation investigations may constitutionally find its way into CODIS.[38] Since DNA collected from abandoned tissue samples is not the fruit of a Fourth

Amendment search, its use remains unregulated by constitutional principles. At present, "CODIS collects DNA profiles provided by local laboratories taken from arrestees, convicted offenders, and forensic evidence found at crime scenes."[39]

In the case of suspects, those who have not yet been arrested, direct DNA collection from blood or a buccal swab requires a warrant. Drawing blood or swabbing the inside of a suspect's cheek invades the suspect's body in a way that implicates the Fourth Amendment.

> Virtually any "intrusio[n] into the human body" . . . will work an invasion of "'cherished personal security' that is subject to constitutional scrutiny[.]" The Court has applied the Fourth Amendment to police efforts to draw blood, . . . scraping an arrestee's fingernails to obtain trace evidence, . . . and even to "a breathalyzer test, which generally requires the production of alveolar or 'deep lung' breath for chemical analysis[.]"[40]

As a result, probable cause, supported by a warrant, is required to effect such a search of an individual who is not yet in police custody.

The Fourth Amendment analysis becomes murkier, however, when a suspect has been taken into custody and arrested. In *Maryland v. King*, the Supreme Court considered the constitutionality of the Maryland DNA Collection Act, a statute that allows collection of DNA samples from arrestees. The case had legal scholars on the edges of their seats, all anticipating a vigorous debate about the limits of Fourth Amendment protection. They were not disappointed. During the middle of oral arguments, Justice Samuel Alito opined that the King case "'is perhaps the most important criminal procedure case that this court has heard in decades. . . . This is what's at stake: Lots of murders, lots of rapes that can be solved using this new technology.'"[41]

Justice Scalia responded, "I'll bet you if you conducted a lot of unreasonable searches and seizures, you'd get more convictions, too."[42]

The Maryland statute authorizes law enforcement to take a DNA sample from "'an individual who is charged with . . . a crime of violence or an attempt to commit a crime of violence; or . . . burglary or an attempt to commit burglary.'"[43] Upon arraignment, the sample is then submitted to state and national databases. If the court subsequently determines that the arrest lacked probable cause or if the defendant is acquitted, the sample is destroyed. The state of Virginia led the way by first adopting arrestee legislation in 2003,[44] but it was certainly not the last state to do so. As of the issuance of the *King* decision, "[a]ll 50 States require[d] the collection of DNA from felony convicts . . . [and t]wenty-eight States and the Federal Government [had] adopted laws similar to the Maryland Act authorizing the collection of DNA from some or all arrestees."[45]

In April of 2009, Alonzo King was arrested on first- and second-degree assault charges. Approximately three months later, in July of 2009, his DNA was uploaded to the Maryland DNA database. Three weeks later, in August 2009, King's DNA was matched with DNA taken from the scene of an unsolved rape case. King was ultimately convicted of that rape charge and appealed on the grounds that the state took his DNA in violation of the Fourth Amendment. While King's trial for the 2009 rape included significant evidence beyond the DNA match, it was undisputed that the state would not have focused on King as a suspect in the rape case were it not for the DNA match. As with most cases of DNA testing, King's DNA was taken with a buccal swab: a long-handled medical swab rubbed gently along the inside of a person's cheek in order to pick up skin cells for analysis. As the *King* decision notes, buccal swabs are painless and require only a minimal physical intrusion into the suspect's body.

The opinions issued by the Court were every bit as spirited as the oral arguments. Writing for a five-person majority, Justice Kennedy issued an opinion upholding the Maryland law. The opinion of the Court makes occasional reference to the notion of a search incident to lawful arrest. The "incident to lawful arrest" doctrine is not relevant here. "[T]he Court has allowed warrantless searches incidental to a valid arrest on the basis of a need to safeguard arresting officers or to prevent the destruction of evidence."[46] Neither of these concerns is implicated in the collection of DNA from an arrestee. While the Court uses this language, its actual reasoning does not depend on this concept.

Instead, the Supreme Court's analysis rests on its interpretation of reasonableness in the context of an arrest. The Court notes that no warrant was required because the individuals subjected to testing, such as King, had been arrested for a serious crime based on probable cause and because the physical intrusion into the suspect's body was minor, quickly administered, and causing no pain or physical distress. For these reasons, the Court holds, there is no need for a warrant or even for individualized suspicion that the suspect may be guilty of other crimes in the database. Rather, the only concern is whether the collection of DNA evidence from arrestees satisfies the requirement of reasonableness.

Where no warrant is necessary, the reasonableness analysis requires a balancing of "the promotion of legitimate governmental interests" against the "degree to which [the search] intrudes upon an individual's privacy."[47] According to *King*, the collection of DNA evidence from an arrestee promotes the government's interest in correctly identifying the person it has taken into custody. The Court notes, quite rightly, that identification of an arrestee should not be limited to a proffered name. Rather, law enforcement should look further to determine whether the individual is who he or she claims to be. The court invokes cases in which dangerous criminals—such as Oklahoma City bomber Timothy McVeigh and serial killer Joel Rifkin—have been stopped or arrested for minor traffic offenses; in these instances, the offenses for which they were stopped offered no clue as to the dangerousness of the individual in custody.

One reason for correctly identifying an arrestee is to accurately determine the individual's criminal history. This serves an important purpose beyond the normal investigative procedures of law enforcement. To ensure officer safety, it is critical for the police to know who they are dealing with. As the court notes, law enforcement takes other steps—beyond collection and analysis of DNA—to effect that government interest. Comparing an arrestee's booking photo to artist sketches of wanted criminals, using computer-assisted fingerprint analysis, and matching tattoos to known gang symbols allow the police to determine whether the sheep they have arrested is really a wolf. This determination of dangerousness is also critical to the process of setting appropriate bail and preventing such a dangerous offender from committing additional criminal acts while on conditional release.

The court describes another purpose for detailed identification of an arrestee: "'ensuring that persons accused of crimes are available for trials.'"[48] The Court argues that an individual who has been arrested for a known crime but who is actually guilty of prior offenses is more likely to flee beyond the reach of the justice system because of the present arrest. In other words, the Court seems to assert that making sure a person is available to stand trial for a past offense can be distinguished, as a government interest, from the interest in actually finding and convicting a suspect for the past offense.

The Court also points out that collecting and analyzing DNA from arrestees may result in the exoneration of the wrongfully convicted. Imagine that Joe is convicted of crime A. Years later, Ed is arrested for crime B, but analysis of his DNA points to his guilt in crime A. Such a finding may result in Joe being exonerated in relation to a crime for which he was wrongfully convicted.

The Court weighs this important government interest in identifying the arrestee and accurately determining past criminal history against the physically noninvasive way in which DNA is collected. The noninvasive nature is even more significant, given the reduced expectation of privacy for individuals who have been taken into police custody, where they may be subjected to highly intrusive cavity searches. Using the balancing test described in *Wyoming v. Houghton*,[49] the Court concludes that collection of DNA from arrestees is reasonable.

In a (typically) scathing dissent, Justice Scalia calls into question the assumptions of the majority's analysis, comparing the collection of DNA from arrestees to the British general warrants that were the very target of the Fourth Amendment. Scalia takes particular umbrage with the notion that the collection of DNA from arrestees is done for purposes of identifying a person in custody:

> So while the Court is correct to note . . . that there are instances in which we have permitted searches without individualized suspicion, "[i]n none of these cases . . . did we indicate approval of a [search] whose primary purpose was to detect evidence of ordinary criminal wrongdoing." *Indianapolis v. Edmond*, 531 U.S. 32, 38 (2000). That limitation is crucial. It is only when a governmental purpose aside from crime-solving is at stake that we engage in the free-form "reasonableness" inquiry that the Court indulges at length today. To put it another way, both the legitimacy of the Court's method and the correctness of its outcome hinge entirely on the truth of a single proposition: that the primary purpose of these DNA searches is something other than simply discovering evidence of criminal wrongdoing. . . . [T]hat proposition is wrong.[50]

Scalia argues that the facts do not support the notion that the primary purpose of the DNA collection was for identification purposes. First, he points out that the sample was not submitted for processing for three months and that the statute specifically states that the sample cannot be analyzed prior to arraignment. If the purpose of the sample had anything to do with identification, it would be processed before further steps in the criminal justice process were carried out.

The majority argues that the delay in obtaining DNA analysis does not seriously undermine its usefulness as a tool of identification. First, the Court notes that fingerprint analysis in the days before the FBI's Integrated Automated Fingerprint Identification System (IAFIS) took considerable time, yet its importance as a means of identifying suspects allowed its use on arrestees. Second, the Court points out that the delay between collection and analysis of DNA samples is not always as long as in the *King* case and that such delays will surely decline as technology improves, noting that "the FBI has already begun testing devices that will enable police to process the DNA of arrestees within 90 minutes."[51] Finally, the majority argues that law enforcement's interest in identifying an arrestee and accurately understanding that person's criminal history does not end within a short period after arrest. Rather, that interest continues throughout the criminal justice process, with continual updates to identifying information constantly revising the way in which an individual arrestee should be treated.

Scalia goes further, however. He notes:

> If anything was 'identified' at the moment that the DNA database returned a match, it was not King—his identity was already known. . . . Rather, what the August 4 match 'identified' was *the previously-taken sample from the earlier crime*. That sample was genuinely mysterious to Maryland; the State knew that it had probably been left by the victim's attacker, but nothing else. King was not identified by his association with the sample [from the cold case]; rather, the sample was identified by its association with King.[52]

While the majority would argue that the identification ran two ways—that the match further identified King as a violent repeat offender—Scalia takes issue with this notion of identification. He argues that it strains the definition of "identification" to include past criminal history under that umbrella.

Finally, Scalia points out that the Maryland statute only authorized collection of DNA from people arrested for serious offenses (violent crimes, burglary, and attempts to commit such crimes). If the purpose of the DNA collection was identification, Scalia questions, why couldn't it be just as effective for that purpose in comparably minor cases? Indeed, techniques including fingerprinting and photographing of arrestees are routinely carried out for the most minor of offenses. The fact of the limitation suggests that the DNA collection is not a routine part of the identification process but, rather, something special. That something special, Scalia suggests, is routine law enforcement: investigation of past crimes.

> As he states, [s]ensing (correctly) that it needs more, the Court elaborates at length the ways that the search . . . served the special purpose of "identifying" King. But that seems to [Scalia] quite wrong—unless what one means by "identifying" someone is "searching for evidence that he has committed crimes unrelated to the crime of his arrest."[53]

The tension between the majority and the dissenting opinions primarily turns on the question of why, exactly, law enforcement collects DNA from arrestees. Scalia alludes to the Court's conclusion that the physical intrusion involved in a DNA swab is minimal:

> We are told that the "privacy-related concerns" in the search of a home "are weighty enough that the search may require a warrant, notwithstanding the diminished expectations of privacy of the arrestee." . . . But why are the "privacy-related concerns" not also "weighty" when an

intrusion into the body is at stake? (The Fourth Amendment lists "persons" first among the entities protected against unreasonable searches and seizures.) And could the police engage, without any suspicion of wrongdoing, in a "brief and . . . minimal" intrusion into the home of an arrestee—perhaps just peeking around the curtilage a bit? . . . Obviously not.[54]

Because his conclusion that the purpose of DNA collection is normal law enforcement is dispositive, however, he spends little time exploring the majority's argument about the minimally invasive quality of DNA collection.

However, the majority's statement deserves further consideration. The majority makes much of the fact that a cotton swab to the inside of the cheek is a quick and gentle intrusion into the bodily integrity of the arrestee, certainly less invasive than the search to which the arrestee will be subjected before being placed in a cell. Yet the collection of DNA is not merely a search; it is also a seizure. It is the collection of biological material from the arrestee.

While most of us would not consider the loss of a few skin cells a major deprivation, the seizure is not only a seizure of the cells themselves but also a seizure of the detailed genetic information contained within those cells. Short of the contents of our hearts, that is the most personal information one could obtain from us.

The majority is careful to note that the only information extracted for entry into state and federal indices is the arrangement of alleles at the 13 specific loci identified by CODIS for consistent comparison. Those 13 loci offer only limited information about a person's genetic makeup, and there is some question whether those loci provide any probative medical information.[55] In other words, the amount of personal information contained in the profile derived from the sample is relatively small; it is enough to distinguish one individual from another but not enough to actually reveal something meaningful—beyond identity—of the donor subject.

The *King* decision indicates that the Court is enchanted by the power of DNA as a law enforcement tool. It is hard to argue for limitations on the collection and use of DNA, given the tremendous public good that it can provide. That said, the Court appears to engage in legal backflips in order to ascribe some use beyond normal law enforcement to DNA collected from arrestees, and it dramatically minimizes the scope of the intrusion involved in DNA collection. As both technology and this area of the law develop, it will be interesting to see whether traditional Fourth Amendment jurisprudence in general and the *King* logic in particular prove adequate to the task of walking the fine line between public interest and individual privacy.

CONCLUSION

Existing case law weighs strongly in favor of the ability of law enforcement to use DNA to solve crimes. So far, the courts have placed no limits on the ability of law enforcement to collect abandoned DNA during the course of their investigations. Moreover, the Supreme Court has explicitly found that police do not need a warrant, probable cause, or even particularized suspicion to test the DNA from an arrestee against a nationwide database of DNA samples from unsolved crimes. The current state of affairs accords little value to an individual's privacy interest in his or her genetic blueprint, yet the possibility for further erosion of that privacy interest is certainly possible if the courts do not halt or reverse the current momentum of their decisions.

First, the question of whether abandoned DNA can be uploaded into the state and national DNA indices remains open. Given that the genetic material has been abandoned, the courts have held that its collection is not a search or seizure within the meaning of the Fourth Amendment, so it is unclear what constitutional basis could exist for limiting the use of such material to investigation of a specific crime. Short of legislative intervention, there is no obvious mechanism that prevents police from gathering abandoned DNA without any constitutional check whatsoever and submitting it to state and national DNA databases, putting an individual's genetic profile into a vast pond in which law enforcement officers from across the country may go fishing for suspects. While most Americans would find this notion appalling, it is the logical trajectory of the current state of the law.

A second unresolved (and troubling) question relates to the collection of DNA from people arrested for violent crimes. The Maryland statute specifically provided for the destruction of the DNA sample if the arrestee was released because of a lack of probable cause to support the arrest or in the event of an acquittal. But is such a limitation constitutionally mandated? In other words, would collection statutes of arrestees be rendered unconstitutional if the State of Maryland allowed law enforcement to maintain those records even after an acquittal?

Based on the rationale in the *King* decision, the answer is most likely no. The Court's decision to uphold Maryland's statute did not depend on this legislative check on the process. Moreover, the Court's decision rests on its comparison of DNA collection to the collection of fingerprints and the taking of mug shots. Both fingerprint cards and mug shots are maintained by police even after a person has been acquitted. Given the Court's reasoning, there is no obvious constitutional reason why DNA samples taken from arrestees would ever need to be destroyed or removed from state and national databases. Again, this gives the government tremendous power to expose individuals to criminal liability through trawling of existing databases.

Finally, the decision in *King* raises a question of whether it would be constitutionally permissible to collect DNA from people arrested or detained for lesser offenses, including traffic stops (potentially). Policy analysts have suggested that extending arrestee DNA policies from serious crimes to all crimes would increase the associated cost savings (by increasing the number of known subject samples) and decrease the administrative costs associated with distinguishing between cases in which arrestee DNA evidence should and should not be collected.[56]

Again, the Court's decision in *King* leaves the door open for policies extending routine collection of DNA to lesser offenses. Because the Court frames the collection of DNA as a way to identify a suspect, similar to fingerprinting, and as a way for law enforcement to appropriately handle individuals with violent criminal histories, the next step in the analogy—to the use of driver's licenses to determine both the identity and criminal history of someone at a traffic stop—seems likely. Indeed, two of the examples offered by the Court to demonstrate the importance of law enforcement having accurate criminal history information were cases in which a violent criminal was stopped for a traffic violation. To the extent that law enforcement needs adequate criminal history information to protect themselves in the custodial setting, as the Court suggests, that need is certainly heightened in the context of a traffic stop during which only one or two officers engage with a suspect in an uncontrolled environment.

Perhaps the one fact that would weigh against the use of DNA collection during traffic stops is the length of time it takes to obtain results. In the custodial context, according to the Court, the delay in results does not undermine the identification function of the DNA collection because continual updates to that information serve an ongoing purpose and because the nature of an arrestee's interaction with law enforcement is itself ongoing; in contrast, traffic stops are fleeting encounters with law enforcement. However, the Court notes that the technology of DNA collection and analysis is changing at a rapid rate, making the process quicker and more streamlined. As the technology improves, we could well find ourselves in a situation in which genetic profiles can be generated rapidly enough to justify their use as an identification tool during a traffic stop. At present, such a use may seem like pure science fiction, but 35 years ago, the average person would have been astonished by the notion that law enforcement would be able to chemically distinguish between every human being on earth with nothing more than a handful of skin cells.

The *King* decision is the first salvo in what will likely become a long and contentious legal debate over the appropriate collection and use of forensic DNA analysis. These opening volleys give law enforcement a heavy advantage in the battle between the government's interest in solving crime and protecting the public, on the one hand, and citizens' interest in their genetic privacy, on the other. Ultimately, the people may need to turn to their legislatures rather than their courts to protect them from highly intrusive DNA policies. In the alternative, the Supreme Court is going to need to consider DNA analysis as its own animal, rather than attempt to analogize to other forms of evidence, when it considers future Fourth Amendment challenges to this critical law enforcement tool.

NOTES

1. *Katz v. United States*, 389 U.S. 347 (1967).
2. *J.D.B. v. North Carolina*, 564 U.S. ____ (2011).
3. Elizabeth Canter, "A Fourth Amendment Metamorphosis: How Fourth Amendment Remedies and Regulations Facilitated the Expansion of the Threshold Inquiry," *Virginia Law Review* 95, no. 1 (2009): 156.
4. *District Attorney's Office for the Third Judicial District, et al., v. Osborne*, 557 U.S. 55, 8 (2009) (slip op.).
5. *Maryland v. King*, 569 U.S. ____, 4 (2012) (slip op.) (citation omitted).
6. Peter Donnelly and Richard D. Friedman, "DNA Database Searches and the Legal Consumption of Scientific Evidence," *Michigan Law Review* 97, no. 4 (1999): 935–936.
7. Seth F. Kreimer and David Rudovsky, "Double Helix, Double Bind: Factual Innocence and Postconviction DNA Testing," *University of Pennsylvania Law Review* 151, no. 2 (2002): 547–617.
8. 293 F. 1013 (D.C. Cir. 1923).
9. 509 U.S. 579 (1993).
10. 545 N.Y.S.2d 985 (Sup. Ct. 1989).
11. James Pinkerton, "Backlog prompts Lykos to call for emergency DNA lab," *Houston Chronicle*, July 18, 2010, http://www.chron.com/news/houston-texas/article/Backlog-prompts-Lykos-to-call-for-emergency-DNA-1713568.php.
12. Benjamin Vetter, "Habeaus, Section 1983, and Post-Conviction Access to DNA Evidence," *The University of Chicago Law Review* 71, no. 2 (2004): 588.
13. Donnelly and Friedman, "DNA Database Searches," 935.
14. The DNA Identification Act of 1994, Pub. L. No. 103–322, §§ 210301-210306 Stat. 108 (1994): 2065.
15. Anthony Pearsall, "The Unexamined 'Witness' in Criminal Trials," *California Law Review* 77, no. 3 (*Symposium: Law, Community, and Moral Reasoning*) (1989): 665–703.
16. Pearsall, "The Unexamined 'Witness,'" 679. *See also*, Clare M. Tande, "DNA Typing: A New Investigatory Tool," *Duke Law Journal* 1989, no. 2 (1989): 474–494.
17. Anthony Pearsall, "The Unexamined 'Witness,'" 679.
18. *Maryland v. King*.

19. Elizabeth E. Joh, "Reclaiming 'Abandoned' DNA: The Fourth Amendment and Genetic Privacy," *Northwestern University Law Review* 100, no. 2 (2006): 857–884.

20. Holly K. Fernandez, "Genetic Privacy, Abandonment, and DNA Dragnets: Is Fourth Amendment Jurisprudence Adequate?," *Hastings Center Report* 35, no. 1 (2005): 21.

21. California v. Greenwood, 486 U.S. 35 (1988).

22. Elizabeth E. Joh, "Reclaiming 'Abandoned' DNA," 861.

23. Ibid., 861–862.

24. 389 U.S. 347 (1967).

25. 425 U.S. 435 (1976).

26. Christopher Slobogin, "Government Data Mining and the Fourth Amendment," *The University of Chicago Law Review* 75, no. 1 (2008): 329.

27. Robert C. Power, "Technology and the Fourth Amendment: A Proposed Formulation for Visual Searches," *The Journal of Criminal Law and Criminology* 80, no. 1 (1989): 5.

28. Elizabeth E. Joh, "Reclaiming 'Abandoned' DNA," 867 (internal citations omitted).

29. 533 U.S. 27 (2001).

30. 489 U.S. 602 (1989).

31. Skinner v. Ry. Labor Executives Ass'n, 489 U.S. 602, 616 (1989).

32. Leigh M. Harlan, "When Privacy Fails: Invoking a Property Paradigm to Mandate the Destruction of DNA Samples," *Duke Law Journal* 54, no. 1 (2004): 179–219.

33. Slobogin, "Government Data Mining and the Fourth Amendment," 318.

34. Ibid., 319.

35. Jay Siegel and Susan D. Narveson, "Why Arrestee DNA Legislation Can Save Indiana Taxpayers Over $50 Million Per Year," accessed June 28, 2013. http://dnasaves.org/files/IN_DNA_Cost_Savings_Study.pdf.

36. Julie E. Samuels, Elizabeth H. Davies, and Dwight B. Pope, "Collecting DNA at Arrest: Policies, Practices, and Implications," *Urban Institute Justice Policy Center*, accessed July 3, 2013, http://www.urban.org/uploadedpdf/412831-Collecting-DNA-at-Arrest-Policies-Practices-and-Implications-Report.pdf.

37. Pearsall, "The Unexamined 'Witness,'" 680 (quoting Address by John Van de Kamp, Attorney General of California, California Criminalistics Institute Seminar on D.N.A. Identification (Jan. 7, 1988) 3–4).

38. Joh, "Reclaiming 'Abandoned' DNA," 857–884.

39. *Maryland v. King*, 6.

40. Ibid., 8 (internal citations omitted).

41. Adam Liptak, "Justices Wrestle Over Allowing DNA Sampling at Time of Arrest," *New York Times*, February 26, 2013, http://www.nytimes.com/2013/02/27/us/supreme-court-hears-arguments-on-dna-sampling.html?_r=0.

42. Ibid.

43. *Maryland v. King*, 4 (quoting Md. Pub. Saf. Code Ann. §2-504(a)(3)(i) (Lexis 2011)).

44. Siegel and Narveson, "Why Arrestee DNA Legislation Can Save."

45. *Maryland v. King*, 7.

46. "Criminal Procedure: Searches Incident to Lawful Arrest Limited to the Area Within Reach of the Arrestee," *Duke Law Journal*, no. 5 (1969): 1085.

47. *Maryland v. King*, 10 (quoting Wyoming v. Houghton, 526 U.S. 295, 300 (1999)).

48. Ibid., 14 (quoting Bell v. Wolfish, 441 U.S. 520, 534 (1979).

49. 526 U.S. 295 (1999).

50. *Maryland v. King*, 3–4 (Scalia, dissenting).

51. Ibid., 22.

52. Ibid., 9 (Scalia, dissenting).

53. Ibid., 5 (Scalia, dissenting).

54. Ibid., 4 (Scalia, dissenting).

55. Ibid., 27 (citing S. H. Katsanis and J.K. Wagner, "Characterization of the Standard and Recommended CODIS Markers," *Journal of Forensic Science* 58 (2013): S169, S171).

56. Siegel and Narveson, "Why Arrestee DNA Legislation Can Save."

4

Biometric Identification as a Requirement for Work Access and Forced Surrendering of Private Information

Pamela LaFeber

The use of biometric technology is not new. We have freely handed over our most personal and private information to others for decades. Even when our information is taken from us, it is not without our knowledge, and the purpose of its use is generally understood. However, as with any technology, biometric technology has become more sophisticated. Today, information gathering, storage, dissemination, and search capability has blurred the line between personal protection and invasion of privacy. This chapter will broadly describe biometrics, discuss why the use of biometric technology is growing, give an overview of the application of biometrics by the government and private-sector organizations, consider the privacy implications of the use of biometrics, discuss court rulings on the use of biometrics in the workplace, and detail a standard set of principles when employing this type of technology.

WHAT IS BIOMETRICS?

At the simplest level, a biometric is a physical or behavioral characteristic that is unique to an individual. The ideal characteristic does not change over the course of a person's lifetime, cannot be altered, and is extremely difficult to forge. A fingerprint, signature, voice, palm print, iris, retina, and face are all examples of biometric types, or modalities, which can be used for identification and authentication purposes. Positive identification and authentication can protect individuals from identity theft, replace the need for passwords, and allow appropriate access to restricted computer networks and other secure information and facilities. This technique for human recognition was first introduced in the 1970s; however, since the events of September 11, 2001, the proliferation of web-based services, and an increasingly mobile society, the use of biometrics has become more common in both the public and private sectors. Systems have advanced from simply allowing access to personal computers and now have a place in many aspects of work and home life.

Biometric application can be divided into three categories: commercial, government, and forensic.[1] Examples of commercial applications include, but are not limited to: mobile phone, facility access, electronic purchases, personal banking, and even entry into Disneyland. Corpse identification and criminal investigation are examples of biometric forensic application. The government can employ biometrics as a requirement to obtain a driver's license and to receive welfare and other public grants. Biometrics is also regularly deployed by the federal government in response to the threat of terrorism, and in 2002, Congress passed the Enhanced Border Security and Visa Entry Reform Act, which mandates biometrics be used for all U.S. visas. It also required that biometric scanners be installed at all U.S. ports of entry.

Over the years, Congress has repeatedly requested that a similar exit program be implemented; however, the U.S. Department of Homeland Security has struggled to successfully move a program past the pilot stage. On June 27, 2013, the Senate approved major changes to America's immigration policies when it passed the Border Security, Economic Opportunity, and Immigration Modernization Act by a vote of 68–32. This act, or Senate Bill 744, is a bipartisan effort to reform the U.S. immigration system. The bill is broad and divided into four categories: (1) border security, (2) immigrant visas, (3) interior enforcement, and (4) reforms to nonimmigrant visa programs.[2] Senators in support of including biometrics in Bill 744 stressed that the September 11 attacks were committed by individuals who had overstayed their visas. Those opposing the inclusion of a biometrics amendment focused on the cost and said that the great expense would only further delay immigration reform. The biometrics amendment died by a 36–58 vote.

Regardless of which of the three biometric application categories is in question, when a system is used for the first time, the individual must enroll. To enroll, the initial fingerprint, optical scan, voice recording, or other characteristic is taken and entered into the database as a template.[3] This template can be linked to user information such as name, employee number, Social Security number, address, phone number, e-mail address, emergency contacts, and so forth. The biometric and associated information is verified and stored in the database. Once all information is entered, the user must attempt a log-in. The user input is compared to the initial template via a statistical algorithm, and the search for a match begins. If a match is found, user access is accepted; if not, access is rejected. If access is successful, the individual is enrolled.[4] The time it takes to enroll someone is directly related to the type of technology being used. In organizations that require higher security, such as nuclear facilities and financial institutions, the enrollment process can be lengthy given the amount of data that must be entered and the verification that must occur. One may think that by working with such unique characteristics, biometrics would deal in absolutes. But the science is still only based on probabilities. Even with a high probability of a match, the evaluation is still statistical and there is still a chance that there is no match or the match delivered is incorrect.

CATEGORIES OF BIOMETRIC TECHNOLOGY

Identification and authentication are the two broad categories of biometric technology. The identification process compares a sample physical biometric to a database and searches for a match. This is referred to as a "one-to-many" search.[5] The second category of biometric technology application is authentication. This is referred to as a "one-to-one" search and involves collecting a person's biometric for comparison to, and verification against, records of that person already stored in a database.[6] The accuracy of authentication is greater than the one-to-many application because the identifier is only compared against known information of the person.[7] Authentication is also seen as less invasive and more respectful of privacy than identification.

Law enforcement agencies have been using the identification process in the investigation and resolution of crimes for years each time a fingerprint is taken and compared to those in a database. Today, the use of fingerprints has been expanded from investigative purposes to gaining access to personal computers and bank information. Facial recognition software can also be used for activities not related to law enforcement. Individuals that have accessibility challenges can experience improved communication when using this type of technology. Voiceprint identification is another biometric that has been used for facility access, bank transaction security, and time and attendance monitoring in the workplace.[8] Because the use of this technology is convenient, can increase security, and assists with fraud detection, it is attractive to organizations in the private and public sectors. Also, timely identification results and the elimination of the need for passwords or documents that may be lost or stolen are just some of the benefits of employing biometric technology.

However, its application is not without challenges and disadvantages. First, biometrics is not a one-size-fits-all concept. Before a biometric modality is selected for a workplace, several factors must be considered. For example, the costs associated with obtaining and storing data, the required speed of identification, population of the workforce, integration into existing organizational systems, and fit and appropriateness for the staff.[9] For example, installing a retinal scan device for employees of a small public library may not be the most cost-effective option. Similarly, a fingerprint reader would not be suitable in an environment where employees would have dirty or greasy hands, such as a municipal fleet services department.

A disadvantage of biometrics is that false positives can occur when one characteristic is compared against the thousands in a database. Fingerprinting is a well-known biometric and can be used for a plethora of noncriminal identification circumstances; however, some individuals may associate fingerprint collection with criminal behavior, consider it an invasion of privacy, and believe that finger scans will be used to track their activities.[10] Facial recognition software is two-dimensional and can be altered by eye glasses, lighting, image resolution, and the angle of a photo.[11] Data on facial characteristics can also be gathered in public places without a person's knowledge. While not an invasion of privacy, this can make people uneasy. Voiceprint technology is less secure than other biometrics because a person's voice changes over the course of a lifetime, can be recorded and used by someone else, can be altered with an illness, and can be distorted by ambient noise. Because of these possibilities, voice biometrics should be used in conjunction with another form of identification or in a low-security-risk environment.[12]

Security is at the heart of any type of identification or authentication tool. In many workplaces, in government agencies, and on many websites, traditional means of access such as a driver's license or password have given way to biometric technology. Because a biometric is unique to the individual, these measurements are thought to be more reliable than conventional verification methods. However, as has been demonstrated, this isn't always the case. How the data is collected, accessed, and used has raised questions regarding the security of the information and whether these methods constitute a threat to privacy. The next section will discuss privacy in general and its evolution through the centuries, detail the types of privacy concerns that are associated with the use of biometrics, introduce the implementation of biometrics in the workplace, describe how the courts have viewed biometrics in the workplace, and review how government and private-sector employers are held accountable when implementing biometric technology.

BIOMETRICS AND THE RIGHT TO PRIVACY
The Individual and the Government

The way Americans view privacy has changed over the years. In the 16th century, people were concerned with privacy of place. James Otis said, "A man's house is his castle; and while he is quiet, he is well guarded as a prince in his castle."[13] In the 1800s, people wanted the right to privacy to grow. In 1890, Samuel Warren and Louis Brandeis reinforced this belief when they authored a *Harvard Law Review* article on privacy and penned that people have the "right to be let alone."[14] Smith also highlighted this shift in his book where he said privacy "is the desire by each of us for physical space where we can be free of interruption, intrusion, embarrassment, or accountability and the attempt to control the tie and manner of disclosures of personal information about ourselves. In the first half of our history, Americans seemed to pursue the first, physical privacy; in the second half—after the Civil War—Americans seemed in pursuit of the second, information privacy."[15]

While individuals may have had the desire to be let alone and coveted privacy of person and information, nothing in the law clearly defended this notion. The highest law of the land, the U.S. Constitution, does not overtly mention privacy. However, scholars believe that the framers were still concerned about protecting individual privacy interests.[16] According to some, citizens should expect protection "from the disturbance of one's seclusion, solitude, and private affairs; from embarrassing public disclosures about private facts; from publicity that places one in a false light; and from appropriation of one's name or likeness for someone else's gain."[17]

The U.S. Supreme Court also recognizes that the Constitution does not specifically guarantee privacy but has ruled that citizens are to be protected from certain government actions and have an inferred right to privacy.[18] Indeed, the Court and its members have interpreted the Constitution to protect individual's privacy in some situations.[19] After his appointment, Brandeis used the phrase "right to be let alone" in his dissent in *Olmstead v. United States*, 277 U.S. 438 (1928). The *Olmstead* case involved a law enforcement agency that used a telephone wiretap on a public phone. The Court found that its use did not constitute a violation of the Fourth Amendment's protection against unreasonable searches and seizures. The Court's rationale was that because the tap was not on Olmstead's actual premises, no physical trespass to the property had occurred.[20] Brandeis disagreed with the 5–4 majority and wrote that the framers "conferred, as against the Government, the right to be let alone, the most comprehensive of rights and the right most valued by civilized men."[21] However, later in *Katz v. United States*, 389 U.S. 347 (1967), the Court overturned *Olmstead* and said that, in fact, a violation of the Fourth Amendment had occurred. It held that persons using public telephones were protected from wiretaps unless the law enforcement agency had secured a warrant.[22] In *Griswold v. Connecticut*, 381 U.S. 479 (1965), the Court heard arguments regarding a state statute that said the sale of contraceptives to married couples was illegal. In his opinion, Justice Douglas wrote that the Constitution protects "zones of privacy," and the Court held that the statute was unconstitutional.[23] One year later, in 1966, Justice Douglas also addressed privacy in *Osborn v. United States*, 385 U.S. 323. Here, he wrote of a fear that the privacy of citizens was being eroded, and its continuation may allow the government to invade the most personal areas of someone's life, unchecked.[24]

So, while the topic is not specifically mentioned in the Constitution, the courts have relied on the Constitution for decades to support a person's expectation of privacy and to prohibit government from going places it shouldn't. The roots of Douglas's zones of privacy are said to be found in the First Amendment's rights of freedom of speech, press, religion, and association; the Third Amendment's prohibition against the quartering of soldiers in one's home; the Fourth Amendment's right for individuals to be secure in their persons, houses, papers, and effects from unreasonable searches and seizures without a warrant issued upon proof of probable cause; the Fifth Amendment's protection from being compelled to testify against oneself in a criminal trial; the Ninth Amendment's provision that "the enumeration in the Constitution, of certain rights, shall not be construed to deny or disparage others retained by the people;" the Tenth Amendment's provision that "the powers not delegated to the United States by the Constitution, nor prohibited by it to the States, are reserved to the States respectively, or to the people;"[25] and the Fourteenth Amendment's provision that no person shall be deprived of life, liberty, or property without due process of law.[26] However, when the government can demonstrate a legitimate interest, such as a threat to public safety, the courts have erred on the side of safety and more loosely interpreted the Fourth Amendment's term "reasonable."[27] For example, in the wake of 9/11, society has tolerated increased security measures. Pat-downs and body scans are now routine in U.S. airports. Since the Boston Marathon bombing in April 2013, municipal governments have added police resources, including bomb-sniffing dogs, police K-9 units, and video surveillance to special events such as races and festivals. Furthermore, increased security protocols and specialized rapid-response training courses have been taught by law enforcement agencies and implemented in school facilities due to elementary, high school, and university shootings.

To add clarity to what is meant by "zones of privacy," the Supreme Court created three categories of privacy: decisional, physical, and informational.[28]

Decisional privacy is defined as the "freedom of the individual to make private choices without undue government interference."[29] An example of a decisional privacy case is *Eisenstadt v. Baird,* 405 U.S. 438 (1972).[30] Here, William Baird was charged with a felony for distributing birth control after lectures on contraceptives at Boston University. When an unmarried 19-year-old student accepted the contraceptive, Baird violated Massachusetts's law regarding crimes against chastity. This law stated that only doctors or pharmacists could dispense birth control, and they could only give the items to married couples. The Supreme Court applied *Griswold,* and in a 6–1 decision held that regardless of marital status, an individual has the right to be free from government interference when deciding whether to have a child. The Court struck down the crimes against chastity law, saying it violated the Fourth Amendment's equal protection clause. The Court went on to say that the law failed to meet the rational basis test and ultimately discriminated against nonmarried individuals by denying them access to contraceptive protection. No case better epitomizes the category of decisional privacy better than *Roe v. Wade,* 412 U.S. 113 (1973). Justice Blackmun issued the opinion for the 7–2 majority and explained that the Court relied on privacy inferred from the Ninth and Fourteenth Amendments to rule that it is a woman's decision whether or not to terminate her pregnancy.[31] This case divided a nation at the time and is still being debated 40 years later. The freedom to make private choices is tantamount to being an American. Because government has been kept at bay during the course of individual decision making, it is unlikely that this type of privacy will be affected by biometric technologies.

The second privacy category is physical. Physical privacy is guaranteed under the Fourth Amendment and is defined as the "freedom from contact with other people or monitoring agents."[32] Governments using biometric technologies must acknowledge that a privacy right is created depending on how and what data is collected and should therefore be cognizant of Fourth Amendment expectations. Likewise, citizens must be mindful that the use of such technology and its resulting data, whether in a criminal or noncriminal context, is not a violation of the Fourth Amendment when it advances a legitimate government interest such as public safety or the more efficient provision of services.[33]

When determining whether a Fourth Amendment violation has occurred, the court will evaluate the nature of the physical intrusion. Actual bodily intrusions can trigger Fourth Amendment protection; however, data can still be gathered in this manner in criminal contexts with a warrant. An example of when a person's Fourth Amendment rights were not violated by biometric technology can be found in the 1986 case of *Christopher Ann Perkey v.*

Department of Motor Vehicles, 42 Cal. 3d. 185. In this case, Plaintiff claimed that the requirement to provide a fingerprint as a condition to receive a California driver's license violated her right against unreasonable search and seizure. She further argued that her privacy rights were violated because the Department of Motor Vehicles (DMV) disseminated fingerprint data to other governmental agencies and private businesses for use not related to vehicular safety. The California Supreme Court concluded that it is unlawful for the DMV to distribute personal information to third parties. However, the Court held that the fingerprint requirement alone did not violate her right to privacy. Specifically, it stated that unlike previously disallowed actions such as forcible stomach pumping or the forced taking of a semen sample, fingerprinting does not require penetration beyond the body's surface and does not oppose the Fourth Amendment's principles of reasonableness.[34]

In *Thomas v. New York Stock Exchange*, 306 F.Supp.1002 (1969), a group of stock exchange employees challenged the constitutionality of a New York statute that mandated that all employees of stock exchange firms be fingerprinted as a condition of employment.[35] They also feared that the fingerprints would be used for future criminal investigations. The district court disagreed and said that the statute did not invade their right of privacy or deny them due process. With regard to Plaintiff's second concern, the court held that the employees did not know the breadth of how the state of New York intended to use the fingerprints in the future. Furthermore, the use of the prints in future investigations still did not constitute a violation of privacy, because the state had proven a rational reason for initially requiring employees to be fingerprinted.[36] The court concluded that "the submission of a fingerprint to an employer was no more an invasion of privacy than the submission of a photograph or signature."[37]

In *Iacobucci v. City of Newport*, 785 F.2d 1354 (6th Cir. 1986), Plaintiffs argued that a city ordinance that mandated employees of liquor establishments to be fingerprinted violated their right of privacy. The California district court did not agree. It ruled that the fingerprint requirement was reasonable, not arbitrary, and bore a rational relationship to a permissible state objective."[38] Specifically, it was noted that fingerprinting would assist law enforcement and prohibit minors and convicted felons from serving alcohol.

Courts have regularly ruled this way when deciding the constitutionality of similar employment mandates. In *People v. Stuller*, 28 L.Ed2d 327 (1971), a bartender's fingerprints were taken and later admitted as evidence in a rape trial. Stuller argued the ordinance requiring his fingerprints to be taken violated his right of privacy. The court disagreed. A federal court also found no invasion of privacy when cabaret employees were mandated to submit to fingerprinting in *Friedman v. Valentine*, 42 N.Y.S.2d 593 (1943). The Friedman logic was followed when another federal court heard arguments about a New York statute requiring the fingerprinting of all employees and members of national security exchanges and all employees of affiliated cleaning companies. The court in *Thom v. N.Y. Stock Exchange*, 306 F.Supp.1002 (1969), rejected the violation of privacy argument and stated that fingerprinting "is only a means of verifying the required information as to the existence or non-existence of a prior criminal record . . . [t]he actual inconvenience is minor; the claimed indignity, nonexistent; detention, there is none; nor unlawful search; nor unlawful seizure."[39]

In 1987 the federal District Court for the Southern District of New York heard a case brought by a utility worker union. In this case, the workers were challenging a federal statute that required them to be fingerprinted to work in a nuclear power plant. The union argued that the fingerprinting mandate violated their Fourth Amendment and right to privacy. Again, the court disagreed and said that the judiciary has consistently upheld fingerprinting for purposes of employment.[40]

In addition to fingerprint cases, the Supreme Court has heard cases that argued that providing a voice exemplar is an unreasonable search. In *United States v. Dionisio*, 410 U.S. 1 (1973), the Court ruled that because a person's voice is constantly exposed in public, it is reasonable to conclude that others will know the sound of his or her voice and that there cannot be an expectation of privacy.[41] It is for this reason—the constant exposure to the public—that facial recognition systems in public places are not a violation of individual privacy.

While some people may still be offended, and feel violated, by submitting to a fingerprint scan, it is unlikely that the courts will find that this action is so objectionable that it violates one's privacy. The rational basis test dictates that a relationship to a legitimate government objective or interest must exist for a fingerprinting requirement to be upheld.[42] Individual privileges and employment types requiring fingerprint scans that meet a rational basis test include, but are not limited to, obtaining a Firearm Owners Identification (FOID) card, holding a liquor license, securing a taxi drivers' license, becoming a police officer, working in a nuclear power facility, being an employees of the Securities and Exchange Commission, and working as a school teacher.

The final category of privacy is informational privacy. Personal information is considered to be any information that could identify an individual. Even data that may not seem to be personal can become personal when used or combined with other data.[43] In this category of privacy, a person seeks to be given the ability to limit the dissemination of information about him or her. Informational privacy has been defined as the right of individuals "to determine for themselves when, how, and to what extent information about them is communicated to others."[44] The concern here is not about the possible intrusive, objectionable collection techniques but rather the ability to control the type of information collected or provided and its use, retention, and disclosure.

A case regarding informational privacy was heard in 2011, when a group of NASA contractors disagreed with information requested during the course of preemployment background checks and argued that the background check questions were too broad and invasive and therefore violated the Constitution. The Supreme Court ultimately decided the case and opined that while the Constitution can be interpreted to protect the privacy of personal information, it did not protect information gleaned from the background checks provided to NASA. The Court held that the checks were reasonable, employment-related requirements and that no violation had occurred.[45]

In 1977, the Supreme Court heard *Whalen v. Roe*, 429 U.S. 589, another informational privacy case, and ruled that a database hosting sensitive medical information did not violate the Constitution. In response to New York's growing drug problem, the state legislature created a commission to evaluate current drug-control laws. The commission's recommendation was that prescriptions for all schedule II drugs be prepared by a physician on state-provided forms that asked for information on the dispensing pharmacy, the drug and prescribed dosage, and name, address, and age of the patient.[46] The forms were then forwarded to the New York State Department of Health, where the information was entered into a database. Any suspicious prescription trending could be analyzed via the data that was entered. In the end, the majority of the Court found that the database holding personal medical information did not constitute an invasion of privacy. The Court reached its decision by focusing on what had been done to secure the database. Specifically, it favored that "1) the forms and the records were kept in a physically secure facility; 2) the computer system was secured by restricting the number of computer terminals that could access the database; 3) employee access to the database was strictly limited; and 4) there were criminal sanctions for unauthorized disclosure."[47]

As can be seen, different government entities employ biometric technology in noncriminal scenarios. The technology is exercised for preemployment requirements, facility access, public safety and security, and streamlined service provision. In sum, the litany of court decisions have held that when a federal, state, or municipality requires fingerprinting and background checks for employment, or they require personal information that otherwise advances government purposes, an individual has negligible constitutional protection of privacy.[48]

The Individual and Information Safeguards

With the growing use of computers and information databases, Congress passed the Privacy Act in 1974. This act regulates the collection, storage, use and dissemination of personal information.[49] Like many laws, this act balances the government's need to gather the data against the rights of those individuals providing the information. In short, the Privacy Act outlines four practices designed to safeguard individuals against unwarranted invasions of privacy by the federal government: (1) to restrict disclosures of personally identifiable records maintained by agencies; (2) to grant individuals increased rights of access to agency records maintained on themselves; (3) to grant individuals the right to see amendment of agency records maintained on themselves upon a showing that the records are not accurate, relevant, timely, or complete; and (4) to establish a code of fair information practices that requires agencies to comply with statutory norms for collection, maintenance, and dissemination of records.[50]

The requirement to establish a code of fair information practices was born of a U.S. Department of Health, Education, and Welfare (HEW) report. In 1972, the Department of HEW created an advisory committee to make recommendations on protecting citizens and applying safeguards to electronic record keeping of personal information. In July 1973, the committee released its report, "Records, Computers and the Rights of Citizens—Report of the Secretary's Advisory Committee on Automated Personal Data Systems." The report establishes five principles upon which the Code of Fair Information Practices (CFIP) is based: (1) there shall be no personal data record-keeping systems whose very existence is secret. An organization must be transparent with its intent to collect biometric information; (2) there must be a way for a person to access the information that has been collected about him or her and to learn how the information is being used. Similarly, the data collector is expected to openly communicate how the information will be protected; (3) the individual must consent to any information being disclosed to a third party. This consent may not apply when information is used for academic research, national security, and law enforcement purposes; (4) the person must be given a mechanism to correct or make edits to information that has been collected; (5) any organization creating, maintaining, using, or disseminating records containing personal data must keep it safe, guarantee the reliability and integrity of the data for its intended use, and take precautions to prevent misuses of the data.[51] The CFIP is thorough, and its intent is to make people more comfortable about submitting personal information to the federal government. However, the CFIP does not apply to non-U.S. citizens, and it does not regulate state and local governments or private-sector organizations' use of personal information. Until there is a mandate to provide a unified approach to securing personal information, each state and private business can, and does, exercise different standards.

For example, the Biometric Information Privacy Act ("Act") in Illinois does address the use of biometrics in private business.[52] Enacted in 2008, this law regulates private entities' collection, use, storage, retention, and destruction of biometric information. The Act does not distinguish between the type of intended use and applies the same regulations whether biometrics are gathered for employment, security, facility access, financial information, or something else. Similar to CFIP, the Illinois law is made up of five components.[53] First, the organization must create a written policy that explains how the biometric information will be used. This policy must address retention and destruction schedules and be made available to the public. The second component addresses parameters around data collection. The business must provide written notification that information is being collected; for what purpose it is being collected; how long it will be collected, kept, and accessed; and receive written authorization from the subject to collect and use the information. The third component prohibits the data from being sold, leased, or traded. The fourth component regulates disclosure. Organizations cannot disclose or disseminate information unless they have the subject's consent, the information is required by law, or the disclosure is mandated pursuant to a valid warrant or subpoena. The final component requires the entity to store biometric data in the same or better manner that it stores other sensitive information. Individuals whose data is compromised are provided relief from the Act and can file suit for damages, injunctive relief, and attorneys' fees.

Other states, including Alaska, Florida, Louisiana, and South Carolina, have granted constitutional and statutory privacy rights.[54] In California, the Database Security Breach Notification Act requires that customers of private- or public-sector services be notified of intrusions that may result in the loss or release of personal information. Recognizing the explosion of e-services and an ever-increasing mobile society, many states have enacted statutes governing electronic transactions. The Government Paperwork Elimination Act of 1998 focuses on improving service delivery by developing procedures to protect electronic signatures and use and accept electronic documents. Undoubtedly, physical paper will not be extinct for some time; however, it is safe to assume that even paper transactions are ultimately stored electronically.

The Court has been consistent in finding that informational privacy is not jeopardized when safeguards are in place to protect valuable personal data. So, according to *Whalen*, *Perkey*, *Nelson*, and other cases, as long as the information gathered has passed the rational basis test, it advances government interests, and the agency has procedures in place to restrict access to and prevent unauthorized dissemination of the information, there has been no violation of privacy.

The Individual and the Private Sector

Just as government agencies can argue that the collection and use of biometric data is necessary to further their interests, so can private businesses. However, individuals have fewer defenses here than in the public sector. They cannot rely on statutes from the state or the Constitution for protection of their privacy rights. Indeed, when it comes to the private sector's use of biometric information, the Constitution is basically hands-off and does not provide for relief when individuals voluntarily give information to private parties. Researchers point to *Smith v. United States*, 442 U.S. 735 (1979) as an example.[55] In this case, Patricia McDonough was robbed. After the robbery, she began receiving obscene and threatening phone calls from a man claiming to be the robber, Michael Lee Smith. Smith argued that the numbers he dialed from his home telephone number could not be given to the police without a warrant. The Court disagreed and opined that Smith had voluntarily provided the numbers to the telephone company via key strokes and, as such, had no reasonable expectation of privacy. It would have been reasonable for Smith to expect to keep the conversation private but not the number he dialed. Furthermore, the Court stated that pen registers cannot prove that a conversation was even held, just that the number was dialed.

Another case, *United States v. Miller*, 425 U.S. 435 (1976), delivered a similar message.[56] Mitch Miller had failed to pay liquor tax on distilling equipment he owned. The Bureau of Alcohol, Tobacco, and Firearms (ATF) issued subpoenas for Miller's bank accounts. The bank dutifully complied and provided the requested information. Miller claimed that the bank records were illegally seized and were obtained in violation of the Fourth Amendment. The Court of Appeals in the Fifth Circuit agreed with Miller.[57] Justice Powell and the majority of the Supreme Court reversed the Fifth Circuit decision and ruled that the bank records that were provided to the ATF were not Miller's private papers. The documents belonged to the bank as part of its business, and Miller had voluntarily given the papers to the bank.

Both *Miller* and *Smith* suggest that individuals should not, and cannot, rely on the Constitution to protect the privacy of the information they voluntarily provide as a matter of doing business. While neither of these cases deal with biometrics, it is logical to deduce from the Court's reasoning that if a fingerprint or other form of personal data is given by an individual during the course of a private-sector transaction, there should be no reasonable expectation of privacy.[58]

Individuals can still find relief outside of the judiciary. After the *Miller* case was decided, Congress passed the Right to Financial Privacy Act (RFPA).[59] This act created procedures that banking institutions must follow when disclosing information to government agencies. Individuals can also look to unions, management, agreements, and laws that outline appropriate actions regarding personal information. One example is the 1986 Electronic Communications Privacy Act, which creates both civil and criminal penalties for the unauthorized interception and disclosure of wire, oral, or electronic communications.[60] Electronic communications is broadly defined as any transfer of signs, signals, writing, images, sounds, data, or intelligence of any nature transmitted in whole or in part by a wire, radio, electromagnetic, photoelectric, or photo-optical system that affects interstate or foreign commerce.[61] The Fair Credit Reporting Act is another statute that individuals can rely on to guarantee their information is protected. This act prevents credit agencies from disclosing personal data such as financial status and medical information to third parties.[62] Clearly, the private sector is given more latitude than government agencies, and courts have consistently found that business actions fall outside of constitutional safeguards. Generally, as a matter of law, a private party in possession of information and related biometric data has the right to disclose it. Accordingly, the private sector enjoys great leeway as far as what it can do with an individual's information. However, organizations can still be held accountable. When businesses do run afoul of morality and what citizens expect, Congress may step in and begin regulating collection and use. While the individual is afforded some protection by these acts, one must be mindful that in the private sector, in the majority of instances, nothing can be done about the use of information that this person has given or that has been collected about him or her.

The Individual and the Data

In addition to the well-documented issues surrounding violation of privacy, there are a myriad of concerns with the collection, use, retention, and disclosure of data in employment situations and day-to-day business scenarios. One emerging concern has been termed "function creep."[63] Function creep occurs when the information that has been collected for a specific reason is used for another. When data is used for unintended or unauthorized purposes, it erodes the trust in a methodology that is supposed to enhance trust and security. The Social Security number (SSN) is often cited as an example of function creep. The original Social Security cards were not intended to be used as identification. However, over the years, the Internal Revenue Service began using SSNs to verify the identity of taxpayers, and now the Social Security card and number is the foremost identity document of most U.S. citizens.[64]

Function creep is essentially created when too much data is gathered. It involves three components but can occur when any one of the three is present. The components are: (1) a policy vacuum, (2) an unmet demand for a specific job, and (3) a slippery slope or covert application.[65] The risk of function creep is greatest when there is a policy vacuum. Every day, public- and private-sector organizations implement new technological solutions. However, the use of the technology can go horribly awry in the absence of specific policies that identify and reinforce the intended purpose of the technology. It is critical that the policy makers are knowledgeable of the specific innovation to craft the most applicable guidelines. The second component occurs when information is gathered for one purpose, such as preemployment screening or employee authentication, but because a need somewhere else is not being met, the information is used to fill the void. The function creep of the use of the Social Security Number occurred because the federal government did not have an acceptable method to identify taxpayers. So, instead of developing a new identifier specifically for the Internal Revenue Service, information already on file was tapped, and the use of the SSN was expanded. Finally, there is the slippery slope, or covert application. Here, function creep is the result of minor, imperceptible changes to the use of the technology over time. Quite often, the technology is only slightly misused; however, the incremental effect of those indiscretions in the aggregate can drastically alter the intended purpose. Warrantless cell phone tracking by law enforcement agencies has been cited as an example of this component of function creep.[66] The potential misuse of the superfluous data was one of the drivers for the creation of the Code of Fair Information Practices. The excess data that results from extraction lends itself to the problem of function creep, and it is one of the reasons why only the data necessary for the stated purpose should be mined.

A second emerging issue is "informatization of the body."[67] Biometrics is touted as a solution to personal and national security. Because the information gathered is unique to an individual, the risk of identity confusion is low. The information that is gleaned from the body can be, and is, used for good. For example, otherwise unrecognizable persons can be incontrovertibly identified via biometric data. Also, biometric information can eliminate job applicants with criminal records. But using such information to identify people and get a glimpse into their lives can be seen as a negative use of this technology.[68]

In 2012 the Federal Trade Commission reported more than 350,000 known cases of identity theft.[69] Over a billion dollars was paid out in 2012 for identity theft and other types of fraudulent behavior.[70] This staggering statistic is leading more and more companies to change how they store their customers' information and how the customers can access their information. Employers are also increasing security via double authentication procedures for data and facility access. The fear that information will be accessed creates the desire to lock down all information. This can result in unnecessary data collection. In certain instances, the implementation of biometric technology is appropriate and warranted; however, as organizations nationwide increase security, there is a risk that biometrics will be deployed in situations where there is little to no benefit to strong user authentication or identification, and thus, there is no need to gather data.[71]

Unnecessary collection leads to the next concern, unauthorized use of biometric technology. This can be considered the greatest risk to an individual's right, or expectation, of privacy. As discussed above, people are not wary of the stated purpose of the collection; rather, it is the undisclosed and unpredicted uses that are the foundation of this concern. However, the courts have ruled that a violation of privacy has not occurred if the data being collected legitimately advances a government interest and if safeguards are put in place to protect the data from unauthorized access or dissemination. Researchers have classified unnecessary collection into two categories, forensic usage and usage as a unique identifier.[72]

Fingerprints are the primary method of forensic identification. Submitting to a fingerprint scan often has a negative connotation because as a common, long-standing part of the criminal booking process, some individuals may associate being fingerprinted with illegal activity. According to Woodward, that number is not compelling. In a survey of 1,000 adults, 75 percent of those polled said they are comfortable having their fingerprints made available to the government or the private sector for identification purposes.[73] The survey indicated that more than half had been fingerprinted at some point in their lives, and only 20 percent thought that fingerprinting stigmatizes a person as a criminal.[74] People do fear that once their fingerprints have been obtained for employment or public-benefits administration, they will be accessed by law enforcement agencies to facilitate investigations.

Individuals are also afraid that biometric data will be used to track their activities, purchases, and general whereabouts. Biometric technology can provide a stronger sense of security than traditional measures, because data is unique to the individual. This unique-identifier characteristic causes concern for some; because it is unique, they fear that it will provide a common link to information stored in disparate databases. From those links, the government, or private-sector organizations, will analyze the web of data that is created and develop a subject profile.[75]

Unauthorized collection is also a concern, however de minimis in nature. The reason that individuals are not as threatened by this is because, currently, there are few technologies that allow the collection of data without the subject's knowledge.[76] Furthermore, those technologies are most often placed in public locations, where the court has ruled an individual should have minimal expectation of a right to privacy. However, increased deployment of certain types of biometric technologies in areas that are not public—for example, work environments—does bring a risk that information is being gathered, and most likely used for some purpose, without consent.

Finally, the unauthorized disclosure of personal information is a prevalent concern with the use of biometric technology. Individuals provide information to a person or organization for a specific purpose and have a reasonable expectation that the information will be kept private. With unauthorized disclosure, a person fears that his or her information will be given to a third party whose intention may not be in the best interest of that person. Again, the courts have found that if the government or private agency can prove a rational basis for the data collection and has implemented procedural safeguards for access and dissemination, there is no violation of privacy.

One such procedural safeguard is the privacy assessment. A comprehensive privacy assessment will determine if a biometric system, for employment or other business, is using information appropriately and in accordance with the intended purpose. This assessment should be done as early as possible in the development and conducted routinely during the life of the system.[77] Questions that should be asked during the assessment include the following: Is the data intended to identify the individual or not; is there a rational basis between the intended purpose of the data gathering and the purpose and authority of the organization; does the actual and intended use match; does the technology have the ability to track people without their knowledge or mine data from other systems; and what policies and procedures are in place to audit the system and its use?[78]

THE BENEFITS OF BIOMETRICS

The discussion thus far may be interpreted as a commentary against the use of biometrics. The concerns associated with the collection, use, and potential use of physical or behavioral information and characteristics have been well documented. The guarantee that the Fourth Amendment protects the data gathered by public- and private-sector organizations is tenuous at best. The requirement to submit to fingerprint scans for employment, just like the implementation of facial recognition technology in public places, may be considered an invasion of privacy for some. The courts have ruled that this is not the case. Others may believe that information provided to private-sector businesses is protected from third-party infiltration. Again, courts have found that when individuals voluntarily provide information, they should have no expectation that the information will be kept private. The greatest risk to privacy does not lie with the type of technology, or technologies, implemented, but rather with the poor management of the information that is stored.[79] The risk increases as organizations reduce trained personnel, cut technology budgets, and fail to draft and implement appropriate policies to keep information safe.[80]

People can look to legislation for assistance and resolution, but currently there is minimal government regulation. The Code of Fair Information Principles (CFIP) discussed previously is a step in the right direction for relief; however, the principles do not regulate state and local governments or private-sector organizations. Also, this code does not apply to information provided by non-U.S. citizens. Some believe that even though the myriad of risks (real and perceived) are well known, this limited government regulation and restrained involvement can be interpreted as support for the use of biometric technology.[81]

Given these issues, it is no surprise that people feel exposed by and vulnerable to the use of biometric technology. The goal of biometrics, to provide privacy and security, is constantly juxtaposed to the perception of biometrics. This section will discuss some of the benefits of biometric technology in employment situations and day-to-day business applications.

The need to enhance security measures makes biometrics an attractive, often relatively inexpensive solution to businesses and governments charged with protecting vital information and securing their workforce. Because these characteristics cannot be falsified, there is a high degree of authentication accuracy. Identification cards can be lost or stolen. PINs, passwords, pass phrases, and user IDs can often be easily guessed or forgotten. In contrast, biometric characteristics cannot be shared, lost, or easily stolen. Depending on the modality selected, real-time identification and verification is possible. One benefit that straddles the cost-benefit line is data mining. In other words, mining information that has been entered into the database to build a personal profile is appropriate, depending on the organization and its authority. Having the ability to search multiple systems to find an identity is not an invasion of privacy if such action passes the rational basis examination. When it fails, and the system becomes vulnerable to intrusion, data mining is seen as a negative characteristic of biometric technology. Transparency can also be viewed as a benefit of the use of biometrics. By increasing the probability that a person is who he or she claims to be, crimes of fraud and identity theft can be reduced.

Biometric modalities such as fingerprint scans, signature recognition, facial recognition, and voice recognition are less expensive and considered minimally invasive compared to iris recognition and retinal scanning. According to some, the extensive list of biometrics that can be used is one of the greatest benefits of the technology. Because there are so many options from which to choose, customers can implement one or more for their particular situation. They do not need to force an authorization or identification method where it is not appropriate. The availability of a growing catalogue of techniques for employment and personal use is referred to as biometric balkanization.[82]

For those who are skeptical and fearful of an invasion of privacy, Woodward discusses two advantages balkanization offers for protection.[83] First, biometric balkanization increases the selection of technologies from which to choose. This selection allows an organization the opportunity to select the optimal biometric for a specific mission. For example, a retinal scan may be appropriate for a federal agency or nuclear facility requiring high security and limited access. This would not necessarily be appropriate for a personal bank transaction or employee time-keeping control. Similarly, biometric balkanization supports what Woodward calls compartmentation.[84] Because different agencies can require and employ different biometrics, this compartmentation is comparable to having multiple PINs and passwords for different operating systems. Compartmentation leads to Woodward's second benefit, the synergy of agency interests and individual concern.[85]

This chapter has stressed that government and private organizations want to protect data and secure their facilities via reliable identification technology. So do consumers and employees, with the added benefit of guaranteed protection of privacy information. The ability to select the appropriate biometric retards the notion of an all-encompassing database that can be accessed by all who subscribe to it. Whether used independently or in combination with other forms, this technology is becoming more popular in the workplace as the opportunity to reduce labor costs is being realized by incorporating a biometric time-keeping system. This eliminates the opportunity for coworkers to punch in and out for each other. Worker productivity can also increase by requiring a scan at the start and end of breaks. Similarly, overtime costs can be cut by requiring the biometric approval of a supervisor. Workplace theft can be curtailed as fingerprint readers are installed on cash registers. Building security is also heightened, as there is no risk for lost, shared, or stolen access cards. IT departments are seeing the benefit of eliminating multiple passwords for logging in to different operating systems. Also, there is a risk that passwords can be accessed when they are stored in the computer system they are supposed to protect.

For example, an organization does not have to be the specific target of a breach to be vulnerable to even a casual hacker. In 2012 a midwestern municipality fell victim to one such individual.[86] Once the protective firewalls were penetrated, numerous passwords that granted access to some of the most secure areas of the city's network were retrieved. Although the intrusion was identified and halted before any personal utility billing and employee information could be stolen, to date, the city has spent more than $600,000 to restore and secure the computer network and related infrastructure.[87] The investigation of the intrusion event revealed several architectural shortcomings of the city's network. Two major weaknesses that were identified in the event debrief were the extensive use of passwords and the storage of those passwords on the network. As the city rebuilds its network, the implementation of biometric technology is being investigated.

BIOMETRICS AS THE FINAL SOLUTION

Unquestionably, biometrics has its drawbacks, limitations, and constitutional criticism. It is also highly regarded as the technological answer to many security questions. But is it the ultimate solution? Those on both sides of the argument have fueled the debate. Walker says biometrics is not the final answer and that it should only be a part of an organization's overall risk-management plan.[88] Complementary technologies, processes, policies, and resources must accompany any biometric implementation. Because these additional factors can be overlooked, several misconceptions have occurred and will be discussed next.

First, even with all of its benefits, biometric authentication is not the strongest application available. This is because security best practices suggest that two, preferably three, factors be part of a strong authentication protocol.[89] The combination of something you know (ID, PIN, password), something you have (USB fobs and one-time password tokens), and something you are (biometric characteristics) is recommended in order to have a very secure authentication process.[90]

Second, biometric authentication is reliable; however, that reliability is not foolproof. The chosen system must be configured properly, and data must be entered accurately. Error types such as false rejection and false acceptance can occur if the system is not set up correctly.

The information produced from the system is only as reliable as the information put into the system.

Next, biometrics has been touted as being virtually impenetrable. Unfortunately, "virtually" does not mean "absolutely." Without question, it is secure. It is more secure than a single-factor authentication method such as a password or PIN. But the biometric template that is created upon enrollment is simply algorithms and text sent from one system to another for comparison.[91] This information can be captured while being transferred, and then it can be replayed to forge an authentication. Organizations must install additional safeguards to halt such "network-sniffing" efforts.[92] Biometrics can also be vulnerable outside of the computer network. Fingerprint forgeries have occurred by lifting a print with tape and applying it to a scanner, and photographs have been used to trick iris- and facial-recognition systems.[93] Undoubtedly, as technologies to secure information become more sophisticated, so will the efforts to thwart that security.

Finally, implementation of biometric technology is not the whole solution. As stated previously in the chapter, several factors, including additional back-end technology, policies, and other risk-mitigation procedures, must work in concert with biometrics. Simply installing a fingerprint reader or other system will not guarantee a more secure environment. Biometrics is part of a larger package that must be evaluated for flexibility, appropriateness, and functionality.

CONCLUSION

Both public- and private-sector organizations must address several issues before biometrics will be fully embraced in today's culture. While society has tolerated additional security measures that involve biometrics in the wake of September 11, 2001, and as a result of increased identity theft, the implementation of certain modalities in organizations has raised privacy concerns. The courts have consistently ruled that if the collection and use surpass the rational basis test and advance the business purpose, there is no invasion of privacy. The requirement to submit to biometric identification in the workplace has been evaluated against these criteria and met with similar attitudes by the courts. They have also held that privacy is not violated if an organization has developed policies to regulate the collection, accuracy, access, use, and retention of the data. Criticisms of biometric technology will go on; however, those criticisms must strike a balance with the reality that knowing an individual's identity is often necessary. Until the balance is realized, the chasm will continue to grow between those who argue biometrics is advantageous to conventional forms of identity and security and those who hold that it is nothing more than an invasion of privacy and a camouflaged weapon of a tyrannical government.

NOTES

1. Anil Jain and Arun Ross, *Introduction to Biometrics* (New York: Springer Science+Business Media, LLC, 2011).
2. American Immigration Council, "A Guide to S.744." Accessed June 1, 2013, http://www.immigrationpolicy.org/special-reports/guide-s744-understanding-2013-senate-immigration-bill.
3. Steve Walker, "Biometric Selection: Body Parts Online," Paper submitted to SANS Institute InfoSec Reading Room, 2002, accessed June 20, 2013, http://www.sans.org/reading_room/whitepapers.
4. Ibid.
5. Ibid.
6. Ibid.
7. Ibid.
8. Judith A. Markowitz, "Voice Biometrics," *Communications of the* ACM 43, no. 9 (2000): 66–73.
9. National Science and Technology Council (NSTC), et al. *Privacy and Biometrics: Building a Conceptual Foundation.* 2006, accessed June 26, 2013, http://www.biometrics.gov.
10. Edmund Spinella, "Biometric Scanning Technologies: Finger, Facial and Retinal Scanning," Paper submitted to SANS Institute InfoSec Reading Room, 2003, accessed June 23, 2013, http://www.sans.org/reading_room/whitepapers.
11. Samir Nanvati, *Biometrics: Identity Verification in a Networked World* (New York: Wiley and Sons, Inc., 2002).
12. Lisa Myers, "An Exploration of Voice Biometrics," Paper submitted to SANS Institute InfoSec Reading Room, 2002, accessed June 20, 2013, http://www.sans.org/reading_room/whitepapers.
13. John D. Woodward, Jr., "The Law and the Use of Biometrics," in *Handbook of Biometrics*, ed. Anil K. Jain, Patrick Flynn, and Arun A. Ross (New York: Springer Science+Business Media, LLC, 2008), 357–379.
14. Samuel D. Warren and Louis D. Brandeis, "The Right to Privacy," *Harvard Law Review* 4 (1890): 193–220.
15. Robert Ellis Smith, *Ben Franklin's Web Site: Privacy and Curiosity from Plymouth Rock to the Internet* (Providence, RI: Sheridan Books, 2000), 6.
16. John D. Woodward, Jr., "The Law and the Use of Biometrics," 357–379.
17. Kenneth P. Nugar and James L. Wayman, "Biometrics and the Constitution," in *Biometric Systems: Technology, Design, and Performance Evaluation*, ed. James Wayman, Anil K. Jain, Davide Maltoni, and Dario Maio (New York: Springer Science+Business Media, LLC, 2005), 322.
18. Ibid., 311–333.
19. Ellen Alderman and Caroline Kennedy, *The Right to Privacy* (New York: First Vintage Books Edition, 1997).
20. *Olmstead v. United States*, 277 U.S. 438 (1928).
21. Ibid.

22. *Katz v. United States*, 389 U.S. 347 (1967).
23. *Griswold v. Connecticut*, 381 U.S. 479 (1965).
24. *Osborn v. United States*, 385 U.S. 323 (1966).
25. John D. Woodward, Jr., "The Law and the Use of Biometrics," 357–379.
26. Kenneth P. Nugar and James L. Wayman, "Biometrics and the Constitution," 311–333.
27. Ibid.
28. John D. Woodward, Jr., "The Law and the Use of Biometrics," 357–379.
29. Ibid.
30. *Eisenstadt v. Baird*, 405 U.S. 438 (1972).
31. *Roe v. Wade*, 410 U.S. 113 (1973).
32. John D. Woodward, Jr., "The Law and the Use of Biometrics," 361.
33. Kenneth P. Nugar, "The Special Needs Rationale: Creating a Chasm in Fourth Amendment Analysis," *Santa Clara Law Review* 32, no. 1 (1992).
34. Christopher Ann Perkey v. Department of Motor Vehicles, 42 Cal.3d.185 721 (F.2d.50; 228 Cal.Rptr.169 1986).
35. *Thomas v. New York Stock Exchange*, 306 F.Supp.1002 (S.D.N.Y. 1969).
36. Ibid.
37. Ibid., p. 1009.
38. *Iacobucci v. City of Newport*, 785 F.2d 1354 (6th Cir. 1986).
39. *Thom v. N.Y. Stock Exchange*, 306 F.Supp.1002 (1969) p. 1009.
40. *Utility Workers Union of America v. Nuclear Regulatory Commission*, 664 F.Supp. 136 (S.D.N.Y. 1987).
41. *United States v. Dionisio*, 410 U.S. 1 (1973).
42. John D. Woodward, Jr., "The Law and the Use of Biometrics," 357–379.
43. National Science and Technology Council (NSTC), Committee on Technology, Committee on Homeland and National Security, and Subcommittee on Biometrics (2006), *Privacy and Biometrics: Building a Conceptual Foundation*, Accessed May 17, 2013, www.biometrics.gov.
44. Allan Westin, *Privacy and Freedom* (New York: The Bodley Head, Ltd. 1967).
45. *National Aeronautics and Space Administration, et al. v. Nelson, et al.* 562 U.S. ____ (Docket Number 09-530) (2011).
46. John D. Woodward, Jr., "The Law and the Use of Biometrics," 357–379.
47. *Whalen v. Roe*, 429 U.S. 589 (1977).
48. John D. Woodward, Jr., "The Law and the Use of Biometrics," 357–379.
49. 5 U.S.C. Section 552(a).
50. The FBI Federal Bureau of Investigation. "FBI Records: Freedom of Information/Privacy Act," Accessed June 15, 2013. http://www.fbi.gov/foia/overview-privacy.
51. Willis Ware, "Records, Computers and the Rights of Citizens—Report of the Secretary's Advisory Committee on Automated Personal Data Systems" (1973) Accessed June 3, 2013, http://epic.org/privacy/hew1973report/foreword.htm#TL; John D. Woodward, Jr., "The Law and the Use of Biometrics," 357–379.
52. Biometric Information Privacy Act 740 ILCS 14/1.

53. Ibid.
54. National Science and Technology Council (NSTC), Committee on Technology, Committee on Homeland and National Security, and Subcommittee on Biometrics (2006) *Privacy and Biometrics: Building a Conceptual Foundation*, Accessed May 17, 2013, www.biometrics.gov.
55. *Smith v. United States*, 442 U.S. 735 (1979).
56. *United States v. Miller*, 425 U.S. 435 (1976).
57. *United States v. Miller*, 500 F.2d 751 (5th Cir. 1974).
58. John D. Woodward, Jr., "Comments Focusing on Private Sector Use of Biometrics and the Need for Limited Government Action" (presented for the National Telecommunications and Information Administration, U.S. Department of Commerce meeting on Elements of Effective Self Regulation for the Protection of Privacy and Questions Related to Online Privacy, July 1998).
59. Right to Financial Privacy Act (RFPA) 12 U.S.C. 3401.
60. 18 U.S.C. Sects 2510–2520.
61. Ibid., (1986) p.2512.
62. 15 U.S.C. Sects 1681–1681t (1976).
63. Emilio Mordini, "Ethics and Policy of Biometrics," in *Handbook of Remote Biometrics for Surveillance and Security*, ed. Massimo Tistarelli, Stan Z. Li, Rama Chellappa (London: Springer-Verlag, 2009), 293–309.
64. Ibid.
65. Ibid.
66. Ibid.
67. Ibid.
68. Ibid.
69. Federal Trade Commission, *Consumer Sentinel Data Book for January–December 2012* (Released February 2013).
70. Ibid.
71. Michael Theime, "Privacy Concerns and Biometric Technologies" (2008) Accessed June 29, 2013, http://www.bioprivacy.org/privacy_fears.htm.
72. Ibid.
73. John D. Woodward Jr., "Comments Focusing on Private Sector Use of Biometrics and the Need for Limited Government Action."
74. Ibid.
75. Michael Theime, "Privacy Concerns and Biometric Technologies" (2008) Accessed June 29, 2013, http://www.bioprivacy.org/privacy_fears.htm.
76. Ibid.
77. National Science and Technology Council (NSTC), Committee on Technology, Committee on Homeland and National Security, and Subcommittee on Biometrics (2006), *Privacy and Biometrics: Building a Conceptual Foundation*, Accessed May 17, 2013, www.biometrics.gov.
78. Ibid.
79. John D. Woodward, Jr., "Comments Focusing on Private Sector Use of Biometrics and the Need for Limited Government Action."

80. Ibid.
81. Ibid.
82. John D. Woodward, Jr., "The Law and the Use of Biometrics," 357–379.
83. Ibid.
84. Ibid.
85. Steve Walker CISSP, ABCP, "Biometric Selection: Body Parts Online," SANS Institute 2002.
86. "Hacking of Naperville website still (likely) under investigation," *Naperville Sun*, April 7, 2013, Accessed June 26, 2013, http://napervillesun.suntimes.com/news/19317118-418/hacking-of-napervilles-website-still-likely-under-investigation.html.
87. Ibid.
88. Steve Walker, CISSP, ABCP, "Biometric Selection: Body Parts Online," SANS Institute 2002.
89. Ibid.
90. Simon Liu and Mark Silverman, "A Practical Guide to Biometric Security Technology," Accessed May 29, 2013, www.computer.org/itpro/homepage/Jan_Feb/security3.htm.
91. Steve Walker, CISSP, ABCP, "Biometric Selection: Body Parts Online," SANS Institute 2002.
92. Ibid.
93. Ibid.

5

Employee Expectations of Privacy in the Workplace: Drug Tests, Work Spaces, Computers, and Social Media

R. Craig Curtis

The most important fact in determining the privacy rights of an employee in his or her workplace is the type of employer. In the private sector, the employee's privacy rights are determined by the employment contract and any applicable state law. This means that any private-sector employer may drug test or search the work space of an employee for whatever reason unless prohibited by state or local law or the terms of any personnel manual, employment contract, or collective bargaining agreement. In the public sector, by contrast, the principle of state action applies. Searches must be justified according to the principles of the Fourth Amendment, as they govern administrative searches. Thus, drug testing and searches of premises and electronic devices must be justified under analysis of the jurisprudence of the Supreme Court. This chapter is concerned primarily with the detailed analysis of the extent to which that jurisprudence allows searches of public-sector employees and their work spaces.

The author would like to acknowledge the assistance of his two research assistants during the course of the drafting of this chapter—Scott Canedy and Mollie Sheerin. Their work was stellar and made this task both easier and more fun.

INTRODUCTION: BASICS OF THE FOURTH AMENDMENT AND THE CONCEPT OF STATE ACTION

The actual language of the Fourth Amendment is instructive, because many citizens do not know what it actually says or means. The language is as follows:

> The right of the people to be secure in their persons, houses, papers, and effects, against unreasonable searches and seizures, shall not be violated, and no warrants shall issue, but upon probable cause, supported by oath or affirmation, and particularly describing the place to be searched, and the persons or things to be seized.[1]

This does not mean that the government may not search a person's private spaces. It does mean that any such searches must be reasonable. When it applies to homes, where the Fourth Amendment protections are strongest,[2] there almost always must be a search warrant.[3] A search warrant can only be issued by a "neutral magistrate" and only after evidence has been presented to that court and the judge has found that the warrant is justified in the law.[4] In criminal matters, this means that the court must find that it is more probable than not that a crime has been committed and that evidence of that crime can be found in the locations described by the person giving testimony to the court.[5]

Warrantless searches of places, motor vehicles, and work spaces have been allowed under certain circumstances. There has long been a vehicle-search exception to the warrant requirement for motor vehicles, but probable cause is required.[6] Whenever a person is arrested, a search incident to that arrest is allowed for purposes of officer safety and to collect evidence of the crime for which the arrest was made.[7] In some cases, some types of evidence are such that the Supreme Court has held that there is no reasonable expectation of privacy.[8] In other cases, the Court had held that no warrant is required when the public official's behavior in question does not constitute a search within the meaning of the Fourth Amendment.[9] The Court has held that warrantless searches of public-sector workplaces, based on less than probable cause, are allowed under a balancing test that is essential to understanding this chapter.

Most people think of the Fourth Amendment as applying only to criminal matters, but it also applies to administrative searches conducted pursuant to the implementation of civil or regulatory laws.[10] Administrative searches are conducted by a public official other than a law enforcement officer and for purposes other than the discovery of evidence of crime. An example of an administrative search would be an inspection of a restaurant by a city or county health department. Additionally, when a motor vehicle is impounded by a public agency, an administrative search of the contents of the vehicle, both to make sure that there are no hazardous materials and to create an inventory of the contents for the purpose of protecting the property of the owner of the vehicle, is allowed.[11] Such searches have long been allowed in many circumstances under a less-than-probable-cause standard, but they may not be used as a pretext for a criminal investigation.[12] Regulatory agencies charged with enforcing administrative rules, such as county health offices or federal agencies, such as the Occupational Safety and Health Administration, are allowed to conduct searches of regulated work sites.[13] Whether these searches can be conducted without notice to the business being searched or whether judicial supervision is required depends on the nature of the business, the potential sanctions, the nature of the grant of authority to the agency conducting the search, and the potential risk of harm to the public.[14]

Public-sector employers are allowed to search the official work records—including computer files—of their employees when work-related concerns require that someone other than the employee have access to these records. These searches are administrative searches and must be analyzed according to the standards applicable to such searches. In her plurality opinion in *O'Connor v. Ortega*, Justice O'Connor stated, "There is surprisingly little case law on the appropriate Fourth Amendment standard of reasonableness for a public employer's work-related search of its employee's offices, desks, or file cabinets. Generally, however, the lower courts have held that any "work-related" search by an employer satisfies the Fourth Amendment reasonableness requirement."[15] For example, if a case worker in a child protection agency was on vacation and an emergency arose concerning one of the children in that employee's case load, the employer would certainly be allowed to access the work records of the employee so that the child could be protected. Disputes most often arise when the employee faces some adverse job action, or even criminal sanction, in which the evidence seized during an administrative search is used to justify the adverse action. In the seminal case of *O'Connor v. Ortega*,[16] the employee was being involuntarily separated from employment and was contesting that outcome.

THE CONCEPT OF STATE ACTION

It is hard for beginning students of constitutional law to understand the limits of the constitution itself. It does not apply to interactions between private citizens. It only applies to the actions of governments. This principle is known as "state action." The most famous application of this doctrine is found in the *Civil Rights Cases*: "In this connection, it is proper to state that civil rights, such as are guaranteed by the Constitution against State aggression, cannot be impaired by the wrongful acts of individuals, unsupported by State authority in the shape of laws, customs, or judicial or executive proceedings. The wrongful act of an individual, unsupported by any such authority, is simply a private wrong . . . ."[17] The *Civil Rights Cases* struck down a federal law that purported to mandate that the keepers of inns, theaters, and other public businesses allow everyone, regardless of race, to use their facilities. The opinion in that case made it clear that states could mandate such behavior but that there was no authority in the Constitution for the federal government to control private conduct in this way.

This doctrine was reinforced in the mid-20th century in several cases, including *Shelley v Kraemer*[18] and *Cooper v. Aaron*.[19] In *Shelley v. Kraemer*, the Court stated that a provision in a deed that did not allow for sale of the property to a person of color, often called a "restrictive covenant," was not, in and of itself, a violation of the Fourteenth Amendment. But it would be a violation of that provision of the Constitution for a state court to enforce a restrictive covenant. This is because two private individuals can enter into a contract without any constitutional implications, but when a state actor, such as a court, takes action based on race, that is state action and does have constitutional implications. In *Cooper v. Aaron*, the Court, in requiring that the schools in Little Rock, Arkansas, be desegregated, had to dismiss the argument that the inability to enforce an order for school desegregation was due to the actions of private citizens. The Court, in holding that the actions of these citizens were a product of actions taken by state and local governments, stated, "The significance of these findings, however, is to be considered in light of the fact, indisputably revealed by the record before us, that the conditions they depict are directly traceable to the actions of legislators and executive officials of the State of Arkansas, taken in their official capacities, which reflect their own determination to resist this Court's decision in the *Brown* case and which have brought about violent resistance to that decision in Arkansas."[20]

The result of the application of this doctrine is that a private-sector employer may freely discriminate on the basis of race without running afoul of the equal protection clause of the Fourteenth Amendment. When presented with this statement, most students will say that such discrimination is still against the law, and they are right, but the constitutional basis for those federal laws that prevent racial discrimination in employment is the commerce clause.[21] Two commonly cited United States Supreme Court cases for this principle are *Katzenbach v. McClung*[22] and *The Heart of Atlanta Hotel v.*

United States.[23] In both those cases, the Supreme Court held that Congress did have the power to use the commerce clause as the constitutional basis for legislation regulating racially discriminatory behavior by small businesses. The Civil Rights Act of 1964 applies only to businesses with more than 15 employees.[24] Smaller businesses are not bound by its provisions and may discriminate so long as they do not violate state or local laws.

In similar fashion, searches by private individuals are not searches within the meaning of the Fourth Amendment because they do not constitute state action. Writing for the majority in the case of *United States v. Jacobsen*, Justice Stevens said, "Whether those invasions were accidental or deliberate, [footnote omitted] and whether they were reasonable or unreasonable, they did not violate the Fourth Amendment because of their private character."[25] That case involved the discovery of cocaine in a package being shipped via Federal Express. The package was damaged in the course of handling and, as was standard policy at Federal Express, employees of the company inspected the package to determine its value for insurance purposes. They discovered white powder, reclosed the package, and called the Drug Enforcement Agency (DEA). When DEA agents arrived, they replicated the inspection performed by the employees of Federal Express and found the white powder, which, upon testing, was discovered to be cocaine. Based on this evidence, the DEA obtained a search warrant for the addressee's premises, executed that warrant, and subsequently arrested Jacobsen and his associates. The Court upheld this search on the basis that the search by the private party was of no constitutional significance in and of itself, but by virtue of having been performed, it eliminated any reasonable expectation of privacy that the defendants had in the package.[26]

THE CONSEQUENCES OF THE CONCEPT OF STATE ACTION FOR EMPLOYEES IN THE PRIVATE SECTOR

As a result of the concept of state action, private-sector employees have no federal constitutional protection against searches of their persons, work spaces, or workplace computers and other electronic gadgets. As a consequence, drug testing, which is a search, is very common in the private sector. Recent survey results indicate that more than half of all private-sector employers routinely drug test all job candidates.[27] The Obama Administration actively promotes drug testing in the workplace.[28] There are a variety of state laws in place concerning drug testing of private-sector workers, but few states significantly restrict the ability of private-sector employees to test their employees for drugs. Most of these laws describe the conditions under which drug testing may be a condition of work, the procedures that must be followed when drug testing employees, and listing the conditions under which an employee may be disciplined or terminated for drug use.[29] A few states may limit the ability of an employer to make passing a drug test a condition of work.[30]

THE FOURTH AMENDMENT AND EMPLOYEES IN THE PUBLIC SECTOR

When a public-sector employer conducts a search of the person, workplace, or electronic equipment of a worker, that search constitutes action by an agent of the state. The Fourth Amendment does protect employees' rights in such circumstances. The legal standard used to determine the legitimacy of such searches is that which applies to an administrative search.[31] That means that a court faced with a challenge to a search conducted by a public employer must apply a balancing test in which the privacy expectations of the employee are balanced against the need for the employer to obtain the property, documents, or information in question. The most commonly cited case for this balancing test is O'Connor v. Ortega.[32] While the case produced no majority opinion, Justice O'Connor's plurality opinion is widely cited and is generally considered controlling, "In the case of searches conducted by a public employer, we must balance the invasion of the employees' legitimate expectations of privacy against the government's need for supervision, control, and the efficient operation of the workplace."[33] The Court made it clear that no warrant was required and that the test to be applied was not probable cause, as is required in searches that are conducted as part of a criminal investigation. A lesser standard is to be applied. "We have concluded, for example, that the appropriate standard for administrative searches is not probable cause in its traditional meaning. Instead, an administrative warrant can be obtained if there is a showing that reasonable legislative or administrative standards for conducting an inspection are satisfied."[34]

As a consequence of the decision in O'Connor v. Ortega, and despite the fact that there were only four votes for O'Connor's opinion,[35] the decision has been largely confirmed in the case of City of Ontario v. Quon,[36] and the lower courts have followed its balancing test.[37] Language in several appellate opinions makes it clear that a policy stated in a personnel manual or other document that announces that the employer retains the right to inspect work spaces or computers limits or eliminates any claim of reasonable expectation of privacy by the employee.[38] This creates a strong incentive for public-sector employers to include explicit language in their personnel manuals granting the employer access to work spaces.

Searches of Physical Spaces

The landmark case of *O'Connor v. Ortega*[39] created the standard test to be applied when considering the validity of a search of the physical space commonly used by a public-sector employee. The parties to that case were Dr. Dennis O'Connor, then director of the Napa State Hospital in Napa, California, and Dr. Magno Ortega, then chief of professional education at the hospital. In response to concerns about Dr. Ortega's use of patient funds to purchase a computer and allegations of sexual harassment of female employees, Dr. O'Connor placed Dr. Ortega on administrative leave and conducted a thorough search of Dr. Ortega's office. The dispute concerned the legality of this search.

The Court, while unable to produce a majority opinion, did produce a plurality opinion announcing a ruling in favor of Dr. O'Connor. That opinion, written by Justice O'Connor, laid out a legal standard that is largely followed to this day. When faced with a legal challenge to such a search, the task is to attempt to balance the interests of the employee in the privacy of his or her work space against the legitimacy of the need for the information, documents, or other evidence by the public-sector employer. The expectations of the employee are a key factor in this equation. Where there is a clear provision in the organization's personnel manual that purports to authorize access to these spaces, the claims of violation of privacy by the employee are weakened. In writing about the nature of the dispute between Dr. Ortega and his hospital in the case of *O'Connor v. Ortega*, Justice O'Connor wrote, "There was a dispute of fact about the character of the search, and the District Court acted under the erroneous assumption that the search was conducted pursuant to a Hospital policy."[40] This language clearly creates the impression that if there is an established organizational policy, this bolsters the legitimacy of the government's interest in conducting the search.[41]

The fifth vote for the outcome in favor of the employer in *O'Connor v. Ortega* came from Justice Scalia. Scalia would apply a test commonly called the "presumption of reasonableness" test.

> The government, like any other employer, needs frequent and convenient access to its desks, offices, and file cabinets for work-related purposes. I would hold that government searches to retrieve work-related materials or to investigate violations of workplace rules—searches of the sort that are regarded as reasonable and normal in the private employer context—do not violate the Fourth Amendment.[42]

To the extent that there is a past pattern of the employer gaining access on a regular basis to the area in question, this significantly weakens the case for the complaining employee. Such was the case in *Muick v. Glenayre Electronics*, decided by the U.S. Court of Appeals for the Seventh Circuit. "Glenayre had announced that it could inspect the laptops that it furnished for the use of its employees, and this destroyed any reasonable expectation of privacy that Muick might have had and so scotches his claim."[43] It should be noted that the existence of a policy may not be controlling, as the court balances the interests of the employee and employer. Where the policy that purports to provide notice to the employee that they have no expectation of privacy in a particular place or device is contradicted by actual practices that arguably create an expectation of privacy, the court will consider that in conducting its balancing analysis.[44]

The claims of the employer regarding the need for the evidence in question depends on whether the employer can provide the court with a valid reason why such evidence would be needed for a legitimate organizational purpose. In the *O'Connor v. Ortega* case, Justice O'Connor described the test for determining the relative strength of the organization's interest in obtaining the evidence in question as follows: "Ordinarily, a search of an employee's office by a supervisor will be 'justified at its inception' when there are reasonable grounds for suspecting that the search will turn up evidence that the employee is guilty of work-related misconduct, or that the search is necessary for a noninvestigatory work-related purpose, such as to retrieve a needed file."[45]

The Supreme Court's opinion in *City of Ontario v. Quon*[46] left some doubt as to whether the balancing test announced by Justice O'Connor or the presumption-of-reasonableness standard laid out by Justice Scalia in his concurring opinion in *O'Connor v. Ortega*[47] would be applied. The *City of Ontario v. Quon* case arose out of a dispute over the use of a pager by Jeff Quon, who was employed as a police sergeant by the City of Ontario, California. The dispute arose out of events taking place during calendar years 2001 and 2002. Quon, like several other officers, had been issued an alphanumeric pager capable of sending and receiving text messages to be used in the course of his official duties. The pager had a character limit on its use, and Quon had exceeded the limit more than once, resulting in extra charges from the service provider. Quon had reimbursed the city for the extra charges, but the continued overages resulted in an investigation by the city. In the course of that investigation, the city acquired transcripts of the text messages sent by Quon from the service provider. Some of the text messages were not work related, and some were of a sexual nature. After an internal affairs investigation, Quon was disciplined and sued the city, alleging that the obtaining of the text messages from this pager account was a violation of the Fourth Amendment.

The Court, in an opinion written by Justice Kennedy, held that the search of Jeff Quon's text messages was reasonable, regardless of whether one applies the balancing test laid out by Justice O'Connor in her plurality opinion in *O'Connor v. Ortega* or the presumption-of-reasonableness standard announced by Justice Scalia in his concurrence in that case. The limited lower court cases decided after *City of Ontario v. Quon* was decided commonly cite *O'Connor v. Ortega* as the controlling precedent.[48] Commentators have been critical of the lack of clarity of the *O'Connor v. Ortega* standard[49] and of the failure of the Court to clarify the legal standard when they had the opportunity to do so in the case of *City of Ontario v. Quon*.[50]

These rules for analysis of whether an employer may search desks, lockers, and locker rooms are fairly well settled in that the lower courts are still using the balancing test from *O'Connor v. Ortega*, but the outcomes of individual cases are not always so settled.

Public-sector employers will continue to monitor workplaces using video cameras as well as searching work spaces and lockers, and they will continue to be sued by their own employees when such searches turn up evidence that is used against the employee in a disciplinary action or turned over to prosecutors to be used in a criminal prosecution.

Drug Testing

The basic Fourth Amendment standard for when a public-sector employee may drug test an employee or potential employee was set out in the cases of *National Treasuries Employees Union v. Van Raab*[51] and *Skinner v. Railway Labor Executives' Association*.[52] Drug tests are searches within the meaning of the Fourth Amendment and can only be conducted when they can be justified according to the standards set out in *O'Connor v. Ortega*. This means that you can drug test pretty much anytime the employee is involved in law enforcement or national security activities, carries a gun or operates any dangerous equipment or machinery, drives a bus or flies an airplane, or there is individual suspicion that the individual is using or selling drugs.

While alcohol is the most commonly used recreational drug in the United States, by far, the most commonly abused illegal drugs in the United States are marijuana, prescription pain killers, and cocaine.[53] The drugs most commonly the subject of drug testing are opiates, marijuana, barbiturates, amphetamines, and hallucinogenic drugs such a peyote or lysergic acid diethylamide, or LSD. In the aftermath of an accident, an employer may test for blood alcohol content as well.[54]

The rules to be followed in the federal practice have been in place since 2008.[55] Only urine may be sampled, and only certified and trained collectors may collect urine specimens. Each specimen is split, and all drug tests must be conducted by drug laboratories certified by Health and Human Services (HHS). These extensive rules are not mandated for public-sector employees at the state and local levels, and there is considerable variation from jurisdiction to jurisdiction in how drug testing is done. In most public-sector organizations, in situations where drug testing is allowed, it is done with less formality and fewer legal constraints than in the federal practice. Perhaps one of the biggest differences is that states are not required to use HHS-certified drug laboratories. They can use the commercially available drug-test kits.

The most common drug tests are urine tests designed so that any person can conduct them. The test kits are available at low cost and from a variety of commercial vendors.[56] The first issue is how to obtain the sample. Two common procedures are the observed sample, in which the employee is watched while urinating into a container, and the unobserved sample, for which the employee urinates into a container in private, most often after completely disrobing and putting on a hospital gown. These procedures are followed to prevent the sample of urine from being contaminated and to ensure it comes from the person in question. These drug-test kits may also include a procedure to test for adulteration or contamination of the sample by testing its pH.

Once the sample is obtained, a low-cost reagent test kit is used. A portion of the urine is set aside for confirmation, should the test show positive, and the rest is tested either by submerging a test strip—treated with chemicals designed to change color if the metabolites of any of the drugs for which the test kit has been designed are present in the urine—or by mixing a chemical with the sample that will cause unambiguous changes in the appearance of the sample if the residue of drug use is present.

These kits are cheap, they can be administered by nonexperts, and they produce unambiguous results. They do not directly measure the presence of drugs in the body. Rather, they test for traces left in the system when the body metabolizes these drugs. The metabolites only remain in the system for a limited period of time, with the sensitivity being determined by the type of drug and the frequency and intensity of use.[57] The use of a hair or fingernail sample yields longer detection times,[58] but it also comes at a greater cost per test. These tests are not 100 percent accurate, and critics and vendors offer very different views of the accuracy of the tests. For example, the National Work Rights Institute reports that the false positive rates is anywhere from 10 to 30 percent.[59] The American Civil Liberties Union assumes a 5 percent error rate, an estimate it calls "conservative."[60] The National Institutes of Health concluded that drug testing by these methods is accurate but conceded that errors can occur. "Accuracy also depends on the choice of laboratory, use of proper equipment and methods, quality control, and adherence to high-quality standards by all involved. As in all laboratory testing, human errors, confounding results, a poorly controlled chain of custody for samples, and other problems lower test reliability."[61] Of course, the vendors of such tests promise a very high level of accuracy, either or 98 or 99 percent.[62]

The procedures to be followed in conducting the drug tests, and the appeals procedures to be followed if a positive result is obtained, are usually found in the employee handbook or personnel manual; these procedures will commonly be the subject of collective bargaining if the workforce is unionized. The National Labor Relations Board has been quite clear that whether to drug test is a subject of mandatory bargaining in the private sector,[63] but in New York at least, the specific method to be used for testing of urine or hair samples is not necessarily the subject of mandatory bargaining.[64]

Since these kits are not foolproof and false results can occur, it is standard practice to split the sample. If a positive result is obtained, the remainder of the urine sample can then be retested using a more sensitive method. Often, urine samples are split before the reagent test is performed, with a portion going to the employee so that he or she can have it tested using methods and service providers that person trusts if a positive result occurs. Splitting of samples is routinely done in the federal practice.

A few states have imposed limits on the use of drug testing. Vermont only allows drug testing of employees on probable cause or if there is an employee-assistance program that provides drug treatment and the employee will not be terminated for a positive drug test.[65] Minnesota prohibits random drug testing unless the employees are as such: "(1) they are employed in safety-sensitive positions, or (2) they are employed as professional athletes if the professional athlete is subject to a collective bargaining agreement permitting random testing but only to the extent consistent with the collective bargaining agreement."[66]

Video Surveillance of the Workplace

Two recent cases arising out of a single video surveillance in Los Angeles County illustrate clearly that the method of surveillance is considered with suspicion by the courts. The two cases, *Carter v. County of Los Angeles*[67] and *Richards v. County of Los Angeles*,[68] arose out of the attempt by Los Angeles County to catch an employee engaging in improper sexual behavior in a break room used by dispatchers. Even though the county had reason to investigate the misconduct alleged against Amber Richards, it did not have the authority to record images of her without her knowledge in an area that was used for multiple purposes, including rest during breaks. Ms. Richard's supervisor authorized the installation of a hidden camera but used the recorded images to investigate other matters as well. In both cases, the district court was quite clear that the video surveillance in this case failed the tests as stated in both the *Ortega*[69] and *Quon*[70] cases. The fact that the area under surveillance was used for non-work-related purposes, such as resting and eating during work breaks, and the failure to provide notice that the area was under surveillance were key factors in the decisions. What the two decisions do not determine is under what circumstances a public-sector employer may put a work area under video surveillance. Is notice to the employee required? What rationale for the surveillance is required? Neither of these questions is answered, and this author strongly suspects that video surveillance will become more common as the technology is used in a greater variety of circumstances in the general population.

Recording Phone Calls

Given the extensive controversy surrounding the revelations in June of 2013 that the federal government was making records of millions of Americans' phone conversations and Internet usage,[71] it is incumbent to ask whether a public-sector employer may record the phone calls made from landlines at the place of work. The case of *Walden v. City of Providence*[72] involved a claim by police and fire department employees of Providence, Rhode Island, that the recording of phone calls made on official phone lines violated the Fourth Amendment. After a 21-day jury trial, nominal damages were awarded to the employees, but the United States Court of Appeals for the First Circuit overturned the award and made it quite clear that recording of phone calls on departmental phone lines does not violate the Fourth Amendment.

Given that many public-sector employees engage in official business on cell phones owned either by the employer or the employee, the issue of access to the records of the use of the devices, whether held by the service providers or stored on the devices themselves, will arise. While the Court did not squarely address this issue in *City of Ontario v. Quon*,[73] the facts did include that the carrier of the pager service for the officer in question was asked to provide records without any subpoena or notice to the officer, and the carrier did so voluntarily. The facts in *Quon* as well as the potential for legal challenges involving a wide variety of uses of digital devices present several legal questions that are not yet settled: (1) Who owns the records—the service provider, the employee, or the employer? (2) What legal test will be used to determine whether the employer can access these records over the objections of the service provider or the employee? (3) Is there a legal difference, for purposes of Fourth Amendment analysis, between access to call records and text records and accessing the content of the text messages or conversations themselves? (4) Can the employer or the police use GPS tracking information from the devices to determine the whereabouts of the employee? And (5) in the case of a smartphone or other device capable of making phone calls or sending text messages, is the device to be treated as a computer or as a phone? The law will have to evolve with the technology. The attorney who argued the case for the officer in *Quon* has stated in print that he believes, based on the questions asked of him by the justices during oral argument, that the justices do not understand the technology and that this bodes ill for the law adapting to the new reality created by digital communications.[74] It is true that the Supreme Court is a conservative institution in most matters. Noted Supreme Court author Jeffrey Toobin commented about the great slowness of the justices to adopt e-mail during the last decade.[75]

Computers, Other Electronic Devices, and Data Chips

The basic principles of law that govern the constitutionality of searches of lockers, desks, and other work spaces also apply to searches of computers and other electronic devices. The distinction between computerized devices, physical spaces, and tools used in the workplace is wide. The first issue is that it is clear that stated policies do matter. If the employer has a clearly stated and consistently implemented policy regarding data stored on workplace computerized devices, the employee likely has no reasonable expectation of privacy. The problem is that it is not always clear what the word "computer" means. For example, a smartphone is essentially a small tablet computer that allows the user to surf the web; send text messages; send e-mails; take, store, and share pictures, and make phone calls. All of these functions have potential workplace utility in the public sector. For purposes of analysis, this chapter will treat laptop and desktop computers differently than other mobile computerized devices, such as tablets, smartphones, and e-readers. The word "computer" will refer to a laptop or desktop unit. The term "device" will be used to refer to cell phones, tablets, handheld units such as might be used by a police officer to issue traffic citations, and e-readers. Additionally, a distinction is made between devices and computers and the small, limited-function computer chips commonly contained within automobiles or other capital equipment. These will be referred to as "data chips."

A second issue arises out of the use of privately owned devices for public purposes. Many salaried public-sector workers have home offices with computers that they have bought with personal money and with data plans purchased with personal funds, that are often used for work purposes as well as for private purposes. For example, a college professor at a public university will almost certainly work from his or her home office on a regular basis, perhaps as much as or more than they work from their office. While these employees may have the use of a university-purchased and university-issued computer, it is also common that personal business will be conducted using these computers. These patterns of use of these computers, and the acceptance of using them in a comingled manner, create complications in determining to what areas of memory on these computers the employer would have access. The same type of comingling of use occurs with other devices. In the *Quon* case, the policy in question was that Officer Quon could use his government-issued pager for personal use as long as he paid for any overage charges resulting from that private use.[76] What makes this even more complicated is that stated policies about computer use may not have clear definitions of the range of computers and devices to which the policy applies.

A third issue that arises with mobile devices is the ownership of the data. Data generated by the use of these devices is stored in two different ways: (1) data generated and stored on the device itself, such as GPS location information; the record of calls made, received, and missed; and the actual contents of text messages sent and received; and (2) data generated and stored by the provider of the data services, such as call records or text messages. If the data is considered to be owned by the service provider, then the employer may obtain the data directly from the service provider, and the Fourth Amendment rights of the employee could, potentially, be of little application to the gathering and use of the data. A New York court has held that Twitter, Inc., does not have standing to challenge a subpoena for records of a client's Twitter account on the grounds that tweets are made public by the sender.[77] The decision of who has rights to the data will play a major role in the outcome of the lawsuit filed by the ACLU against the federal government over the National Security Agency's monitoring of cell phone calls and Internet usage because of the government's collection of call records from ACLU's cellular phone service provider, Verizon.[78]

An issue that was sidestepped nicely by the Supreme Court in the criminal case of *U.S. v. Jones*[79] was whether the use of GPS data to track the location of a person over an extended period of time violates the Fourth Amendment. The case was decided based on the holding that the placement of the device itself was a Fourth Amendment violation, thus rendering it unnecessary for the Court to address the issue of whether the use of the device to track the location of a person is a violation of any reasonable expectation of privacy. For public-sector employees who have devices that record location information, the question of whether the employer can track them while they are at work has not come up, but it will, and there is no clear legal standard for determining whether this is allowable under the Fourth Amendment.

For employees who operate motor vehicles, the access to data stored on any data chips installed by the maker of the vehicle is another unanswered question. When such a vehicle is involved in a collision, all sorts of data about the speed and direction of the vehicle can be retrieved. These data could easily be used in a disciplinary proceeding or, for that matter, in determining whether the employer will seek to employ the doctrine of *Monell v. New York City Department of Social Services*,[80] to extricate their company from the lawsuit. To this author, it seems that the employee has no rights to the data, and the employer has every right.

A final issue has to do with information that results from the use of personal or work computers to access social networking sites such as Facebook or LinkedIn. There are public-sector employees who have reason to use these sites as part of their official duties and, of course, millions who use these sites for personal purposes. The content generated on such sites is publicly available and archived by the hosts. Even content that is restricted by the user may be available through reposts by third parties. The legal question that arises, and it is more of a First Amendment question than a Fourth Amendment question, has to do with the use of this content by employers to discipline an existing worker or to deny a potential worker a position. While the concerted action provisions of the 1935 National Labor Relations Act provide protection for private employees who engage in speech concerning workplace issues, this provision of the law is not well known and is rarely invoked by nonunion employees.[81] This law would not apply in the public sector to protect employees, but it is easy to anticipate the kinds of arguments that an employer would use to justify taking action against an employee based on posts on a social media site. Some states do limit the ability of employers to use social media in screening potential employees or in investigating current employees,[82] but these laws are not of constitutional status.

That employees have posted work-related content on social media sites and that the public-sector employers have taken notice of this behavior is not in dispute, but the law has not yet evolved on this issue, in the sense that the Supreme Court has not heard such a case. Until it does, employees of public-sector organizations are advised to post nothing on the web that they would not want their employer to see. In both tort cases and criminal cases, courts have held that the gathering of data from social media accounts via access granted by third parties, or "friends," do not implicate Fourth Amendment rights.[83] Given that gaining information from third parties is not prohibited by the Fourth Amendment in criminal matters, where the courts have applied a more rigorous standard of review with regard to the Fourth Amendment than in the context of workplace searches, it makes sense that the courts will resolve these issues in favor of employers.

CONCLUSION

With regard to searches of physical spaces, the legal landscape is largely unchanged since the O'Connor v. Ortega case was decided in 1987. The word "privacy" is not in the U.S. Constitution, and this author feels safe in predicting that the federal judiciary is not likely to create any such doctrine in a way that could be criticized as activist. Moreover, the post-9/11 environment is one in which privacy rights are valued less, and security concerns are valued more than in the past. Case-by-case decisions will turn on the individual facts and not on any change in the law. From an employee perspective, if there is a policy that says the employer has access, you can assume that the employer indeed has access, and you should act as though the space is not private.

It is also a safe bet to predict that the judiciary will not develop any new, clear doctrines that will clarify the rights of public-sector employees and employers regarding the use of, or access to, computers, devices, or data chips. The outlook for those employees who would like to think that they maintain a significant zone of privacy at work is not good. The uncertainty that arises from the application of the O'Connor v. Ortega test has survived both time and critics. In applying this test to modern cases involving the use of information technology that was not in existence in 1987, judges have a great deal of discretion to decide cases in whatever way their personal views of the workplace dictate. The type of balancing test set out by Justice O'Connor in this case has been called by one commentator "a poor fit for providing much protection on technology such as computers, cell phones, and Blackberries."[84] Given that many judges did not grow up in an environment rich with electronic communication, and they do not exist in one at work or at home today to the extent that younger workers do, it is safe to say that for the time being, the law will lag behind the privacy challenges posed by the use of computers, devices, and data chips.

Employers are well aware that they have the power to monitor computer usage, cell phone activity, and even content posted on social media sites. Employees entering the workforce who are used to living in an information rich environment, almost constantly connected with friends and even strangers via e-mail, text messages, Facebook, Twitter, Tumblr, Vine, Pinterest, Path, Instagram, Snapchat, or whatever social medium they prefer, will need to learn to constrain their activities once they accept a job. Employers are using these tools to screen potential employees.[85] Public-sector managers are savvy enough to craft personnel policies that provide notice to the employee that he or she is being monitored so that the balancing test in case of litigation is skewed toward the employer.

There are a number of state laws that provide greater protection for employee privacy than in the federal practice, and public-sector employees and managers are advised to check with their attorneys about these few exceptions. Eight states—Arkansas, Colorado, Nevada, New Mexico, Oregon, Utah, Vermont, and Washington—passed laws in 2013 that limit the ability of employers to require that employees provide passwords to social media accounts as a condition of work.[86] Summary websites on employee privacy such as that maintained by the Privacy Rights Clearinghouse[87] can provide a start to finding these state laws, but the status of the law in this area, especially with regard to policies to do with the use of computers and electronic devices, is such that laypersons or human resource professionals may not be not well equipped to protect themselves without professional legal advice.

NOTES

1. United States Constitution, Amendment IV.
2. Stephanie M. Stern, "The Inviolate Home: Housing Exceptionalism in the Fourth Amendment," *Cornell Law Review* 95 (2010): 905–908.
3. *Kyllo v. United States*, 533 US 27, 31 (2001), "With few exceptions, the question whether a warrantless search of a home is reasonable and hence constitutional must be answered no."
4. *New York v. Belton*, 453 US 454, 457 (1981), "It is a first principle of Fourth Amendment jurisprudence that the police may not conduct a search unless they first convince a neutral magistrate that there is probable cause to do so."
5. The current standard for determining the sufficiency of the evidence for the issuance of a search warrant, the "totality of the circumstances" test, was set out in *Illinois v. Gates*, 426 US 213, 230–239 (1983).
6. See, *Carroll v. United States*, 267 U.S. 132 (1925); *New York v. Belton*; and, *Arizona v. Gant*, 556 US 332 (2009).
7. *Chimel v. California*, 395 U.S. 752 (1969). See also *Arizona v. Gant*, note 6 supra.
8. See, e.g., *California v. Greenwood*, 486 U.S. 35, 37, 40–41 (1988), holding that a search of garbage left at the curb was not a search.
9. See, e.g., *Walter v. United States*, 447 US 649 (1980), and *United States v. Jacobsen* 466 US 109 (1984) both of which involved a search that had been conducted by law enforcement after a private party had conducted a search and then brought the contraband to the attention of law enforcement. In both cases, the Court held that the private search had effectively eliminated a reasonable expectation of privacy on the part of defendants and that so long as the search by law enforcement did not exceed the scope of the search by the private party, there was not violation of the Fourth Amendment.
10. *Camara v. Municipal Court*, 387 US 523 (1967).

11. *Florida v. Wells*, 495 U.S. 1 (1990); *Colorado v. Bertine*, 479 U.S. 367 (1988); *South Dakota v. Opperman*, 428 U.S. 364 (1976).

12. See, e.g., *Florida v. Wells*, note 11, supra, at p. 4, "Our view that *standardized criteria* . . . must regulate the opening of containers found during inventory searches is based on the principle that an inventory search must not be a ruse for a general rummaging in order to discover incriminating evidence."

13. *Marshall v. Barlow's, Inc.* 436 US 307 (1978).

14. See, *Camara v. Municipal Court*, note 10, supra, at 539, "Since our holding emphasizes the controlling standard of reasonableness, nothing we say today is intended to foreclose prompt inspections, even without a warrant, that the law has traditionally upheld in emergency situations."

15. *O'Connor v. Ortega*, 480 US 709, 720–721 (1987).

16. Id.

17. *Civil Rights Cases*, 108 US 3, 17 (1883).

18. *Shelley v. Kraemer*, 334 U.S. 1 (1948).

19. *Cooper v. Aaron*, 358 US 1 (1958).

20. Id. at 15.

21. Article I, Section 8, of the Constitution says, in relevant part, "The Congress shall have Power . . . to regulate Commerce with foreign Nations, and among the several States, and with the Indian Tribes."

22. *Katzenbach v. McClung*, 379 US 294 (1964).

23. *The Heart of Atlanta Hotel v. United States*, 379 US 241 (1964).

24. 42 USC §2000e(b).

25. *United States v. Jacobsen*, note 9 supra, at 115.

26. Id. at 121, "The fact that, prior to the field test, respondents' privacy interest in the contents of the package had been largely compromised is highly relevant to the reasonableness of the agents' conduct in this respect. The agents had already learned a great deal about the contents of the package from the Federal Express employees, all of which was consistent with what they could see. The package itself, which had previously been opened, remained unsealed, and the Federal Express employees had invited the agents to examine its contents. Under these circumstances, the package could no longer support any expectation of privacy[.]"

27. Society for Human Resource Management. "Drug Testing Efficacy SHRM Poll," posted September 7, 2011, accessed June 11, 2013, (http://www.shrm.org/research/surveyfindings/articles/pages/ldrugtestingefficacy.aspx.

28. Office of National Drug Control Policy, Executive Office of the President. "National Drug Control Strategy, 2013," p. 6, accessed August 1, 2013, http://www.whitehouse.gov//sites/default/files/ondcp/policy-and-research/ndcs_2013.pdf.

29. See, e.g., Florida Statutes, Title XXXI, Chapter 440, sections 101 and 102; Iowa Code Section 730.5; Montana Workforce Drug and Alcohol Testing Act, MCA Sections 39-2-205 through 39-2-211.

30. See, e.g., 21 Vermont Statutes Annotated, sections 511–520; Rhode Island General Laws, Section 28-6.5-1; 2012 Minnesota Statutes, sections 181.950–181.957.

31. See, *Camara v. Municipal Court*, note 10, supra.

32. *O'Connor v. Ortega*, note 15, supra.
33. Id. at 719–720.
34. Id. at 723.
35. Justice Scalia agreed in a separate concurrence that the search of Dr. Ortega's office was a search within the meaning of the Fourth Amendment, but would allow searches of workspaces by public employers in virtually all circumstances on the grounds that employer access to an employee's work space are just part of the normal operations of any organization, private or public (480 US at 729–732).
36. *City of Ontario v. Quon*, 560 US ___, 130 S Ct 2619 (2010).
37. See, e.g., *City of Ontario v. Quon*, note 36, supra; *Wiley v. Department of Justice*, 283 F3d 1426 (2003), striking down a search of an employee's car based on an interpretation of *O'Connor v. Ortega*; *Muick v. Glenayre Electronics*, 280 F3d 741 (2002); *U.S. v. Fernandes, Floyd*, 272 F3d 938 (2001). Sheila A. Bentzen, "NOTE: Safe for Work? Analyzing the Supreme Court's Standard of Privacy for Government Employees in Light of *City of Ontario v. Quon*," Iowa L Rev 97 (2012): 1283, 1289, n31 provides a list of US Courts of Appeals cases that followed Justice O'Connor's *O'Connor v. Ortega*'s balancing test.
38. See, e.g., *Muick v. Glenayre Electronics*, note 37, supra, at 743, "But Glenayre had announced that it could inspect the laptops that it furnished for the use of its employees, and this destroyed any reasonable expectation of privacy that Muick might have had and so scotches his claim," and citing a long list of cases that held that a policy that stated that the employer retains access eliminates the employee's reasonable expectation of privacy. See, also, *City of Ontario v. Quon*, note 36, supra, at 9–10, discussing the reasonableness of Quon's expectation of privacy given that there was a clear employer policy that use of computers was not private.
39. *O'Connor v. Ortega*, note 15 supra.
40. Id at 728.
41. *Muick v. Glenayre Electronics*, note 37, supra, at 743. See also, *United States v. Linder*, 2012 U.S. Dist. LEXIS 112134 (N.D. IL (Eastern Division), August 9, 2012).
42. *O'Connor v. Ortega* note 15, supra, at 732, Justice Scalia concurring.
43. *Muick v. Glenayre Electronics*, note 37, supra, at 743.
44. See e.g., *Carter v. County of Los Angeles*, 70 F. Supp. 2d 1042; 2011 U.S. Dist. LEXIS 17554 (C. D. CA, Feb. 22, 2011); and *Looney v. Washington County, Oregon*, 2011 U.S. Dist. LEXIS 75624 (OR (Portland Division), July 23, 2011).
45. *O'Connor v. Ortega*, note 15, supra, at 726.
46. *City of Ontario v. Quon*, note 36 supra. Sheila A. Bentzen, note 37, supra, at 1291.
47. *O'Connor v. Ortega*, note 15, supra, at 729–732.
48. See, e.g., *True v. Nebraska*, 612 F.3d 676 (8th Cir. 2010); *Carter v. County of Los Angeles*, note 44, supra; *Gwynn v. City of Philadelphia*, 866 F. Supp. 2d 473, LEXIS 43482 (E. D. PA, March 28, 2012); *James v. Hampton*, 2012 U.S. Dist. LEXIS 6542 (E. D. Michigan, Jan. 22, 2012); *Jones v. Houston Community College System*, 816 F. Supp. 2d 418; 2011 U.S. Dist. LEXIS 113524 (2011); *Looney v. Washington County, Oregon*, note 44, supra; *Richards v. County of Los Angeles*, 775 F. Supp. 2d 1176; 2011 U.S.

Dist. LEXIS 308 (C. D. CA, March 1, 2011); *Tangredi v. City of New York Department of Environmental Protection*, 2012 U.S. Dist. LEXIS 36788 (S. D NY, Feb. 16, 2012); *United States v. Johnson*, 871 F. Supp. 2d 539; 2012 U.S. Dist. LEXIS 67275 (2012); *United States v. Linder*, 2012 U.S. Dist. LEXIS 112134 (N.D. IL (Eastern Division), August 9, 2012); *Washington v. Fischer*, 2012 Wash. App. LEXIS 2724 (2012).

49. See, e.g., E. Miles Kilburn, "CASENOTE: Fourth Amendment—Work-Related Searches by Government Employers Valid on "Reasonable" Grounds." *J Crim L & Criminology* 78 (1988): 792, 826; Heather Hanson, "NOTE: The Fourth Amendment in the Workplace: Are We Really Being Reasonable?" *Va L Rev* 79 (1993): 243; Michelle Hess, "NOTE: What's Left of the Fourth Amendment in the Workplace: Is the Standard of Reasonable Suspicion Sufficiently Protecting Your Rights?" *Fed Cir B J* 15 (2005): 255; Dieter C. Dammeier, "Symposium: Privacy and Accountability in the 21st Century: Fading Privacy Rights of Public Employees," *Harv L & Pol'y Rev* 6 (2012): 297, 300–302.

50. Sheila A. Bentzen, note 37, supra, at 1291–1292. Dieter C. Dammeier, note 49, supra, at 302–304.

51. *National Treasuries Employees Union v. Von Raab*, 489 U.S. 656 (1989).

52. *Skinner v. Railway Labor Executives' Association*, 489 US 602 (1989).

53. Substance Abuse and Mental Health Services Administration. 2012. *State Estimates of Substance Use and Mental Disorders from the 2009–2010 National Surveys on Drug Use and Health*, NSDUH Series H-43, HHS Publication No. (SMA) 12-4703. Rockville, MD: Substance Abuse and Mental Health Services Administration, accessed August 1, 2013, http://www.samhsa.gov/data/NSDUH/2k10State/NSDUHsae2010/.

54. See, e.g., City of Chicago Personnnel Rules. 2010, p. 48, dated November 18, 2010, accessed June 10, 2013, http://www.cityofchicago.org/dam/city/depts/dhr/supp_info/Personnel_Rules_Revised11_26_2010.pdf; Boston Police Rules and Procedures. 1998. Substance Abuse Policy, p. 4, dated December 17, 1998, accessed August 1, 2013 (http://www.cityofboston.gov/Images_Documents/rule111_tcm3-9570.pdf.

55. United States Department of Health and Human Services, Substance Abuse and Mental Health Services Administration, "Mandatory Guidelines for Federal Workplace Drug Testing Programs," Federal Register 73 (228): 71858, Tuesday, November 25, 2008, accessed June 10, 2013, http://www.gpo.gov/fdsys/pkg/FR-2008-11-25/pdf/E8-26726.pdf.

56. Example online companies include Medical Disposables (http://www.medicaldisposables.us/default.asp); Uritox Medical (http://www.uritoxmedicaltesting.com/index.html); and Craig Medical (http://craigmedical.com/drugtests.htm).

57. Alain G. Verstraete, "Detection Times of Drugs of Abuse in Blood, Urine, and Oral Fluid," *Ther Drug Monit* 26 (2004): 200; Mayo Clinic, "Approximate Detection Times Table," last updated, January, 2011, accessed June 10, 2013, http://www.mayomedicallaboratories.com/articles/drug-book/viewall.html.

58. Medical Disposables.com, "Drug Detection Periods in Drug Tests," accessed June 10, 2013, http://www.medicaldisposables.us/Drug_Detection_Periods_s/1835.htm.

59. National Work Rights Institute, "Drug Testing in the Work Place" copyright 2010, accessed August 2, 2013, http://workrights.us/?products=drug-testing-in-the-workplace.

60. American Civil Liberties Union, "Work Place Drug Testing," posted March 12, 2002, accessed August 2, 2013, http://www.aclu.org/racial-justice_womens-rights/workplace-drug-testing.

61. Center for Substance Abuse Treatment. "Medication-Assisted Treatment for Opioid Addiction in Opioid Treatment Programs." Rockville, MD Substance Abuse and Mental Health Services Administration (US); 2005. (Treatment Improvement Protocol (TIP) Series, No. 43.) Chapter 9. Drug Testing as a Tool. Copyright 2005, accessed August 2, 2013, http://www.ncbi.nlm.nih.gov/books/NBK64151/.

62. Medical Disposables promises a 99% accuracy rate, http://www.medicaldisposables.us/home-drug-test-kits-onsite-drug-testing-s/1818.htm; Uritox Medical says, "Up to 99% accurate," http://uritoxmedicaltesting.com/aboutus.html; Craig Medical asserts that "Results are obtained within 8 minutes and are 98% accurate when instructions are followed precisely," http://craigmedical.com/drugtests.htm.

63. See, Mark M. Rabuano, "An Examination of Drug Testing as a Mandatory Subject of Collective Bargaining in Major League Baseball," *U Pa Journal of Labor and Employment Law* 4 (2002): 439, 452–453.

64. See, *In the Matter of City of New York v Patrolmen's Benevolent Association of the City of New York, Inc.*, 14 N.Y.3d 46; 924 N.E.2d 336; 897 N.Y.S.2d 382; 2009 N.Y. LEXIS 4486; (2009). This case was shepardized on June 10, 2013, and no subsequent citations to jurisdictions outside of New York were found.

65. 21 Vermont Statutes Annotated, section 512.

66. Minnesota Statutes, section 181,951, subdivision 4.

67. 770 F. Supp. 2d 1042; 2011 U.S. Dist. LEXIS 17554 (C. D. Ca, Feb 22, 2011).

68. 775 F. Supp. 2d 1176; 2011 U.S. Dist. LEXIS 308 (C. D. CA, March 1, 2011).

69. *O'Connor v. Ortega*, note 15, supra.

70. *City of Ontario v. Quon*, note 36, supra.

71. Barton Gellman and Laura Poitras, "U.S., British Intelligence Mining Data from Nine U.S. Internet Companies in Broad Secret Program," Washington Post Online, posted June 6, 2013, accessed June 13, 2013, http://www.washingtonpost.com/investigations/us-intelligence-mining-data-from-nine-us-internet-companies-in-broad-secret-program/2013/06/06/3a0c0da8-cebf-11e2-8845-d970ccb04497_story.html?hpid=z1.

72. *Walden v. City of Providence*, 596 F.3d 38; 2010 U.S. App. LEXIS 3652 (1st Cir 2010).

73. *City of Ontario v. Quon*, note 36, supra.

74. Dieter C. Dammeier, note 49 supra, p. 304, at note 56.

75. Jeffrey Toobin, *The Nine: Inside the Secret World of the Supreme Court* (New York: Anchor Books 2007), at 57.

76. Dieter C. Dammeier, note 40 supra, p. 302. The facts of the case in *City of Ontario v. Quon*, note 36, supra, are set out in considerable detail in Justin Conforti, "Somebody's Watching Me: Workplace Privacy Interests, Technology Surveillance,

and the Ninth Circuit's Misapplication of the *Ortega* Test in *Quon v. Arch Wireless,*" *Seton Hall Circuit Review* 5 (2009): 461.

77. *New York v. Harris,* Criminal Court of the City of New York, County of New York, Decision and Order, Docket No: 2011NY080152, denial of motion to quash subpoena (2012).

78. Associated Press, "Lawsuits Challenging Government Surveillance Face Daunting Legal Obstacles," posted June 12, 2013, accessed August 1, 2013, http://www.foxnews.com/us/2013/06/12/lawsuits-challenging-government-surveillance-face-daunting-legal-obstacles/; American Civil Liberties Union, "ACLU v. Clapper – Complaint," posted June 11, 2013, accessed June 13, 2013, http://www.aclu.org/national-security/aclu-v-clapper-complaint.

79. *United States v. Jones,* 565 US ____, 132 SCt 945 (2012).

80. *Monell v. New York City Department of Social Services,* 436 US 658 (1978).

81. Ariana C. Green, "Privacy Law: Using Social Networking to Discuss Work: NLRB Protection for Derogatory Employee Speech and Concerted Activity," *Berkeley Tech L J* 27 (2012): 837.

82. Dara Kerr, "Six States Outlaw Employer Snooping on Facebook," C|NET, January 2, 2013, accessed August 1, 2013, http://news.cnet.com/8301-1023_3-57561743-93/six-states-outlaw-employer-snooping-on-facebook/.

83. See, e.g., *United States v. Meregildo* (United States District Court for the Southern District of New York, 2012), accessed June 25, 2013, http://nysd.uscourts.gov/cases/show.php?db=special&id=204; *Romano v. Steelcase, Inc.,* 2010 NY Slip Op 20388 [30 Misc 3d 426] (2010), accessed June 25, 2013, http://www.courts.state.ny.us/Reporter/3dseries/2010/2010_20388.htm.

84. Justin Conforti, note 76, supra, at 464–465.

85. Science Daily, "Many Employers Use Facebook Profiles to Screen Job Applicants," posted July 23, 2012, accessed August 1, 2013, http://www.sciencedaily.com/releases/2012/07/120723095208.htm.

86. National Conference of State Legislatures, "Employer Access to Social Media Usernames and Passwords 2013," last updated July 31, 2013, accessed August 1, 2013, http://www.ncsl.org/issues-research/telecom/employer-access-to-social-media-passwords-2013.aspx.

87. Privacy Rights Clearinghouse, "Fact Sheet 7: Workplace Privacy and Employee Monitoring," updated June 2013, accessed August 1, 2013, https://www.privacyrights.org/fs/fs7-work.htm.

6

The Privacy Rights of Minors in a Digital Age

Gardenia Harris

Carrie S., a straight A student and captain of the high school cheerleading squad, knowingly violated the school policy prohibiting cell phones by bringing her cell phone to school. Between classes, she furtively sent text messages to her friend who attends another high school. After being reported by a student hall monitor, the assistant principal confiscated her cell phone and communicated her transgression to the cheerleading coach. While the phone was in his possession, the assistant principal scanned the call log, text messages, Internet bookmarks, and photos stored on Carrie's cell phone to see if she had violated any other school policies. One photo showed Carrie drinking alcohol, an action that is a violation of the athletic policy. Carrie was suspended from school for three days and was later kicked off the cheerleading squad. It is obvious that school officials' searches of cell phones raise Fourth Amendment concerns due to the highly personal nature of the contents contained on cell phones.[1] Does Carrie have any legal recourse under Fourth Amendment protections?

MINORS AND THE BILL OF RIGHTS

The Supreme Court has long acknowledged that the Constitution and the Bill of Rights apply to minors. Yet, the Court has also determined that minors' particular vulnerabilities, reduced decision-making capacity, and the "special needs"[2] contexts they occupy (i.e., schools), warrant a restriction of their rights compared to adults. For example, the Supreme Court's finding in *In re Gault* was that juveniles facing adjudication for delinquency and incarceration are only entitled to some of the procedural safeguards under the due process clause of the Fourteenth Amendment, and there are legitimate reasons for treating juveniles and adults differently.[3] In a similar vein, in *Tinker v. Des Moines Independent Community Schools*, the Supreme Court confirmed that students and teachers have First Amendment rights, but the special characteristics of the school environment necessitates some restriction of their rights in certain contexts. For example, school administrators may regulate free speech if they can demonstrate constitutionally valid reasons for doing so. The Court specifically cited the need for school discipline and the protection of the rights of others as valid reasons for curtailing students' rights.[4]

The Supreme Court has likewise confirmed that Fourth Amendment protections apply to minors, but under certain contexts, minors have reduced Fourth Amendment rights compared to adults.[5] This chapter reviews the principles underlying Fourth Amendment protections in the digital age, and then it examines minors' privacy rights within the home and school settings.

PRE-DIGITAL AGE

The Fourth Amendment was adopted in reaction to English and Colonial authorities' arbitrary use of general searches and seizures.[6] Its fundamental purpose is the protection of citizens from unreasonable government intrusion[7] by requiring governmental authorities to obtain a search warrant that specifies what they intend to search or seize.[8] The warrant requirement is designed to ensure that authorities have sufficient cause to justify a search[9] and reflects the notion that judges, rather than law enforcement officials, are in the best position to determine if the necessary cause or suspicion exists to conduct a search.[10]

The Fourth Amendment uses physical language in that it protects people's rights to be secure in regard to material effects such as "their persons, houses, papers, and effects."[11] Warrants have traditionally authorized law enforcement officials to enter a location, such as a home, and seize physical objects, such as paper files and printouts. These types of objects contain a relatively narrow range of information about an individual, or they contain information about a limited number of persons.[12] Therefore, prior to the digital age, officials executing a search warrant could limit their investigation to the items identified in the warrant.

THE DIGITAL AGE

The Court has demonstrated a willingness to extend Fourth Amendment protections to intangible interests and nonphysical intrusions.[13] Notably, in the digital age, the government can obtain evidence from computer hard drives, web-based e-mail applications, cell phones, Internet searches, and so forth. Digital data differ from the information contained in physical files in that there is no separate compartmentalized storage container or set of papers that authorities can examine to view the specified information. Instead, authorities must scan through a large amount of information in order to locate the material identified in the search warrant.[14] Therefore, those conducting computer searches are likely to see data and information not directly identified in the search warrant.[15] As a result, the examination of evidence contained on computer hard drives and similar devices is problematic under the Fourth Amendment. The search of these devices may reveal evidence that is relevant to the investigation and the individual or individuals under investigation, but it is also likely to expose information that is not directly related to the matters or persons identified in the search warrant.[16] Also, some of that irrelevant information could implicate the individual in regard to policy violations or criminal activity that were not initially the source of the investigation.[17] For example, computer spreadsheets generally contain information on a large number of individuals and also contain some additional information on each individual that is not related to the focus of the investigation.[18]

The increased scope of a cell phone's context and functions potentially subjects students to a greater intrusion upon their privacy expectations.[19] Smartphones store considerably more information than early generation cell phones. Much of it is the type of information an individual would generally not carry or would not be within that person's immediate reach (i.e., photos, e-mail applications, audio and video files, and Internet browser histories).[20] Finally, early generation cell phones and pagers did not require in-depth searches to uncover evidence on them. Previously, there was no need to utilize passwords, unlock devices, scroll through the home page, open applications, and so on. Instead, evidence could generally be found in two steps—opening the phone and looking through the phone numbers.[21]

Therefore, in the digital age, citizens' privacy interests no longer center on the physical contents inside the device's casing but on information contained in its memory.[22] Currently, defendants are less concerned about intrusion upon their property and are more concerned about intrusions into their informational stores.[23] Ashley Snyder notes that the potential vast amount of information that can be recovered from a cell phone raises the issue of whether cell phones and smart phones should be treated differently under the law.[24] She contends that the case law demonstrates that when courts consider it significant that cell phone technology has dramatically changed the volume of personal information individuals carry with them, they will consider warrantless cell phone searches to be an unreasonable intrusion of privacy.[25]

Thus, in the digital age, intrusion upon real property is not the primary concern of modern Fourth Amendment complaints.[26] The law regarding how the Fourth Amendment treats digital information stored on computers, cell phones, and similar devices is far from settled, particularly as it pertains to the rights of minors. Only a small number of Fourth Amendment cases dealing with modern technology have been decided by the Court, and the Court has never ruled on a Fourth Amendment case dealing with a search of information on the Internet.[27]

THIRD-PARTY DOCTRINE

In a series of cases, the Court upheld the principle that the Fourth Amendment does not protect information that has been voluntarily disclosed to a third party. This was later referred to as the "third-party doctrine."[28] According to this doctrine, once an individual voluntarily discloses information to a third party, Fourth Amendment protections no longer apply.[29] In other words, people do not have a reasonable expectation of privacy after they intentionally reveal information to a third party,[30] even if that information was disclosed to the third party for a definite purpose.[31]

Based on this principle, the Supreme Court has ruled that Fourth Amendment protections do not apply to phone numbers dialed via a telephone, because telephone users cannot have a reasonable expectation of privacy. They know that making a phone call requires them to convey the telephone number to the phone company. Further, the caller is knowledgeable that telephone companies regularly record such transactions.[32] The Court further ruled that the phone numbers were disclosed to an automated machine, instead of a human being, was of no consequence. The Court conceived of the machine as a substitute for a traditional telephone operator.[33] According to the Court, all that mattered was that the defendant voluntarily exposed the phone numbers to a third party's machine.[34]

Although the Court has never directly ruled on a Fourth Amendment case dealing with a search of information on the Internet,[35] several scholars associate Internet servers with telephone switchboards. They observe that almost all Internet communications are stored for some time on third-party servers, or Internet Service Providers (ISPs).[36] Lower courts are not in agreement as to how or whether the third-party doctrine should apply to advancing technology.[37] Although the Supreme Court has yet to determine how the third-party doctrine applies to the Internet,[38] Bedi contends, "The implication is that an individual will likely loose Fourth Amendment protections regarding any information he or she exposes to a third party's machine in the normal course of business regardless of whether a human observes the information."[39] Minors are likely to be covered under this doctrine, too. Erin Smith Dennis notes that "as the information we share with third parties via technology increases exponentially the Third-Party Doctrine mandates that public exposure to warrantless government surveillance increases in tandem."[40]

PRIVACY DEFINED

In *Whalen v. Roe*,[41] the U.S. Supreme Court recognized the right to two types of privacy: "decisional" and "informational." Decisional privacy involves the interest in being able to make certain decisions, free from government influence.[42] The Court has recognized minors' decisional privacy rights to a limited extent in the context of abortion and access to contraception,[43] but it also recognizes that children lack the maturity to make certain decisions. For example, it has upheld parental notification laws for certain activities.[44] Information privacy, the "confidentiality" dimension of privacy, refers to the individual's interest in securing personal information and avoiding its disclosure.[45] This right can also be limited for minors. For example, children are considered to have no constitutionally protected right to conceal information from their parents.[46]

REASONABLE EXPECTATION OF PRIVACY

The warrant requirement only comes into play if there has been a search within the meaning of the Fourth Amendment.[47] In order for the Fourth Amendment to be implicated, an individual must have a protected interest.[48] The Supreme Court has a long tradition of protecting U.S. citizens' reasonable expectation of privacy.[49] The underlying mandate of the Fourth Amendment is that searches be reasonable;[50] thus, the Fourth Amendment only protects individuals against searches that violate their "reasonable expectation of privacy."[51] To distinguish a legitimate, reasonable search and seizure from an illegal, unreasonable search, Courts utilize the two-pronged reasonable-expectation-of-privacy test created in *Katz v. United States*.[52] First, the subject of the search must believe the place or thing being searched is private. Second, that expectation of privacy must be reasonable in terms of societal standards.[53] For example, it is reasonable to expect privacy in your home but less reasonable to expect privacy in your yard, especially if an individual knows his or her actions are observable by others.[54] Like an adult, a child has a reasonable expectation of privacy—of being free from unreasonable searches.[55]

CONTEXT AND EXCEPTIONS

Context is important in Fourth Amendment analyses for both minors and adults. Context encompasses the physical context as well as the nature of the privacy interest under consideration.[56] Citing state interests such as safety, practicality, and evidence collection, the Supreme Court recognizes a number of exceptions to the warrant requirement that allow officials to conduct searches without a warrant and in some instances even without probable cause.[57] One such exception is the "search incident to arrest doctrine."[58] Under this doctrine, officers are entitled to search the suspect's body to ensure he or she is not armed and to prevent him or her from destroying evidence. This incident is automatic and allows authorities to open containers found on the person—even when there is no probable cause to believe anything illegal is inside.[59]

The Fourth Amendment rights of children have been adjusted in certain special needs contexts, particularly the public school environment. For example, in public school settings, state officials are responsible for the supervision of a large number of youth.[60] The Court has held that public school officials' searches of students and their belongings are exceptions to standard Fourth Amendment protections.[61] Hudson observes that in a post-Columbine environment, safety concerns often trump other considerations, and the issue is "whether school officials' actions are a reasonable way of protecting students from danger or whether they violate students' Fourth Amendment rights."[62]

MINORS' SPECIAL STATUS

In *Prince v. Massachusetts*, the Court again ruled that while children share many of the rights of adults, the rights of children can be restricted to a greater degree than the rights of adults, because children face different potential harms from similar activities. The Court further determined that parental authority is not absolute and can be restricted by the state if doing so is in the child's best interests.[63] *Bellotti v. Baird* (Bellotti II) identified three justifications for limiting the scope of minors' constitutional rights: (1) the particular vulnerabilities of minors, (2) minors' lesser ability to make good decisions, and (3) parents' role in rearing children.[64]

Children, like adults, have a legitimate expectation of privacy and of being free from unreasonable searches of their person, home, papers, and effects; however, minors' rights can be restricted in both the home and the school settings. The parental right to raise children generally takes precedent over children's rights to privacy,[65] and similarly, the Court has allowed school authorities to place limits on children's rights.[66]

MINORS' PRIVACY IN THE FAMILY

The Supreme Court has long maintained that privacy in the family context occupies a special place in society.[67] It has, therefore, shown a strong hesitation in recognizing children's right to privacy within the family unit, partially due to the nature of the parent-child relationship.[68] On the one hand, children require a close, intimate relationship with their parents, and yet the establishment of a child's individual personhood is essential for his or her eventual psychological individuation and independence.[69]

Conversely, courts have recognized parents' constitutional right to control children and raise them in the manner they see fit with minimal government intrusion.[70] The Court presumes that parents' legal obligation to act in their children's best interests rests on their ability to monitor their children's communications.[71] For these reasons, among others, parents are considered to be the primary guarantors of their children's privacy.[72]

Since the Fourth Amendment only applies to representatives of the state, the Fourth Amendment provides no protection from parental intrusion into children's bedrooms or other personal space. Thus, the Fourth Amendment does not provide children with any recourse if a parent enters a room, retrieves evidence, and delivers it to the police.[73] In effect, children have no constitutionally protected right to conceal information from their parents.[74] Few cases have addressed minors' expectation of privacy within the home without the consent or involvement of their parents,[75] and no Supreme Court case directly addresses the issue of children's privacy in their relationship with their parents. However, when the court has decided cases where parents have recorded their children's phone calls, the court has decided in the parents' favor.[76]

VICARIOUS CONSENT RULE

The phrase "vicarious consent" refers to the concept that parents are considered to be the primary gatekeepers of their children's privacy;[77] therefore, they can consent on behalf of their children. The idea of considering parental consent equal to minor consent is based on the notions that due to their age, minors cannot give legal consent, and the notion that parents need to be able to monitor their children's conversations in order to act in their children's best interests.[78] Cases involving wiretapping demonstrate the Court's attempt to balance children's privacy rights with the parental duty to protect their children.[79] A parent can record a child's conversation even if the child objects, because the parent, in effect, consents for the child. In this context, the recording of a child's phone conversation would be legal if the parent provided the consent instead of the child. However, for this action to be protected by the Court, a parent must demonstrate a "good faith and reasonable belief that the interception was necessary to safeguard the best interests of the child."[80]

MINOR'S RIGHTS AND PARENTS—COMMENTARY

The Court's treatment of minor's privacy rights has garnered substantial criticism. Professor Benjamin Schmueli and Professor Ayelet Blecher-Prigat contend there is a "privacy problem" in American jurisprudence around minor's privacy rights, in that the proper balance between child privacy and parental authority has not been met. They believe too little attention has been paid to the significance and value of privacy for children in relation to their parents.[81] Schmueli and Blecher-Prigat further note that there is an assumption of unity of interests between parent and child[82] that, in reality, does not always exist. While acknowledging that children's privacy rights need to be balanced against children's need for protection, nurturing, and care, Schmueli and Blecher-Prigat call for an increased recognition of children's privacy in terms of the parental relationship. They maintain that there must be further examination of whether protection is dependent on an invasion of privacy and whether an invasion actually contributes to the child's safety.[83]

MINORS AS SOLO ACTORS

Although children have no expectation of privacy or constitutional authority to ward off parents' inspection of their bedroom, e-mail accounts, computer, and so forth, minors retain an important interest and protected right in avoiding the state's intrusion into their private living space and private property.[84] Absent urgent circumstances, a child's expectation of privacy exists in his or her personal living space. For example, a child may deny law enforcement entry into his or her bedroom or personal property. Similarly, minors living alone have the right to exclude the state, absent some pressing circumstances justifying entry without consent.[85] Therefore, when at home, a minor can deny police the right to search his or her digital devices.

MINORS' PRIVACY IN THE SCHOOL SETTING

Under the Constitution, "students are persons" in school as well as out of school. Supreme Court cases such as *Tinker v. Des Moines* confirmed that students do not forfeit their constitutional rights when they attend school.[86]

However, public school students are not automatically entitled to the same rights that adults would have in a similar situation.[87] In analyzing students' civil rights in a public school setting, the Court also considers the special characteristics of the school setting.[88] For example, the Court has long recognized the right of the state to limit children's rights in order to ensure their safety.[89]

Although searches generally require a search warrant, the Supreme Court has upheld particular exceptions to the warrant requirement that allow government officials to conduct searches without a warrant and, at times, without probable cause. The Court has held that public school officials' searches of students and their belongings is one such exception to standard Fourth Amendment protections.[90] The Court sought to balance students' legitimate expectations of privacy against the school's interest in conducting searches to maintain control of the school environment (i.e., to maintain order, enforce discipline, ensure safety, find evidence of wrongdoing, etc.). With these competing interests in mind, the Court allowed the elimination of the warrant requirement as well as the probable cause requirement in regard to school searches.[91]

The Supreme Court case *New Jersey v. TLO* is a landmark case in students' privacy rights. TLO, the initials of a minor female student whose name was not released due to her age, was observed, by a teacher, violating school rules by smoking cigarettes in the bathroom. TLO was taken to the principal's office and questioned by the assistant vice principal. During questioning, she denied the allegations. However, when the assistant vice principal demanded to look through her purse, he saw cigarettes and cigarette rolling papers commonly used by marijuana smokers. Based on his initial observations, he then conducted a more thorough search of her purse and found items associated with marijuana dealing, including marijuana, a pipe, plastic bags, and letters implicating her as marijuana dealer. School officials notified TLO's mother and also turned the evidence over to the police. Based on this evidence, TLO was charged by the local police and subsequently adjudicated as a juvenile delinquent. TLO moved to suppress the admission of the evidence found in her purse, based on the principle that the first search of her purse for cigarettes and then the second search for evidence of marijuana dealing violated her Fourth Amendment rights.[92] The state of New Jersey argued that school officials were acting "in loco parentis," a doctrine that allows persons acting in the parental role to receive the same exemptions to the Bill of Rights that parents enjoy. However, the Court rejected that argument, noting that "public officials act not from parents' permission but from state-imposed duties and rules."[93]

The Supreme Court ruled in favor of the state of New Jersey, deeming the search reasonable under the Fourth Amendment and noting that the search was based on a justified suspicion that the student had broken a school rule.

In this ruling, the Court relaxed the standard Fourth Amendment requirement that government officials obtain a warrant based on probable cause before searching an individual's person or belongings.[94] The Court considered the warrant requirement to be too impractical for the school environment, maintaining that requiring school officials to obtain a search warrant prior to taking disciplinary action would reduce their ability to control the school environment.[95] The Court also relaxed the standard requirement for probable cause[96] (having knowledge of sufficiently trustworthy facts to believe that an infraction has been committed),[97] considering it to be too inflexible for the school setting.[98] Instead, the Court imposed the "reasonable suspicion" standard. Unlike probable cause, "reasonable suspicion" is not derived from the constitution but was created by the court to describe a level of suspicion lower than probable cause.[99] Reasonable suspicion means that the authority has "sufficient knowledge to believe" that illicit activity is at hand. This threshold of knowledge is less than that of probable cause.[100]

When applying the standard of reasonable suspicion to determine the reasonableness of a school search, the Court first considers whether the search was justified at its inception. A search is justified at its inception if school officials had reasonable grounds to believe that the search would yield evidence of a crime or a violation of a school rule.[101] Next, the court considers whether the search as it was conducted was "reasonably related to the circumstances which justified the interference in the first place."[102] A search would be reasonably related in scope as long as the search was confined to the objectives of the search and did not use excessive methods "in light of the age and sex of the student and nature of the infraction."[103]

Two Court cases have applied the TLO test to events involving the possession and use of cell phones in violation of school rules prohibiting their use.[104] The first case, *Klump v. Nazareth Area School District* involved Nazareth High School student Christopher Klump. Nazareth Area High School's cell phone policy allowed students to carry their cell phones on campus, but it did not allow them to use them during school hours. Klump's cell phone inadvertently fell out his pocket and landed on his knee. A nearby teacher confiscated his cell phone to enforce the policy. While the phone was in their possession, school administrators called other Nazareth Area High School students listed in Klump's phone directory to determine if they were also violating school policy. In addition, the school officials accessed Christopher's text messages and voice-mail messages and used the instant messaging application to hold a conversation with Christopher's brother without bothering to identify themselves.[105] In deciding *Klump*, the Court ruled that school officials were justified in seizing a student's cell phone when its use violated school policy. However, the search of the phone was not justified at its inception, because school officials had no reason to believe the student was in violation of another school policy.[106]

The second school case regarding a cell phone, *JW v. Desoto County School District*, involved Richard Wade, a 12-year-old African American student attending Southaven Middle School, who was observed by a school employee reading a text message from his father during class. Pursuant to the school's policy, the school employee confiscated his cell phone. School officials later viewed photos stored on the phone and concluded the photos were gang-related. Richard was later suspended for the remainder of the school year. Richard filed suit, claiming that his Fourth Amendment rights had been violated by the unreasonable search and seizure of his cell phone. He further claimed that his suspension based on the photos uncovered from his cell phone violated his due process rights.[107]

In deciding the case, the Court upheld both the search and seizure of Richard's cell phone. The Court considered the seizure of the phone to be reasonable because it was consistent with school policy. The Court considered it a reasonable search, in light of a violation of school policy, for school officials to confiscate the cell phone and examine it to determine how the student was using the cell phone.[108] The cases differed in outcome because Klump unintentionally violated school policy when the phone inadvertently fell out of his pocket, and school officials went on a fishing expedition. In contrast, Wade had intentionally violated the cell phone policy.[109]

The emergence of cell phone technology has prompted calls for heightened safeguards to protect students' Fourth Amendment rights against a TLO standard that many consider to be insufficient to protect students' privacy interests.[110] Spung argues that the distinctive characteristics of students' cell phones necessitates a reexamination of the TLO standard that guides the searches public school officials make through the contents of these devices.[111]

The Court's adoption of less stringent standards for school searches has met with strong criticism. Spung laments that the TLO standard released school officials from the Fourth Amendment's warrant and probable cause requirement.[112] Others, like Barry Feld, believe the Court failed to provide an adequate justification for the application of a lower search threshold for students in schools. They charge that the reasonable suspicion standard provides only minimal protections, because school officials are not likely to characterize their own actions as unreasonable.[113] For example, officials could cite body language or physical appearance as reasons for being suspicious of a student's actions. Barry Feld further maintains that TLO increased the number of searches that school authorities conducted and that it allowed the state to introduce evidence against juveniles that would be inadmissible if seized from an adult under comparable circumstances.[114]

Amy Vorenberg argues that school officials' searches of students' cell phones should require a warrant—unless there are urgent safety concerns. From her perspective, if there are no urgent safety concerns, then warrantless searches of students' cell phones violate students' Fourth Amendment rights.[115]

CHILDREN'S ONLINE PRIVACY PROTECTION ACT

Congress has made it clear that it also harbors serious concerns regarding Internet threats to children's informational privacy. To address this issue, Congress and the Federal Trade Commission (FTC) have taken steps to protect minors' informational privacy rights. Adopted in 1998, and effective in 2000, the Children's Online Privacy Protection Act (COPPA), is designed to protect the privacy of children under age 13, who presumably do not know how to protect themselves while online.[116]

COPPA was enacted to reduce the risk of harm posed by e-commerce,[117] the practice of purchasing and receiving personal information about children (including their name and location)[118] and sexual solicitation of minors.[119] COPPA places parents in control of what online information is collected from their young children[120] to prevent the irresponsible disclosure of children's personal information.[121]

To achieve these goals, Congress and the FTC took steps to assure that youngsters under 13 years of age don't share their personal information on the Internet without the express approval of their parents. More specifically, they established guidelines for operators of websites or online services (including mobile apps) that "collect, use, or disclose personal information from children, and operators of general audience Web sites or online services with actual knowledge they are collecting, using or disclosing information from children under 13."[122]

More specifically, COPPA mandates that children's personal information (name, address, e-mail address, etc.) cannot be visible anywhere on the website. In addition, site operators must take steps to notify parents of a child's use of the site and must receive verifiable parental consent prior to the collection, utilization, or disclosure of the child's personal information. Web site operators are further required to keep such information secure following its collection. Parents may request details about the type of information being collected. Enforced by the Federal Trade Commission, violation of this law can result in legal action as well, as fines up to $11,000.[123]

As the digital age progressed, critics questioned whether COPPA, written in the late 1990s when the Internet was in its early stages, was still relevant in the face of the changing Internet landscape and advancing technologies. Following two years of agency review and public input, the revised COPPA went into effect on July 1, 2013. It gives parents greater control over the online collection of their children's personal information.[124] The amended rule added the following four protected categories: (1) geolocation information; (2) photographs, videos, and audio files; (3) usernames; and (4) "persistent identifiers" (cookies) that can be used to recognize users over time or across different websites.[125] COPPA also closes loopholes that had allowed child-directed apps and websites to permit third parties to collect personal information from children without parental consent or notice.[126]

Some critics charge that COPPA unfairly closes off a significant portion of the Internet to children under 13, because many websites ban youth from accessing them or some of their features to avoid the costs and effort it takes to comply with COPPA. For example, Deborah Gray and Linda Christiansen question why Congress established the cutoff age at 13, despite the FTC's formal recognition that the disclosure of online information is problematic for people of all ages.[127] They argue that, in general, children under 18 are not mature enough to understand the difficulties associated with sharing private information—both now and in the future.[128] These authors advocate for extending the coverage of the act through age 17.[129]

CONCLUSION

An examination of Carrie's situation reveals that Carrie has no recourse under the Fourth Amendment to rectify her situation. In the school setting, the standard warrant requirement would automatically be waived, due to its impracticality. Since Carrie intentionally violated the school policy, it is likely the Court would uphold school officials' confiscation of her cell phone, because it was consistent with school policy. Such a search would be considered reasonable at its inception, because it was prompted by a rule violation. Finally, an analysis of the search's scope would examine whether the search was commensurate with the alleged offense. Carrie's intentional violation of a school policy would open her up to searches to determine her level of compliance with other school rules. Using *JW v. Desoto County School District* as a precedent, it seems the Court would uphold the scope of the search because the search was limited to the cell phone. Therefore, Carrie, as a minor, would probably have little success in suppressing evidence uncovered from her cell phone and would be released from the cheerleading squad. Only time will tell if the Court takes minors' privacy jurisprudence in another direction.

NOTES

1. A. James Spung, "From Backpacks to Blackberries: (Re)Examining *New Jersey V. T.L.O.* in the Age of the Cell Phone," *Emory Law Journal* 61, no. 1 (2011): 121. See for a general discussion of the impact of the pervasiveness of cell phones in modern schools.

2. See Kristin Henning, "The Fourth Amendment Rights of Children at Home: When Parental Authority Goes Too Far," *William and Mary Law Review* 53, no. 1 (2011): 65.

3. In re *Gault*, 387 U.S. 1, 20 (1967).

4. *Tinker v. Des Moines Independent Community School District*, 393 U.S. 503 (1969).

5. Amy Vorenberg, "Indecent Exposure: Do Warrantless Searches of a Student's Cell Phone Violate the Fourth Amendment?" *Berkeley Journal of Criminal Law* 17, no. 1 (2012): 63.

6. Thomas K. Clancy, "What Is a "Search" within the Meaning of the Fourth Amendment?" *Albany Law Review* 70, no. 1 (2006): 4.

7. Ibid., 3.

8. A. James Spung, "From Backpacks to Blackberries: (Re)Examining *New Jersey V. T.L.O.* in the age of the Cell Phone," *Emory Law Journal* 61, no. 1 (2011): 124.

9. Ibid.

10. Ashley B. Snyder, "The Fourth Amendment and Warrantless Cell Phone Searches: When Is Your Cell Phone Protected? *Wake Forest Law Review* 46, no. 1 (2011): 160–62. See for an overview of Fourth Amendment Principles Applied to Cell Phones.

11. U.S. Const. amend. IV.

12. Josh Goldfoot, "The Physical Computer and the Fourth Amendment," *Berkeley Journal of Criminal Law* 16, no. 1 (2011): 114–122. See for a general discussion of the unique aspects of computer searches.

13. Thomas K. Clancy, "What Is a "Search" within the Meaning of the Fourth Amendment?" *Albany Law Review* 70, no. 1 (2006): 35. See for a discussion of *Kyllo*, 533 U.S. at 34.

14. Josh Goldfoot, "The Physical Computer and the Fourth Amendment," *Berkeley Journal of Criminal Law* 16, no. 1 (2011): 21. See for a general discussion of the unique aspects of computer searches.

15. Ibid., 138.

16. Ibid., 139.

17. Ibid., 112.

18. Ibid., 138.

19. A. James Spung, "From Backpacks to Blackberries: (Re)Examining *New Jersey V. T.L.O.* in the Age of the Cell Phone," *Emory Law Journal* 61, no. 1 (2011): 147.

20. Adam M. Gershowitz, "The iPhone Meets the Fourth Amendment," *UCLA Law Review* 56, no. 1 (2008): 41–42.

21. Ibid., 41.
22. Nita A. Farahany, "Searching Secrets," *University of Pennsylvania Law Review* 160, no.5 (2012): 1249.
23. Ibid.
24. Ashley B. Snyder, "The Fourth Amendment and Warrantless Cell Phone Searches: When Is Your Cell Phone Protected?" *Wake Forest Law Review* 46, no. 1 (2011): 163.
25. Ibid., 173.
26. Nita A. Farahany, "Searching Secrets," *University of Pennsylvania Law Review* 160, no. 5 (2012), 1249.
27. Daniel T. Pesciotta, "I'm Not Dead Yet: *Katz, Jones*, and the Fourth Amendment in the 21st Century," *Case Western Reserve Law Review* 63, no. 1 (2012): 215.
28. Monu Bedi, "Facebook and Interpersonal Privacy: Why the Third Party Doctrine Should Not Apply," *Boston College Law Review* 54, no. 1 (2013): 9–10. See for a discussion of the history of the Third Party Doctrine.
29. *United States v. Miller*, 425 U.S. 435, 443 (1976).
30. Daniel T. Pesciotta, "I'm Not Dead Yet: *Katz, Jones*, and the Fourth Amendment in the 21st Century," *Case Western Reserve Law Review* 63, no. 1 (2012): 202.
31. Monu Bedi, "Facebook and Interpresonal Privacy: Why the Third Party Doctrine Should Not Apply," *Boston College Law Review* 54, no. 1 (2013): 9–10. See for a discussion of the history of the Third Party Doctrine.
32. See *Smith v. Maryland*, 442 U.S. at 744–46 (1979).
33. Monu Bedi, "Facebook and Interpresonal Privacy: Why the Third Party Doctrine Should Not Apply," *Boston College Law Review* 54, no. 1 (2013): 13. See for a discussion of *Smith*, 442 U.S. 744–45 (1979) (citation omitted) the history of the Third Party Doctrine.
34. Ibid., 14. See *Smith*, 442 U.S, at 744–45 (1979); Matthew Tokson, "Automation and the Fourth Amendment," *Iowa Law Review* 96, no. 2 (2011): supra note 23, at 600.
35. Daniel T. Pesciotta, "I'm Not Dead Yet: *Katz, Jones*, and the Fourth Amendment in the 21st Century," *Case Western Reserve Law Review* 63, no. 1 (2012): 215.
36. See Monu Bedi, "Facebook and Interpersonal Privacy: Why the Third Party Doctrine Should Not Apply," *Boston College Law Review* 54, no. 1 (2013): 2. See for a discussion of Court cases dealing with digital communication.
37. Monu Bedi, "Facebook and Interpersonal Privacy: Why the Third Party Doctrine Should Not Apply," *Boston College Law Review* 54, no. 1 (2013): 18. See for a discussion court cases dealing with digital communication.
38. Ibid.
39. Ibid., 14.
40. Erin Smith Dennis, "A Mosaic Shield: *Maynard*, The Fourth Amendment, and Privacy Rights in the Digital Age," *Cardozo Law Review* 33, no. 2 (2011):753.
41. Whalen v. Roe, 429 U.S. 589 (1977).
42. Helen L. Gilbert, "Minors' Constitutional Right to Informational Privacy," *University of Chicago Law Review* 74, no. 4 (2007): 1378.

43. Benjamin Shmueli and Ayelet Blecher-Prigat, "Privacy for Children," *Columbia Human Rights Law Review* 42, no. 3 (2011): 778–79.

44. Charlene Simmons, "Protecting Children While Silencing Them: The Children's Online Privacy Protection Act and Children's Free Speech Rights," *Communication Law and Policy* 12, no. 2 (2007): 134.

45. Helen L. Gilbert, "Minors' Constitutional Right to Informational Privacy," *University of Chicago Law Review* 74, no. 4 (2007): 1378.

46. Benjamin Shmueli and Ayelet Blecher-Prigat, "Privacy for Children," *Columbia Human Rights Law Review* 42, no. 3 (2011): 780.

47. Ashley B. Snyder, "The Fourth Amendment and Warrantless Cell Phone Searches: When is your Cell Phone Protected?" *Wake Forest Law Review* 46, no. 1 (2011): 163.

48. Thomas K. Clancy, "What is a "Search" within the Meaning of the Fourth Amendment?" *Albany Law Review* 70, no. 1 (2006): 2.

49. Daniel T. Pesciotta, "I'm Not Dead Yet: *Katz, Jones*, and the Fourth Amendment in the 21st Century," *Case Western Reserve Law Review* 63, no. 1 (2012): 189.

50. Barry C. Field, "T.L.O. and Reddings Unanswered (Misanswered) Fourth Amendment Questions: Few Rights and Fewer Remedies," *Mississippi Law Journal* 80, no. 3 (2011): 855.

51. Daniel T. Pesciotta, "I'm Not Dead Yet: *Katz, Jones*, and the Fourth Amendment in the 21st Century," *Case Western Reserve Law Review* 63, no. 1 (2012): 199.

52. Barry C. Feld, "T.L.O. and Reddings Unanswered (Misanswered) Fourth Amendment Questions: Few Rights and Fewer Remedies," *Mississippi Law Journal* 80, no. 3 (2011): 855.

53. Daniel T. Pesciotta, "I'm Not Dead Yet: *Katz, Jones*, and the Fourth Amendment in the 21st Century," *Case Western Reserve Law Review* 63, no. 1 (2012): 199.

54. Ibid., 207.

55. Kristin Henning, "The Fourth Amendment Rights of Children at Home: When Parental Authority Goes Too Far," *William and Mary Law Review* 53, no. 1 (2011): 69.

56. Ibid., 62.

57. See Katz v. United States, 389 U.S. 347, 357 (1967).

58. Adam M. Gershowitz, "The iPhone Meets the Fourth Amendment," *UCLA Law Review* 56, no. 1 (2008): p. 32.

59. Ibid., 28.

60. Kristin Henning, "The Fourth Amendment Rights of Children at Home: When Parental Authority Goes Too Far," *William and Mary Law Review* 53, no. 1 (2011): 65.

61. A. James Spung, "From Backpacks to Blackberries: (Re)Examining *New Jersey v. T.L.O.* in the age of the Cell Phone," *Emory Law Journal* 61, no. 1 (2011): 126.

62. David L. Hudson, Jr., "The Fourth Amendment in Public Schools," *Insights on Law & Society* 11, no. 2 (2011): 14.

63. *Prince v. Massachusetts*, 321 U.S. 158 (1944).

64. *Bellotti v. Baird*, 443 U.S. 622 (1979).

65. Benjamin Shmueli and Ayelet Blecher-Prigat, "Privacy for Children," *Columbia Human Rights Law Review* 42, no. 3 (2011): 778–79.

66. Charlene Simmons, "Protecting Children While Silencing Them: The Children's Online Privacy Protection Act and Children's Free Speech Rights," *Communication Law and Policy* 12, no. 2 (2007): 131.

67. Benjamin Shmueli and Ayelet Blecher-Prigat, "Privacy for Children," *Columbia Human Rights Law Review* 42, no. 3 (2011): 772.

68. Ibid., 762.

69. Ibid., 775–76. See for a discussion of the conflict between family privacy and individual privacy.

70. Kristin Henning, "The Fourth Amendment Rights of Children at Home: When Parental Authority Goes Too Far," *William and Mary Law Review* 53, no. 1 (2011): 73.

71. Benjamin Shmueli and Ayelet Blecher-Prigat, "Privacy for Children," *Columbia Human Rights Law Review* 42, no. 3 (2011): 781.

72. Ibid., 777.

73. Kristin Henning, "The Fourth Amendment Rights of Children at Home: When Parental Authority Goes Too Far," *William and Mary Law Review* 53, no. 1 (2011): 70.

74. Benjamin Shmueli and Ayelet Blecher-Prigat, "Privacy for Children" *Columbia Human Rights Law Review* 42, no. 3 (2011): 780.

75. Kristin Henning, "The Fourth Amendment Rights of Children at Home: When Parental Authority Goes Too Far," *William and Mary Law Review* 53, no. 1 (2011): 72.

76. Benjamin Shmueli and Ayelet Blecher-Prigat, "Privacy for Children," *Columbia Human Rights Law Review* 42, no. 3 (2011), 780–781. See for a general discussion of vicarious consent.

77. See Benjamin Shmueli and Ayelet Blecher-Prigat, "Privacy for Children," *Columbia Human Rights Law Review* 42, no. 3 (2011), 781.

78. *Thompson v. Dulaney*, 838 F. Supp. 1535 (D. Utah 1993) District Court of Utah.

79. Benjamin Shmueli and Ayelet Blecher-Prigat, "Privacy for Children," *Columbia Human Rights Law Review* 42, no. 3 (2011): 782.

80. Ibid., 781.

81. Ibid., 790.

82. Ibid., 781.

83. Ibid., 791.

84. Kristin Henning, "The Fourth Amendment Rights of Children at Home: When Parental Authority Goes Too Far," *William and Mary Law Review* 53, no. 1 (2011): 108.

85. Ibid., 70.

86. Charlene Simmons, "Protecting Children While Silencing Them: The Children's Online Privacy Protection Act and Children's Free Speech Rights," *Communication Law and Policy* 12, no. 2 (2007): 131.

87. See *Bethel School District No 403 v. Fraser*, 478 U.S. 675, 682 (1986).

88. See *Tinker v. Des Moines Independent Community School District*, 393 U.S. 503, 506 (1969).

89. Charlene Simmons, "Protecting Children While Silencing Them: The Children's Online Privacy Protection Act and Children's Free Speech Rights," *Communication Law and Policy* 12, no. 2 (2007): 130.

90. A. James Spung, "From Backpacks to Blackberries: (Re)Examining New Jersey V. T.L.O. in the age of the Cell Phone," *Emory Law Journal* 61, no. 1 (2011): 126.

91. Ibid., 128.

92. See *New Jersey v. T.L.O.*, 469 U.S. 325 (1985).

93. David L. Hudson, Jr., "The Fourth Amendment in Public Schools," *Insights on Law & Society* 11, no. 2 (2011): 15.

94. *New Jersey v. T.L.O.*, 469 U.S. 325, 339–45 (1985).

95. Ibid.

96. Ibid.

97. The Free Legal Dictionary, "Probable Cause and Reasonable Suspicion," accessed August 18, 2013, http://legal-dictionary.thefreedictionary.com/Probable+Cause+and+Reasonable+Suspicion.

98. David L. Hudson, Jr., "The Fourth Amendment in Public Schools," *Insights on Law & Society* 11, no. 2 (2011): 15.

99. Devallis Rutledege, "Probable Cause and Reasonable Suspicion: These Familiar Terms Are Often Confused and Misused," *Police Patrol: The Law Enforcement Magazine*, June 7, 2011, accessed August 18, 2013, http://www.policemag.com/channel/patrol/articles/2011/06/probable-cause-and-reasonable-suspicion.aspx.

100. The Free Legal Dictionary, "Probable Cause and Reasonable Suspicion," accessed August 18, 2013, http://legal-dictionary.thefreedictionary.com/Probable+Cause+and+Reasonable+Suspicion.

101. *New Jersey v. T.L.O.*, 469 U.S. at 341–42 (1985).

102. *New Jersey v. T.L.O.*, 469 U.S. at 342 (1985). (quoting Terry v. Ohio, 392 U.S. 1, 20 (1967).

103. *New Jersey v. T.L.O.*, 469 U.S. at 342 (1985).

104. Naomi Harlin Goodno, "How Public Schools can Constitutionally Halt Cyberbullying: A Model Cyberbullying Policy that Considers First Amendment, Due Process, and Fourth Amendment Challenges," *Wake Forest Law Review* 46, no. 4 (2011): 672.

105. *Klump v. Nazareth Area High School*, 425 F. Supp.2d 622 (2006).

106. *Klump v. Nazareth Area School District*, 425 F. Supp. 2d 622, 627 (E.D. Penn, 2006).

107. *J.W. v. Desoto County School District*, No 2:09-cv-00155-MPM-DAS, 2010, WL 439 4059 at 1-2 (N.D. Miss 2010).

108. Naomi Harlin Goodno, "How Public Schools Can Constitutionally Halt Cyberbullying: A Model Cyberbullying Policy that Considers First Amendment, Due Process, and Fourth Amendment Challenges," *Wake Forest Law Review* 46, no. 4 (2011): 673.

109. Ibid. 674.

110. A. James Spung, "From Backpacks to Blackberries: (Re)Examining *New Jersey v. T.L.O.* in the age of the Cell Phone," *Emory Law Journal* 61, no. 1 (2011): 114.

111. Ibid., 115.

112. Ibid., 134.

113. Barry C. Feld, "T.L.O. and Reddings Unanswered (Misanswered) Fourth Amendment Questions: Few Rights and Fewer Remedies," *Mississippi Law Journal* 80, no. 3 (2011): 855.

114. Ibid., 898.

115. Amy Vorenberg, "Indecent Exposure: Do Warrantless Searches of a Student's Cell Phone Violate the Fourth Amendment?" *Berkeley Journal of Criminal Law* 17, no. 1 (2012): 64.

116. Charlene Simmons, "Protecting Children While Silencing Them: The Children's Online Privacy Protection Act and Children's Free Speech Rights," *Communication Law and Policy* 12, no. 2 (2007): 126.

117. Anita L. Allen, "Minor Distractions: Children, Privacy and E-Commerce," *Houston Law Review* 38, part 3 (2001): 752.

118. Sasha Grandison, "The Child Online Privacy Protection Act: The Relationship between Constitutional Rights and the Protection of Children," *University of the District of Columbia Law Review* 14, no. 1 (2011): 210. See for a general discussion of the e-commerce exploitation of children that prompted this act.

119. Ibid., 221.

120. Business Center, "Complying with COPPA: Frequently Asked Questions. A Guide for Business and Parents and Small Entity Compliance Center," *Federal Trade Commission*, revised June 2013, http://business.ftc.gov/documents/Complying-with-COPPA-Frequently-Asked-Questions.

121. Anita L. Allen, "Minor Distractions: Children, Privacy and E-Commerce," *Houston Law Review* 38, part 3 (2001): 752.

122. Business Center. "Complying with COPPA: Frequently Asked Questions. A Guide for Business and Parents and Small Entity Compliance Center," *Federal Trade Commission*, revised June 2013, http://business.ftc.gov/documents/Complying-with-COPPA-Frequently-Asked-Questions.

123. Sara Grimes, "Revisiting the Children's Online Privacy Protection Act" (blog), *Joan Ganz Cooney Center*, March 25, 2013, http://www.joanganzcooneycenter.org/2013/03/25/revisiting-the-childrens-online-privacy-protection-act/.

124. Ibid.

125. Business Center, *Complying with COPPA: Frequently Asked Questions. A Guide for Business and Parents and Small Entity Compliance Center, Federal Trade Commission*, revised June 2013, http://business.ftc.gov/documents/Complying-with-COPPA-Frequently-Asked-Questions.

126. Federal Trade Commission, "FTC Strengthens Kids' Privacy, Gives Parents Greater Control over Their Information by Amending Children's Online Privacy Protection: Rule Being Modified to Keep Up with Changing Technology," December 19, 2012. http://www.ftc.gov/opa/2012/12/coppa.shtm.

127. Deborah M. Gray and Linda Christiansen, "A Call to Action: The Privacy Dangers Adolescents Face through Use of Facebook.com," *Journal of Information Privacy and Security* 62, no. 2 (2010): 23.

128. Ibid., 22, 26.

129. Ibid., 25.

7

Library Patrons and the National Security State

Chad Kahl

Imagine visiting your local library; you have just heard about a terrorist attack on American interests abroad, and you want to learn more, especially because the suspected people involved in the attack share your ancestry—one even shares a derivation of your surname—and you visited that area the past summer. You get access to a library laptop; sit down; and explore more about where, why, and how the attacks occurred. You also read information about the suspected terrorists and their beliefs, what happened to the region of the world that your ancestors left to find a new life in the United States, and so on. You are particularly interested when you read about how the terrorist organization recruits participants from around the world using social media, so you explore a number of the Internet chat forums, Facebook sites, and web pages to find the tools used to attract attention from around the world. After you use the computer, you look to see if the library has any books on the history of that area of the world as well as more general information about terrorism to answer the question, "What goes through their minds?" Now, imagine doing that same search knowing that government agents are going to review your Internet search history and check on what sites you have visited, see what books you have checked out from the library and speculate why you wanted all the information in the first place. Why all this interest, especially for someone who visited there this last summer? Also imagine this monitoring taking place without your knowledge, despite your being a U.S. citizen, and the library staff being prevented by federal law to say a word to you about it happening. This scenario is completely plausible, given the current state of governmental powers after the horrific September 11, 2001, attacks and the passage of the USA PATRIOT Act.[1]

Given the scope of libraries and their broad use—there are over 120,000 libraries in the United States,[2] 58 percent of Americans have public library cards, and nearly 6.6 million reference questions are answered weekly[3]—it's not surprising that governmental access to library patron information is a much-debated topic. This chapter will examine the development of the professional ethos of libraries and librarians; the historical arc of Supreme Court decisions on the Fourth Amendment, especially those related to search and seizure; how governmental efforts increasing information gathering have impacted library patrons; and how libraries responded to governmental intrusions over the past century, with a special emphasis on the protection of library patrons' rights.

LIBRARY AND LIBRARIAN PROFESSIONAL ETHOS

The professional organization for libraries and librarians, the American Library Association (ALA), has established a strong professional ethos for promoting intellectual freedom and free expression, fighting censorship and the abuse of government power, and protecting library patrons' free access to information and their privacy and confidentiality. This can be seen through a perusal of the profession's foundational documents and the development of specialized organizational units.

The Code of Ethics of the American Library Association was first adopted in 1939, and it demonstrates these values with a number of the code's principles:

> II. We uphold the principles of intellectual freedom and resist all efforts to censor library resources.
> III. We protect each library user's right to privacy and confidentiality with respect to information sought or received and resources consulted, borrowed, acquired or transmitted . . .
> VII. We distinguish between our personal convictions and professional duties and do not allow our personal beliefs to interfere with fair representation of the aims of our institutions or the provision of access to their information resources.[4]

The profession also recognizes the special role that libraries play in society and a commitment to the library patron with the Library Bill of Rights that was originally passed in 1939. A number of the bill's policies reflect the professional ethos:

> III. Libraries should challenge censorship in the fulfillment of their responsibility to provide information and enlightenment.

IV. Libraries should cooperate with all persons and groups concerned with resisting abridgment of free expression and free access to ideas.
V. A person's right to use a library should not be denied or abridged because of origin, age, background, or views.[5]

The very act of reading and exposing differing points of view was recognized by the ALA and the American Book Publishers Council (now called the Association of American Publishers) when they passed The Freedom to Read Statement in 1953. The statement begins with the powerful statement, "The freedom to read is essential to our democracy. It is continuously under attack." Among the seven propositions was the sixth, which said, "*It is the responsibility of publishers and librarians, as guardians of the people's freedom to read, to contest encroachments upon that freedom by individuals or groups seeking to impose their own standards or tastes upon the community at large; and by the government whenever it seeks to reduce or deny public access to public information* (italics in the original)."[6] The timing of the statement was in response to the political times, as Joan Starr noted that "the ALA planned and executed a successful publicity campaign, marking its first significant confrontation with McCarthy, and giving librarians 'a high profile as defenders of an essential democratic freedom, the freedom to read.'"[7]

The ALA Office for Intellectual Freedom was established in 1967 and was charged with advanced ALA efforts in promoting intellectual freedom.[8] Its creation showed an organizational commitment to sustained activities that put the foundational documents into daily use.

The Policy on Confidentiality of Library Records was adopted by the ALA Council in 1971:

The Council of the American Library Association strongly recommends that the responsible officers of each library, cooperative system, and consortium in the United States:

1. Formally adopt a policy that specifically recognizes its circulation records and other records identifying the names of library users to be confidential.
2. Advise all librarians and library employees that such records shall not be made available to any agency of state, federal, or local government except pursuant to such process, order, or subpoena as may be authorized under the authority of, and pursuant to, federal, state, or local law relating to civil, criminal, or administrative discovery procedures or legislative investigative power.

3. Resist the issuance of enforcement of any such process, order, or subpoena until such time as a proper showing of good cause has been made in a court of competent jurisdiction.[9]

The policy not only advises a course of action but also calls for resistance until a proper request has been made by a proper court.

In 1973, the ALA Council also expressed a point of view regarding government intrusion[10] with its Policy on Government Intimidation, which states, "The American Library Association opposes any use of governmental prerogatives that lead to the intimidation of individuals or groups and discourages them from exercising the right of free expression as guaranteed by the First Amendment to the U.S. Constitution. ALA encourages resistance to such abuse of governmental power and supports those against whom such governmental power has been employed."[11]

More specifically, the profession pays careful attention to the importance of privacy. In 2002, the ALA Council adopted the following statement, Privacy: An Interpretation of the Library Bill of Rights. The statement notes:

Privacy is essential to the exercise of free speech, free thought, and free association . . . Further, the courts have upheld the right to privacy based on the Bill of Rights of the U.S. Constitution . . . In a library (physical or virtual), the right to privacy is the right to open inquiry without having the subject of one's interest examined or scrutinized by others. Confidentiality exists when a library is in possession of personally identifiable information about users and keeps that information private on their behalf. Protecting user privacy and confidentiality has long been an integral part of the mission of libraries.

When users recognize or fear that their privacy or confidentiality is compromised, true freedom of inquiry no longer exists . . . Lack of privacy and confidentiality has a chilling effect on users' choices. All users have a right to be free from any unreasonable intrusion into or surveillance of their lawful library use. Users have the right to be informed what policies and procedures govern the amount and retention of personally identifiable information, why that information is necessary for the library, and what the user can do to maintain his or her privacy. Library users expect and in many places have a legal right to have their information protected and kept private and confidential by anyone with direct or indirect access to that information.

The library profession has a long-standing commitment to an ethic of facilitating, not monitoring, access to information . . . [T]he collection of personally identifiable information should only be a matter of routine or policy when necessary for the fulfillment of the mission of the library. Regardless of the technology used, everyone who collects or accesses personally identifiable information in any format has a legal and ethical obligation to protect confidentiality.[12]

The ALA has stepped up its resistance to governmental intrusions on its professional ethical standards, and this will be seen in a later examination of the effect of the Patriot Act on library patrons.

FOURTH AMENDMENT DEVELOPMENTS

The Fourth Amendment to the Constitution reads, "The right of the people to be secure in their persons, houses, papers, and effects, against unreasonable searches and seizures, shall not be violated, and no Warrants shall issue, but upon probable cause, supported by Oath or affirmation, and particularly describing the place to be searched, and the persons or things to be seized."[13]

The inclusion of these protections into the Bill of Rights occurred in response to the British use of general warrants that allowed governmental officials large leeway in terms of on whom, where, and when searches could be done. The Fourth Amendment allowed the constitutional framers to specifically prevent the use of the "unreasonable searches and seizures" and required warrants to be specifically detailed by an agent of the court.[14]

The historical examination of the development of Fourth Amendment rights, especially in the areas of search and seizures, demonstrates the continual conflict between the state—primarily through the actions of law enforcement—and the private citizen.[15] Starting with *Boyd v. United States* (1884),[16] the Supreme Court began to better define how evidence could be legally gathered through the use of proper searches. In general, greater rights were afforded private citizens with the establishment of the federal exclusionary rule with *Weeks v. United States* (1914)[17] and its extension to the states in *Mapp v. Ohio* (1961),[18] the establishment of the (later named) "fruit of the poisonous tree" doctrine in *Silverthorne Lumber Co. v. United States* (1921)[19] and culminating in *Katz v. the United States* (1967)[20] that established the reasonable-expectation-of-privacy standard.

Since the late sixties and early seventies, the balance between the conflicting sides has trended toward the rights of the state. By examining Supreme Court cases since *Katz*, one can begin to see how Fourth Amendment rights have been eroded in many different circumstances in roughly the last half century.

This can be seen with the establishment of exceptions that allow warrantless searches, such as "hot pursuit" (*Warden v. Hayden*, 1967);[21] "plain-view" allowances (*Chimel v. California*, 1969);[22] "murder scene" (*Mincey v. Arizona*, 1978);[23] "inevitable discovery" (*Nix v. Williams*, 1984);[24] "independent source" (*Segura v. United States*, 1984);[25] "good faith" (*United States v. Leon*, 1984);[26] clerical errors that do not obviate warrant (*Arizona v. Evans*, 1995);[27] and prevention of destruction of evidence (*Kentucky v. King*, 2011).[28] There has been expansion of police techniques: "stop and frisk" is permitted (*Terry v. Ohio*, 1968).[29] Drivers have seen a number of reductions to their former Fourth Amendment rights with the "plain-view" allowances extension to vehicles (*New York v. Belton*, 1981);[30] probable cause allowing searching of stopped vehicles (*United States v. Ross*, 1982);[31] permission of fixed sobriety highway checkpoints (*Michigan State Police v. Sitz*, 1990);[32] containers and packages within vehicles becoming searchable (*Florida v. Jimeno*, 1991,[33] and *California v. Acevedo*, 1991);[34] greater leeway to frisking driver and passengers (*Maryland v. Wilson*, 1997);[35] expansion of searches to all personal belongings that could contain contraband (*Wyoming v. Houghton*, 1999);[36] reasonable suspicion, rather than probable cause, as a test for vehicle search (*United States v. Arvizu*, 2002);[37] traffic checkpoints being utilized to gather information about a crime (*Illinois v. Lidster*, 2004);[38] vehicles searched after arrest, even if the arrested is no longer in vehicle (*Thornton v. United States*, 2004);[39] and evidence being utilized at routine traffic stop as long as police activity does not unusually extend the stop (*Illinois v. Caballes*, 2005).[40] Students have continually had their protections eroded, as seen with allowance of warrantless searches with standards lower than probable cause (*New Jersey v. TLO*, 1985);[41] permission of random drug testing of athletes (*Vernonia School District v. Acton*, 1995);[42] and permission of random drug testing for students in extracurricular activities (*Board of Education v. Earls*, 2002).[43]

In addition, the third-party doctrine was established after *Katz* with *United States v. Miller* (1976)[44] and *Smith v Maryland* (1979),[45] which asserted "that there is no expectation of privacy in bank records or the phone numbers called . . . The question of library records . . . is whether or not you have an expectation of privacy, or whether the Court would use the third-party doctrine. With few exceptions (e.g., *Kyllo v. United States*),[46] *Katz* cases have ended up favoring the state."[47] If the doctrine were used for library records, as famous Fourth Amendment treatise author Wayne LaFave notes in the quote taken from *Miller*: "This Court had has held repeatedly that the Fourth Amendment does not prohibit the obtaining of information revealed to a third party and conveyed to him by Government authorities, even if the information is revealed on the assumption that it will be used for a limited purpose and the confidence placed in the third party will not be betrayed."[48]

The scope and frequency at which the state has regained power has accelerated in recent years toward the end of the Burger Court (1969–1986), through the Rehnquist Court (1986–2005), and into the Roberts Court (2005–present). Having established this shift in the Supreme Court's views on Fourth Amendment rights, one also should review how governmental efforts have affected the rights of library patrons.

GOVERNMENTAL EFFORTS AND THE LIBRARY PATRON

While issues related to Fourth Amendment rights, especially privacy and warrantless searches, have been debated in the Supreme Court for decades, there has been little done to protect the library patron's right to privacy. In 2006, Stacey Bowers noted, "While today federal legislation grants privacy rights regarding video rental records, cable records, banking records, and health records, there is still no federal legislation that provides protection for library records."[49] This statement was made even before the public began to find out about additional warrantless searches being done by the government in the name of national security.

One notable legislative attempt was made in 1988 after a newspaper published the video rentals of Supreme Court nominee Robert Bork. Bills were introduced in both houses of Congress. The Senate introduced the Video Privacy Protection Act and the House of Representatives introduced the Video and Library Privacy Protection Act. The House bill "provided protection for library records, including circulation records, reference interview records, database search records, interlibrary loan records, and other personally identifiable records regarding a patron's use of library services and materials."[50] However, the Federal Bureau of Investigation opposed the House version because of its inclusion of protection for library records. "The FBI worked behind the scenes to persuade various Senators and Congressmen that the only way the FBI could support this proposed bill was to use a National Security Letter (NSL) to obtain information regarding library records ... the NSL exemption would allow the FBI to gain access to library records without judicial review or notification and would essentially allow them to bypass all the requirements of proposed legislation."[51] The negotiations on the bills stalled, so the library privacy protections were removed from the bill, and it eventually passed into law with privacy protections for video purchases and rentals only. "Using its high-level influence, the FBI successfully thwarted Congress' attempt to provide for federal legislative protection of library records."[52]

Protections for library records that do exist are at the state level. The ALA notes that "forty-eight of 50 states have [state confidentiality] law on the books, but the language varies from state to state."[53] The two states without statutes do have attorney general opinions that state that those library records are not subject to open-records laws.[54]

The legislation with the most impact on libraries, especially patron records, has been the USA PATRIOT (Uniting and Strengthening America by Providing Appropriate Tools Required to Intercept and Obstruct Terrorism) Act (USAPA). The 342-page bill became law on October 26, 2001, just six weeks after the September 11 attacks, in an atmosphere of utter fear.[55] It consists of ten titles and numerous sections:

Title I: Enhancing domestic security against terrorism (6 sections)
Title II: Surveillance procedures (25 sections)
Title III: Anti-money-laundering to prevent terrorism (46 sections)
Title IV: Border security (21 sections)
Title V: Removing obstacles to investigating terrorism (8 sections)
Title VI: Victims and families of victims of terrorism (8 sections)
Title VII: Increased information sharing for critical infrastructure protection (1 section)
Title VIII: Terrorism criminal law (17 sections)
Title IX: Improved intelligence (8 sections)
Title X: Miscellaneous (16 sections)

Reviewing the nearly 150 sections is beyond the scope of this chapter, so the focus will be on sections that most seemingly impact Fourth Amendment matters and the privacy of library records.[56]

Section 203: Requires sharing of foreign intelligence information between federal agencies, even prior to grand juries. Issues with Fourth Amendment impact: "The broad definition of 'foreign intelligence' includes virtually anything related to national defense, national security, or foreign affairs."[57] No court orders are required during sharing, and sharing of information is not limited to the scope of investigation.[58]

Section 206: "Roving wiretap" authority was incorporated into the Foreign Intelligence Security Act (FISA).[59]

Section 207: Extends the length of FISA warrants

Section 208: Expands the number of judges available for FISA warrant reviews.

Section 209: Lessens the legal rules for warrant access to voice mail; makes it equivalent to email.

Section 211: Adds cable companies to telephone and electronic communication providers that "may be required to provide law enforcement officials with customer identifying information without notifying their customers."[60]

Section 212: Allows electronic communications services to share records of customers if there is belief it will "protect life and limb" and allow services to share "non-content" information of customers.

Section 213: Allows delayed notice of "sneak and peek" warrants—"one that authorizes officers to secretly enter (either physically or electronically), conduct a search, observe, take measurements, conduct examinations, smell, take pictures, copy documents, download or transmit computer files, and the like; and depart without taking any tangible evidence or leaving notice of their presence."[61]

Section 214: Expands scope of the use of trap and trace devices and pen registers to the broader definition of "foreign intelligence" as broadly redefined in Section 203.

Section 215: "Allows federal investigators to seize 'any tangible thing' in FISA type investigations by showing only that the items sought are 'relevant' to an ongoing investigation related to terrorism or intelligence activities. The court *must* grant the warrant to the requesting agency, even if the judge believes the request is without merit."[62] Also expands the classifications of FBI agents able to apply for court order.

Section 216: Expands use of trap and trace devices and pen registers (and their modern software equivalents) to the "modern communication technologies" rather than only telephone technology. Authorization was formerly limited to jurisdiction of the target phone; this was relaxed to multiple jurisdictions anywhere in the United States. Federal agents have up to 30 days to report their use.

Section 217: Makes it easier for law enforcement investigators to access computer communication.

Section 218: Lessens standard for attaining a FISA warrant from "the purpose for the surveillance is to obtain foreign intelligence information" to a "significant purpose" in obtaining foreign intelligence information. In doing so, it encourages greater participation by domestic law enforcement in what was formerly the domain of foreign intelligence efforts.[63]

Section 219: Allows magistrates to issue warrants in and out of their jurisdiction.

Section 220: Allows for the issuance of nationwide warrants for electronic evidence so that an investigation is not required to also get a warrant of jurisdiction where target is located.

Section 358: "Allows governmental investigators access to consumer records without a court order."[64]

Section 505: Lessens threshold for FBI use of a National Security Letter: "Prior to section 505, the FBI was required to assert that the information sought was related to a foreign power, foreign agent, an international terrorist, or an individual engaged in clandestine intelligence activities . . . and [now] allows a NSL to be issued when the FBI certifies, the information sought is 'relevant to an authorized foreign counterintelligence investigation.'"[65]

Section 507: Makes educational records available to the government without court order.

In reviewing the highlighted sections, one can see a substantial reduction in constitutional protections previously enjoyed by private citizens. The scope of what information can be gathered increased immensely with the broader definitions given to "foreign intelligence" in Section 203 and what can be seized and "any tangible thing" as seen in Section 215. The widespread sharing of information between intelligence agencies (with an external focus) and law enforcement agencies (with an internal focus) may have improved the consolidation of information gathering and analysis, but it also invited increased surveillance of U.S. citizens. Surveillance tools (e.g., roving wiretaps, trap and trace devices, and pen registers) were newly allowed or less restricted. Warrants were extended in terms of length and scope (e.g., FISA warrants) and who can grant them (e.g., granting out-of-jurisdiction magisterial privileges for national electronic warrants). The amount of information available to the government without notice was expanded to include cable company customers for "noncontent" information as well as consumer and educational records.[66] Foerstel also pointed out that "the loss of judicial oversight of executive power is one of the more troubling effects of the Patriot Act. Under many of the act's provisions, the court exercises not review function whatsoever. For example, the court is often required to grant government access to sensitive personal information upon the mere request by a government official."[67]

Some of the aspects of USAPA were deliberately given sunset privileges that would require Congressional reauthorization. For instance, in the sections reviewed above, Sections 203 (in part), 206, 207, 209, 212, 214, 215,

217, 218, and 220 were all set to expire January 1, 2006. The reauthorizations have allowed for renewed debate between civil libertarians and those recommending additional governmental powers.

When the first reauthorization neared, at the end of 2005, the Security and Freedom Ensured Act (SAFE Act) was introduced into the Senate;[68] it aimed to limit the use of roving wiretaps and sneak and peek warrants, return FISA to pre–Section 218 requirements, prevent use of National Security Letters to access library records, and expand the number of sections with sunset provisions. It died in the Committee on the Judiciary.[69] USAPA was reauthorized with the passage of the USA PATRIOT Improvement and Reauthorization Act of 2005.[70] It extended 14 of the 16 expiring sections; enacted new sunset dates of December 31, 2009 for Sections 206 and 215; provided greater judicial and legislative oversight of Sections 206 and 215 and National Security Letters; required higher level approval of Section 215 FISA warrants for certain types of records, including library records; enhanced procedures and oversight on Section 213; wiretaps were made available for greater number of federal crimes; created a new federal crime for "special event of national significance"; and so forth.[71] The PATRIOT Act Additional Reauthorizing Amendments Act of 2006[72] soon followed in February 2006 and amended the first reauthorization act. Recipients of Section 215 orders are subject to a nondisclosure requirement (i.e., a gag order). This second reauthorization act allowed the recipient to petition a FISA judge to modify or remove the gag order. It also allowed recipients of Section 215 orders or National Security Letters to consult with an attorney without prior notification to the government. Lastly, it clarified that libraries are not subject to National Security Letters unless they function as an electronic communication service provider. An additional reauthorization on "lone wolf" surveillance from FISA reauthorization was included.

Before the sunset provisions could lapse at the end of 2009, the Judicious Use of Surveillance Tools In Counterterrorism Efforts (JUSTICE) Act of 2009 was introduced to the Senate;[73] it proposed changes to FISA to impose limits on roving wiretaps, trap and trace devices, and pen registers; repealed retroactive immunity to telecommunication providers for illegal disclosure of customer records; prohibited of warrantless collection of some American citizen communications; allowed a FISA order recipient with the ability to challenge it; and redefined the term, "domestic terrorism," but died in committee.[74] Instead, President Obama signed HR 3961[75] into law, and this extended the sunset provisions for an additional year.[76]

The latest reauthorizations, the FISA Sunsets Act of 2011,[77] and the PATRIOT Sunsets Extension Act of 2011,[78] meant that the roving wiretaps,

Section 215 ("tangible things" and "lone wolf" provisions) were extended until June 1, 2015.[79]

USAPA also provided additional latitude for the use of another governmental tool, the National Security Letter (NSL). An NSL "is a form letter signed by an FBI agent that can direct third parties to provide customer account information and transactional communication records."[80] NSLs existed before USAPA but were initially designed for use in financial investigations when they were enacted in 1986. They were expanded in the mid-1990s to offer access to governmental employees' financial and communication records as an investigatory tool after the Aldrich Ames case.[81] However, USAPA greatly expanded the scope of the NSLs. NSLs were given the expanded scope of Section 203's foreign intelligence definition and were no longer limited to records of a foreign power, their agents, or those that communicated with either. The FBI recognized the new powers it received when its general counsel send a memo that warned, "NSLs are powerful investigative tools in that they can compel the production of substantial amounts of relevant information. However, they must be used judiciously . . . In deciding whether to or not to re-authorize the broadened authority, Congress certainly will examine the manner in which the FBI exercised it . . . The greater availability of NSLs does not mean they should be used in every case."[82] Similar to Section 215 orders, they also had nondisclosure requirements.

Despite the FBI general counsel's message, the use of NSLs absolutely exploded. According to a 2007 report by the U.S. Department of Justice's Office of the Inspector General (OIG),[83] the number of NSL requests grew from approximately 8,500 in 2000 to 39,346 in 2003; there were 56,507 in 2004 and 47,221 in 2005. The percentage that focused on investigations of Americans grew from 39 percent in 2003 to 53 percent in 2005.[84]

The 2007 report also noted other problems with the reported use of the NSL powers by the FBI. The FBI underreported its use; the OIG reviewed a sample of 77 case files that supposedly contained 293 NSLs, when actually they contained 17 percent more NSLs than reported to the FBI database, which in turn is used in reporting data to Congress. The FBI was required to self-report any violations in its use of NSLs; it reported only 26 of the 143,074 requests, but in the sample case files, OIG found 22 potential violations, a rate of 7.5 percent, versus the rate of 0.00018 percent reported by the bureau. OIG also found over 700 exigent letters that were not authorized by statute and were granted even when there seemed to be no exigency.[85] The 2008 OIG report chronicled that in 2006, the number of requests had grown to 49,425, and those focused on U.S. citizens increased to 57 percent. However, "based on our review, we believe the FBI and the Department have taken significant steps to address the findings of the OIG's first report on NSLs and have made significant progress in implementing corrective actions. However, we also believe it is too soon to state with full confidence whether the steps the FBI and the Department have taken will eliminate fully the problems we identified in our first report on NSLs."[86] The OIG report from 2010[87] focused on the use of "exigent letters and other informal requests for telephone records" from 2003 to 2007. It noted that "IG's Office discovered that 'the FBI's use [of] exigent letters became so casual, routine, and unsupervised that employees of all three communications service providers sometimes generated exigent letters for FBI personnel to sign and return to them.' . . . Although critical of the FBI's initial response and recommending further steps to prevent reoccurrence, the IG's Report concluded that the 'FBI took appropriate action to stop the use of exigent letters and to address the problems created by their use.'"[88]

The American Civil Liberties Union, working with other organizations, has filed a number of lawsuits challenging the NSLs. In 2004, in *Doe v. Ashcroft*,[89] the ACLU—on behalf of an Internet Service Provider—challenged an NSL and its accompanying gag order as being unconstitutional. A district court found the NSL statute unconstitutional in regard to the First and Fourth Amendments. After the government appealed, a number of changes to NSLs and its gag order provision were made in the 2006 USAPA amendments (described earlier), so the appeals court sent the case back to the district court in May 2006[90] to consider the constitutionality of the changes. In September 2007, the ACLU returned to court to challenge the amended gag provisions of the 2006 USAPA amendments. The district court[91] struck down the NSL provisions of USAPA because of violation of the First Amendment and the separation of powers. The appeals court, in December 2008,[92] agreed in part that portions of the NSL statute were unconstitutional; it also said the government needed to justify its multiyear gag order. In August 2009, the government submitted secret court "attachment," but the appeals court ordered the government to partially disclose the document;[93] in October, the district court said the gag order could continue;[94] and in November, ACLU filed for release of information on the secret attachment. In March 2010, the court ordered a less redacted version of secret attachment to be released and government complies.[95] Finally, the gag order was removed in August 2010 by settlement, and Doe was revealed to be John Merrill after six years of forced anonymity.[96]

In August 2005, in *Doe v. Gonzalez*,[97] the ACLU sued the government on behalf of a "library entity" that was fighting an NSL demanding library records on reading materials checked out and Internet usage by library patrons. At the time of filing, ACLU Associate Legal Director Ann Beeson stated, "Our client wants to tell the American public about the dangers of allowing the

FBI to demand library records without court approval . . . If our client could speak, he could explain why Congress should adopt additional safeguards that would limit Patriot Act powers."[98] Later that month, it was revealed that the "library entity" was the Library Connection, a consortium of 26 Connecticut libraries. In September 2005, a district order ruled that the "gag order" was unconstitutional. In April 2006, the government dropped its demand for the records and the gag order.[99]

In March 2013, in *Re National Security Letter*,[100] an unidentified electronic communication service provider filed a petition that:

> Challeng[ed] the NSL and the FBI's authority to issue the letter. Specifically, the provider argued that (1) the NSL nondisclosure provision is an unconstitutional prior restraint under the First Amendment; (2) the judicial standard of review of NSL nondisclosure requirements violates separation-of-powers principles; and (3) that both the NSL itself and the accompanying nondisclosure requirement do not satisfy strict scrutiny under the First Amendment . . . The lower court held that the nondisclosure provisions were unconstitutional based on the First Amendment standards established in *Freedman v. Maryland* . . . . The court found that (1) the NSL statute did not satisfy *Freedman* because it does not require the Government to institute judicial proceedings; (2) the statute prohibits the mere fact of receipt of an NSL even though disclosure of that fact will not cause harm in many cases; (3) the NSL nondisclosure provisions are indefinite unless the recipient brings a judicial challenge; and (4) the statute "impermissibly attempts to circumscribe a court's ability to review the necessity for nondisclosure orders." The court found that the nondisclosure provisions were not reasonably severable from the substantive NSL provisions, and struck down the entire NSL provision. The court issued an injunction, but stayed its order pending appeal.[101]

In June 2013, the British newspaper *The Guardian*, based on files leaked by Edward Snowden, broke the story about widespread National Security Agency (NSA) spying. Over the following months, more and more information has become known about the global scope of the efforts as well as the breadth of captured information, including call records, e-mails that leave and enter the United States, and so forth.[102] In response, on behalf of other nonprofit civil liberties organizations, *ACLU v. Clapper* was filed in June 2013.[103] The lawsuit noted: "The ACLU is a customer of Verizon Business Network Services, which was the recipient of a secret FISA Court order published by *The Guardian* . . . The lawsuit argues that the government's blanket seizure of and ability to search the ACLU's phone records compromises sensitive information about its work, undermining the organization's ability to engage in legitimate communications with clients, journalists, advocacy partners, and others" in violation of its First and Fourth Amendment rights.[104] A federal judge denied ACLU's call for a temporary injunction and dismissed the case, per the government's motion, in December 2013. ACLU appealed in January 2014.[105]

In response to the controversy over the NSA, President Obama issued a memorandum on August 27, 2013, which established the Review Group on Intelligence and Communications Technologies. They released a final report, "Liberty and Security in a Changing World," on December 12, 2013 that 46 recommendations that would "protect our national security and advance our foreign policy while also respecting our longstanding commitment to privacy and civil liberties, recognizing our need to maintain the public trust (including the trust of our friends and allies abroad), and reducing the risk of unauthorized disclosures." It called for significant reforms on surveillance of both Americans and non-Americans.[106]

Federal legislative efforts have done little to protect the rights of library patrons. With the exception of the 2006 reauthorization bill, efforts to ameliorate the effects of USAPA have all failed. Legislative acts never make it out of committee and are not even discussed in floor debate. Much of the successful efforts to change USAPA have occurred in court cases. Even in those instances, despite relief for those individual parties, substantial changes to the law rarely occur.

LIBRARY RESPONSES TO GOVERNMENTAL INTRUSION

To this point, the focus has been on the judicial and legislative efforts to allow greater access to citizens' information. So, how have libraries responded? How have they done so historically? The interaction of the library profession and the government over national security interests has changed dramatically since World War I. According to Starr, "In previous times of crisis, American authorities have used two security-related strategies that directly involve libraries and librarians: restrictive information management and domestic information gathering."[107] Information restriction has included expanded security classification, removal of sources from physical and online information repositories, enhanced security clearance, and nondisclosure requirements and collection alteration. Governmental recruitment of library staff informants, use of listening and tracking mechanisms, and request for information about patrons are examples of domestic information gathering.[108]

In the case of World War I, the library profession contributed to the war efforts by becoming "one of the 'Seven Sisters,' a group of quasi-official agencies that the government turned to for social services for servicemen... [that] translated into book drives for servicemen, bond sales and food conversation campaigns."[109] Libraries complied with both information management and domestic information gathering, such as removal of books on explosives based on an order from military intelligence, removal of German-language books, and so forth.[110]

Despite the prewar passage of the Code of Ethics of the American Library Association and the Library Bill of Rights, libraries responded similar to how they had done in the previous world war; within a week of Pearl Harbor, the ALA issued a declaration calling on libraries to become war information centers. Once again, libraries complied with War Department information management efforts, when ALA forwarded a War Department removal order to nearly 190 libraries. Archibald MacLeish, the librarian of Congress, also served as head of the Office of Facts and Figures and Office of War Information, the primary propaganda government agency, and recruited librarians into domestic information gathering.[111]

The Cold War saw the beginning of the library profession's greater resistance to government coercion, despite "official actions that effectively created a momentum of fear and an unprecedented pressure to conform, under the guise of ensuring national security. Threats of funding or employment loss intimidated librarians and other public servants."[112] The Library Bill of Rights was strengthened in 1948, and the Freedom to Read statement was passed in 1953 (as detailed previously). "Unlike many other organizations (the National Education Association, labor unions, some bar and medical associations, and even the board of the American Civil Liberties Union), however, [ALA] never required a political test for membership, and it spoke out, through its resolution, against loyalty programs that failed to protect the civil rights of employees. In this ALA differed from the political scientists and educators who approved of forbidding Communists to teach."[113]

In the 1960s and 1970s, borrower records for books on explosives and guerilla warfare were pursued by agents of the Alcohol, Tobacco, and Firearms (ATF) agency, then a division of the Internal Revenue Service. The ALA and the National Education Association passed resolutions against the ATF investigations and asked librarians to report requests. In August 1970, the ALA and ATF released a joint statement on guidelines to govern future interactions between libraries and ATF agents.[114]

The next controversial effort was the Library Awareness Program. First announced in a September 1987 *New York Times* article, the program was designed to have librarians assist FBI agents in the identification of any interest by citizens of the Soviet Union and other countries "hostile to the United States" on information related to national security, such as unclassified technical reports. Despite initial assurance that the effort was limited to three cities, it became clear that agents had approached librarians in cities across the United States. Congressional investigations followed. Professional library organizations reacted strongly against the efforts.[115]

The next substantial conflict between the library profession and the government was USAPA. As seen in the earlier examinations of the act, libraries were impacted by the new rights afforded law enforcement and intelligence agencies, especially with Section 215 orders and National Security Letters. In a response to a House Judiciary Committee letter regarding the number that Section 215 had been used to gain access to library or book store records, Assistant Attorney General Daniel J. Bryant noted that "if the FBI were authorized to obtain the information the more appropriate tool for requesting electronic communication transactional records would be a National Security Letter (NSL)."[116] A 2005 survey by the American Library Association's Office for Information Technology Policy[117] said that 137 requests had been made to academic and public libraries; given the NSL gag-order requirement, the number could be greatly underreported.[118]

The librarian profession responded more fully beginning in January 2003, with the passing of the Resolution on the USA Patriot Act and Related Measures that Infringe on the Rights of Library Users. It recognized the role of libraries in open, democratic societies; opposed "suppression of ideas"; reiterated the essential nature of privacy; called for greater awareness that the USAPA increased the likelihood that library patrons might be surveilled without their knowledge or consent; announced ALA's opposition to governmental suppression of free and open exchange of information. It also stressed the importance of educating library staff and patrons about USAPA and other governmental efforts to curtail privacy and target library records; recommended the adoption and implementation of patron privacy and patron retention records; called sections of USAPA "a present danger to the constitutional rights and privacy rights of library users"; and urged Congress to provide active oversight, hold hearings, and amend the act.[119] On its web site, ALA offered educational materials for staff and patrons. The ALA Office for Intellectual Freedom offered legal assistance to libraries served with warrants but without service of their own attorney. It also worked with other nonprofit advocacy groups to file FOIA requests and file lawsuits to get more information on the scope of law enforcement contacts with libraries.[120] As recommended by ALA, many libraries examined their current practices and some "took more drastic steps, such as shredding daily lists of reference requests and computer access logs, while others stepped back from plans to acquire or upgrade integrated systems, fearing newer networks could be more easily compromised by law enforcement agency investigations."[121] At a September 2003 speech, Attorney General famously referred to librarians' concerns at "breathless reports and baseless hysteria."[122] ALA President Carla Hayden responded by saying, "We are deeply concerned that the Attorney General should be openly contemptuous of those who seek to defend our Constitution,"[123] and the association initiated a public relations campaign centered on "another hysteric librarian for freedom."[124]

CONCLUSION

Over time, we have seen two diametrically opposed trends when it comes to views on the sanctity of library records. On one side, we have seen how the government has nearly continually expanded its capabilities to surveil its citizens, as seen through the evolution of Supreme Court Fourth Amendment cases in the past 50 years and from legislative efforts since September 11. On the other side, we have seen a corresponding development of a professional librarianship ethos that actively opposes these efforts and collaborates with traditional civil liberty advocacy group efforts to unite their efforts.

NOTES

1. USA PATRIOT Act, Pub. L. No. 107-56, 115 Stat. 272 (2001).

2. American Library Association, "Number of Libraries: ALA Library Fact Sheet 1" (accessed 4/26, 2014), http://www.ala.org/tools/libfactsheets/alalibraryfactsheet01.

3. American Library Association, "Quotable Facts 2012," http://www.ala.org/offices/ola/quotablefacts/quotablefacts

4. American Library Association, "Code of Ethics of the American Library Association" (accessed April 27, 2014), http://www.ala.org/advocacy/proethics/codeofethics/codeethics.

5. American Library Association, "Library Bill of Rights" (accessed April 27, 2014), http://www.ala.org/advocacy/intfreedom/librarybill.

6. American Library Association and the Association of American Publishers, "The Freedom to Read Statement" (accessed April 27, 2014), http://www.ala.org/advocacy/intfreedom/statementspols/freedomreadstatement.

7. Joan Starr, "Libraries and National Security: An Historical Review," *First Monday* 9 (2004).

8. American Library Association, "Office for Intellectual Freedom" (accessed April 27, 2014), http://www.ala.org/offices/oif.

9. American Library Association, "Policy on Confidentiality of Library Records" (accessed April 27, 2014). http://www.ala.org/advocacy/intfreedom/statementspols/otherpolicies/policyconfidentiality.

10. Jennifer L. Freer, "The Patriot Act and the Public Library: An Unanticipated Threat to National Security" (unpublished manuscript, last modified October 21, 2001).

11. American Library Association, "Policy on Governmental Intimidation" (accessed April 27, 2014), http://www.ala.org/Template.cfm?Section=otherpolicies&Template=/ContentManagement/ContentDisplay.cfm&ContentID=12454.

12. American Library Association, "Privacy: An Interpretation of the Library Bill of Rights" (accessed April 27, 2014), http://www.ala.org/advocacy/intfreedom/librarybill/interpretations/privacy.

13. U.S. Const. amend. IV.

14. John M. Scheb, John M. Scheb II and Otis H. Stephens Jr., "Fourth Amendment (1791)," in *Encyclopedia of American Civil Rights & Liberties*, ed. Kara E. Stooksbury, Vol. 1 (Westport, CT: Greenwood Press, 2006), 395-398.

15. Otis H. Stephens Jr. and Richard A. Glenn, *Unreasonable Searches and Seizures: Rights and Liberties Under the Law* (Santa Barbara, CA: ABC-CLIO, 2006), 443. LaFave, Wayne R., *Search and Seizure: A Treatise on the Fourth Amendment*, 4th ed. (St. Paul, MN: Thomson/West, 2004); Michael Gizzi, e-mail message to author, April 28, 2014; *WestlawNext* database, last accessed May 1, 2014, Thomson/West.

16. *Boyd v. United States*, 116 U.S. 616 (1886).

17. *Weeks v. United States*, 232 U.S. 383 (1914).

18. *Mapp v. Ohio*, 367 U.S. 643 (1961).

19. *Silverthorne Lumber Co. v. United States*, 251 U.S. 385 (1920).

20. *Katz v. United States*, 389 U.S. 347 (1967).

21. *Warden v. Hayden*, 387 U.S. 294 (1967).

22. *Chimel v. California*, 395 U.S. 752 (1969).

23. *Mincey v. Arizona*, 437 U.S. 385 (1978).

24. *Nix v. Williams*, 467 U.S. 431 (1984).

25. *Segura v. United States*, 468 U.S. 796 (1984).

26. *United States v. Leon*, 468 U.S. 897 (1984).

27. *Arizona v. Evans*, 514 U.S. 1 (1995).

28. *Kentucky v. King*, 131 S.Ct. 1849 (2011).

29. *Terry v. Ohio*, 392 U.S. 1 (1968); Anne Bowen Poulin, "The Plain Feel Doctrine and the Evolution of the Fourth Amendment," *Villanova Law Review* 42, no. 3 (1997), 741–788.

30. *New York v. Belton*, 453 U.S. 454 (1981).

31. *United States v. Ross*, 456 U.S. 798 (1982).

32. *Michigan State Police v. Sitz*, 496 U.S. 444 (1990).

33. *Florida v. Jimeno*, 500 U.S. 248 (1991).

34. *California v. Acevedo*, 500 U.S. 565 (1991).

35. *Maryland v. Wilson*, 519 U.S. 408 (1997).

36. *Wyoming v. Houghton*, 526 U.S. 295 (1999).

37. *United States v. Arvizu*, 534 U.S. 266 (2002).
38. *Illinois v. Lidster*, 540 U.S. 419 (2004).
39. *Thornton v. United States*, 541 U.S. 615 (2004).
40. *Illinois v. Caballes*, 543 U.S. 405 (2005).
41. *New Jersey v. T.L.O.*, 469 U.S. 325 (1985).
42. *Vernonia School District v. Acton*, 515 U.S. 646 (1995).
43. *Board of Education of Independent School District No. 92 of Pottawatomie County v. Earls*, 536 U.S. 822 (2002).
44. *United States v. Miller*, 425 U.S. 435 (1976).
45. *Smith v. Maryland*, 442 U.S. 735 (1979).
46. *Kyllo v. United States*, 533 U.S. 27 (2001).
47. Michael Gizzi, e-mail message to author, April 28, 2014.
48. Wayne R. LaFave, *Search and Seizure: A Treatise on the Fourth Amendment* (St. Paul, MN: Thomson/West, 2004).
49. Stacey L. Bowers, "Privacy and Library Records," *The Journal of Academic Librarianship* 32, no. 4 (2006): 377–383.
50. Ibid., 378.
51. Ibid.
52. Ibid.
53. American Library Association, "State Privacy Laws regarding Library Records" (Accessed April 28, 2014), http://www.ala.org/offices/oif/ifgroups/stateifcchairs/stateifcinaction/stateprivacy. The text of all state privacy laws and attorney general opinions can be found at the ALA web site.
54. Chris Matz, "Libraries and the USA PATRIOT Act: Values in Conflict," *Journal of Library Administration* 47, no. 3/4 (2008): 69–87.
55. For a review of the unseemly process that led to the drafting of the act, read Herbert N. Foerstel, *Refuge of a Scoundrel: The Patriot Act in Libraries* (Westport, CT: Libraries Unlimited, 2004).
56. Ibid. Charles Doyle, *Terrorism: Section by Section Analysis of the USA PATRIOT Act* (Washington, DC: Library of Congress, 2001).
57. Foerstel, *Refuge of a Scoundrel*, 57.
58. Ibid.
59. Foreign Intelligence Surveillance Act, Pub. L. No. 95-511, 92 Stat. 1783 (1978).
60. Doyle, *Terrorism*, 9.
61. Ibid., 10.
62. Foerstel, *Refuge of a Scoundrel*, 57.
63. Doyle, *Terrorism*, 14.
64. Foerstel, *Refuge of a Scoundrel*, 58.
65. Doyle, *Terrorism*, 41.
66. For a comprehensive of how USAPA could expand electronic surveillance at libraries, see Susan Nevelow Mart, "Protecting the Lady from Toledo: Post-USA PATRIOT Act Electronic Surveillance at the Library," *Law Library Journal* 96, no. 3 (2004).

67. Foerstel, *Refuge of a Scoundrel*, 60.
68. SAFE Act of 2003, S. 1709, 108th Cong. (2003).
69. "Analysis of the SAFE Act," *Electronic Frontier Foundation* (accessed April 29, 2014), https://w2.eff.org/patriot/safe_act_analysis.php.; "Patriot Act Overview," *Center for Democracy & Technology* (Accessed April 29, 2014), https://cdt.org/insight/patriot-act-overview/#2.
70. USA PATRIOT Improvement and Reauthorization Act of 2005, Pub. L. No. 109-177, 120 Stat. 192.
71. Brian T. Yeh and Charles Doyle, *USA PATRIOT Improvement and Reauthorization Act of 2005: A Legal Analysis* (Washington, DC: Library of Congress, 2006).
72. USA PATRIOT Act Additional Reauthorizing Amendments Act of 2006, Pub. L. No. 109-178, 120 Stat. 278.
73. JUSTICE Act of 2003, S. 1686, 111th Cong. (2009).
74. Judicious use of Surveillance Tools in Counterterrorism Efforts Act of, S. 1686, 111th Cong. (2009).
75. To extend expiring provisions of the USA PATRIOT Improvement and Reauthorization Act of 2005 and Intelligence Reform and Terrorism Prevention Act of 2004 until February 28, 2011, H.R. 3961, 111th Cong. (2010) became P. Law No: 111-141, officially titled Medicare Physician Payment Reform Act of 2009.
76. The White House, Office of the Press Secretary, "Bills Signed into Law by President" (Accessed April 29, 2014), http://frwebgate.access.gpo.gov/cgi-bin/getdoc.cgi?dbname=111_cong_bills&docid=f:h3961enr.txt.pdf; Rachel Brand, "Reauthorization of the USA PATRIOT Act: New Federal Initiatives Project" (Accessed April 29, 2014), http://www.fed-soc.org/publications/detail/reauthorization-of-the-usa-patriot-act.
77. FISA Sunsets Extension Act of 2011, Pub. L. No. 112-3, 125 Stat. 5.
78. PATRIOT Sunsets Extension Act of 2011, Pub. L. No. 112-14, 125 Stat. 216.
79. U.S. Department of Justice, Office of Justice Programs, "Uniting and Strengthening America by Providing Appropriate Tools Required to Intercept and Obstruct Terrorism (USA PATRIOT) Act of 2001" (Accessed April 29, 2014), https://it.ojp.gov/default.aspx?area=privacy&page=1281.
80. American Association of Law Libraries, Government Relations Office & Government Relations Committee "National Security Letters" (accessed April 30, 2014), http://www.aallnet.org/Documents/Government-Relations/ib022610.pdf.
81. Doyle, *Terrorism*.
82. "US Civil Liberties: National Security Letters (NSLs)," *The Center for Grassroots Oversight* (accessed April 26, 2014), http://www.historycommons.org/timeline.jsp?civilliberties_surveillance=civilliberties_national_security_letters&timeline=civilliberties.
83. U.S. Department of Justice, Office of the Inspector General, "A Review of the Federal Bureau of Investigation's Use of National Security Letters" (accessed April 30, 2014), http://www.justice.gov/oig/special/s0703b/final.pdf.
84. Electronic Privacy Information Center, "National Security Letters" (accessed April 30, 2014), http://epic.org/privacy/nsl/.

85. Ibid.
86. U.S. Department of Justice, Office of the Inspector General, "A Review of the FBI's Use of National Security Letters: Assessment of Corrective Actions and Examination of NSL Usage in 2006" (accessed April 30, 2014), http://www.justice.gov/oig/special/s0803b/final.pdf.
87. U.S. Department of Justice, Office of the Inspector General, "A Review of the Federal Bureau of Investigation's Use of Exigent Letters and Other Informal Requests for Telephone Records" (accessed April 30, 2014), http://www.justice.gov/oig/special/s1001r.pdf.
88. Charles Doyle, *National Security Letters in Foreign Intelligence Investigations: Legal Background* (Washington, DC: Library of Congress, 2014).
89. *Doe v. Ashcroft*, 334 F.Supp.2d 471 (SDNY 2004).
90. *Doe v. Gonzalez*, 449 F.3d 415 (2d Cir. 2006).
91. *Doe v. Gonzalez*, 500 F.Supp.2d 379 (SDNY 2007).
92. *Joe Doe, Inc. v. Mukasey*, 549 F.3d 861 (2d Cir. 2008).
93. *Doe v. Holder*, 640 F.Supp.2d 517 (SDNY 2009).
94. *Doe v. Holder*, 665 F.Supp.2d 426 (SDNY 2009).
95. *Doe v. Holder*, 703 F.Supp.2d 313 (SDNY 2010).
96. "*Doe V. Holder*: Internet Service Provider's NSL," *American Civil Liberties Union* (accessed April 30, 2014), https://www.aclu.org/national-security/doe-v-holder.
97. *Doe v. Gonzalez*, 386 F.Supp.2d 66 (D.Conn. 2005).
98. American Civil Liberties Union, "FBI Uses Patriot Act to Demand Information with no Judicial Approval from Organization with Library Records" (accessed April 30, 2014), https://www.aclu.org/national-security/fbi-uses-patriot-act-demand-information-no-judicial-approval-organization-library-.
99. American Civil Liberties Union, "Librarians' NSL Challenge" (Accessed April 30, 2014), https://www.aclu.org/national-security/librarians-nsl-challenge.
100. Re National Security Letter, 920 F.Supp.2d 1064 (ND Cal. 2013).
101. Electronic Privacy Information Center, "In Re National Security Letter" (accessed April 30, 2014), http://epic.org/amicus/national-security/in-re-nsl/default.html.
102. The Guardian, "The NSA Files" (accessed April 30, 2014), http://www.theguardian.com/world/the-nsa-files.
103. *American Civil Liberties Union v. Clapper*, 959 F.Supp.2d 274 (S.D.N.Y. 2013).
104. American Civil Liberties Union, "ACLU Files Lawsuit Challenging Constitutionality of NSA Phone Spying Program" (accessed April 30, 2014),https://www.aclu.org/national-security/aclu-files-lawsuit-challenging-constitutionality-nsa-phone-spying-program.
105. American Civil Liberties Union, "ACLU v. Clapper—Challenge to NSA Mass Call-Tracking Program" (accessed April 30, 2014), https://www.aclu.org/national-security/aclu-v-clapper-challenge-nsa-mass-phone-call-tracking.
106. *Liberty and Security in a Changing World* (Washington, DC: White House, 2014).
107. Starr, "Libraries and National Security: An Historical Review."

108. Ibid.
109. Ibid., World War I.
110. Ibid.
111. Ibid., World War II.
112. Ibid., The Early Cold War and McCarthyism.
113. Louise S. Robbins, "Champions of a Cause: American Librarians and the Library Bill of Rights in the 1950s," *Library Trends* 45 (1996): 33.
114. Foerstel, *Refuge of a Scoundrel*, 1–3.
115. Ibid., 3–43.
116. Ibid., 64.
117. Abby Goodrum, "Impact and Analysis of Law Enforcement Activity in Academic and Public Libraries" (accessed April 30, 2014), http://www.ala.org/offices/sites/ala.org.offices/files/content/oitp/publications/booksstudies/LawRptFinal.pdf.
118. Ursula Gorham-Oscilowski and Paul T. Jaeger, "National Security Letters, the USA PATRIOT Act, and the Constitution: The Tensions Between National Security and Civil Rights," *Government Information Quarterly* 25 (2008) (accessed April 30, 2014), doi:10.1016/j.giq.2008.02.001.
119. American Library Association, "Resolution on the USA Patriot Act and Related Measures that Infringe on the Rights of Library Users" (accessed April 30, 2014), http://www.ala.org/Template.cfm?Section=ifresolutions&Template=/ContentManagement/ContentDisplay.cfm&ContentID=11891.
120. Starr, "Libraries and National Security," The library profession's response.
121. Matz, "Libraries and the USA PATRIOT Act," 76–77.
122. Katherine K. Coolidge, "'Baseless Hysteria': The Controversy between the Department of Justice and the American Library Association Over the USA PATRIOT Act," *Law Library Journal*, no. 1 (2005): 97.
123. Starr, "Libraries and National Security:" The library profession's response.
124. Matz, "Libraries and the USA PATRIOT Act," 77.

8

Where Is the Suspect? The Potential for the Use of Private Location-Tracking Data by Law Enforcement

R. Craig Curtis

INTRODUCTION: FROM *OLMSTEAD* TO *KATZ* TO *U.S. v. JONES*

Deciphering what the Fourth Amendment protects and does not protect in the context of the use of data generated by the use of electronic communication devices such as cellular phones, tablet computers, and other handheld devices is both difficult and essential. It is difficult because the United States Supreme Court, and the rest of judiciary, is slow to adapt to new challenges. Not only is the judiciary a passive actor in the policy making process, but also, there are traditions of conservatism and restraint that slow down the judicial response to new challenges. It is essential because local, state, and federal law enforcement agencies routinely ask the providers of electronic communication services for data about their customers. There are literally millions of requests made each year, and many are made without any judicial supervision at all.[1] Until the Fourth Amendment rights of the users of these devices and services are made clear by the judiciary, the providers of service will likely follow inconsistent policies, and the customers will not know what they can and can't expect to be private. In other words, the privacy rights expected by citizens will not match the rights granted by the courts.

Because there are no definitive cases on the issues, in order to understand how the Fourth Amendment applies to questions of privacy rights for users of handheld devices of all sorts, it is necessary to trace the history of the Court's jurisprudence on the issue of privacy rights and developing technology. The first major case in which the United States Supreme Court addressed the effect that communication technology would have on the meaning of the Fourth Amendment was *Olmstead v. United States*.[2] In that case, government agents became aware of a very large bootlegging operation in the city of Seattle. The main suspect, Roy Olmstead, and several coconspirators were importing alcoholic beverages on a large scale from Canada, storing them in several large caches, and distributing the liquor widely throughout western Washington. The opinion in the case contained the statement that there were more than 50 employees and that yearly income for the conspiracy would have been an estimated 2 million dollars, a very large sum of money at that time. Federal agents, without seeking a warrant, tapped the office phone and several home phones of the bootleggers. They were able to place the taps along existing phone wires without physical trespass on the office spaces or homes of the conspirators. The data were gathered over the course of several months. Extensive transcriptions of the conversations were made and introduced into evidence at the trial. In holding that the wiretaps did not violate the Fourth Amendment, the majority focused on the lack of physical trespass by the government agents in question. In comparing the status of phone calls and telegraph messages to items sent in the mail, the Court said:

> The United States takes no such care of telegraph or telephone messages as of mailed sealed letters. The Amendment does not forbid what was done here. There was no searching. There was no seizure. The evidence was secured by the use of the sense of hearing, and that only. There was no entry of the houses or offices of the defendants.[3]

In dissent, Justice Brandeis foreshadowed the concerns that led the Court to overrule *Olmstead* in 1967 in *Katz v. United States*.[4]

> The progress of science in furnishing the Government with means of espionage is not likely to stop with wiretapping. Ways may someday be developed by which the Government, without removing papers from secret drawers, can reproduce them in court, and by which it will be enabled to expose to a jury the most intimate occurrences of the home.[5]

The Court had occasion to revisit the *Olmstead* ruling in 1942 in the frequently cited case of *Goldman v. United States*[6] but declined to do so, despite there being at least three justices ready to reconsider the trespass standard.

The change in the approach of the Court from 1928 to 1967 could be said to be due as much to changes in the basic approaches to constitutional interpretation on the part of the justices as to the advances in the ways that private citizens used and conceived of communication technology. The Warren Court is generally considered one of the most activist Courts in our history,[7] and a different attitude toward new developments in technology was only expected because liberal activist judges are known for viewing the Constitution as an evolving document.[8]

The facts in *Katz v. United States*[9] are fairly straightforward. Charles Katz was part of an illegal gambling operation that spanned the coasts. He worked in Los Angeles. His job included communication with associates in Boston and Miami, accomplished via the use of a pay phone in Los Angeles. Federal agents noted that he was using this pay phone on a regular basis and placed a recording device on the outside of the phone booth so that they could record Katz's side of these conversations. They did not seek a warrant to do so, and the federal courts at the district court and in the Ninth Circuit agreed with the government that no warrant was required because there was no intrusion into the physical space occupied by the defendant.

The Supreme Court, in an opinion written by Justice Stewart, held that Katz did have a reasonable expectation of privacy in these phone conversations and that the evidence obtained via warrantless wiretap was not admissible. In so holding, the Court overruled *Olmstead* and changed the underlying test to be used for analysis of cases wherein government agents secure evidence via the interception of electronic communication. Justice Stewart stated that the old common law doctrine of trespass, as laid out in *Olmsted*, is no longer the basis for determining whether a person's expectation of privacy is reasonable with the meaning of the Fourth Amendment:

> Once this much is acknowledged, and once it is recognized that the Fourth Amendment protects people—and not simply "areas"—against unreasonable searches and seizures, it becomes clear that the reach of that Amendment cannot turn upon the presence or absence of a physical intrusion into any given enclosure.
>
> We conclude that the underpinnings of *Olmstead* and *Goldman* have been so eroded by our subsequent decisions that the "trespass" doctrine there enunciated can no longer be regarded as controlling. The Government's activities in electronically listening to and recording the petitioner's words violated the privacy upon which he justifiably relied while using the telephone booth, and thus constituted a "search and seizure" within the meaning of the Fourth Amendment.[10]

The most common citation of language from the *Katz* decision actually comes from Justice Harlan's concurrence. In stating what he thought was the correct legal test to be applied, Justice Harlan wrote, "My understanding of the rule that has emerged from prior decisions is that there is a twofold requirement, first that a person have exhibited an actual (subjective) expectation of privacy and, second, that the expectation be one that society is prepared to recognize as 'reasonable.'"[11]

In more recent times, with the addition of several justices who consider themselves to be originalists to the Court,[12] one would expect that the Court would struggle to apply a legal doctrine written in a time when the powers of the state were far more limited than today. Not surprisingly, in *United States v. Jones*,[13] the Court relied on a common law conception of the law of trespass to find that the search in question was tainted, since the placement of the Global Positioning System (GPS) device on Jones's wife's car was only accomplished by an unwarranted intrusion into a parking lot in another jurisdiction entirely. In doing so, the Court was able to avoid deciding the more difficult issue of what the Fourth Amendment means when applied to the use of GPS technology to track the whereabouts of a suspect.

The facts of the *United States v. Jones* case are fairly simple, even if the case took a long time to make it to final judgment. In 2004, Antoine Jones, owner of a nightclub in the District of Columbia, came under suspicion of selling cocaine. He was investigated by a joint task force of the District of Columbia Metropolitan Police and the Federal Bureau of Investigation. They sought a warrant in 2005 in the Federal District Court for the District of Columbia to place an electronic tracking device on his wife's car, a Jeep Grand Cherokee that was allegedly being used by Jones as part of his drug-dealing activities. That warrant was granted, but it mandated that the device be placed within ten days and within the District of Columbia. The warrant was placed on the undercarriage of the car on day 11 while the car was parked in a public parking lot in Maryland. The police tracked the car for 28 days, gathering a huge amount of data. At trial, the defendant objected to the use of the tracking data, and the motion to exclude evidence was partially granted in that the court kept out data gathered when the car was in the defendant's garage. The rest of the data was deemed admissible.

The first trial, conducted in 2006, resulted in a hung jury. During the second trial, conducted in 2007, the evidence was once again admitted in part, and a conviction was obtained. Jones was sentenced to life in prison. The United States Court of Appeals for the District of Columbia Circuit reversed, holding that the admission of the GPS tracking data was a violation of the Fourth Amendment.[14] The government's motion for rehearing en banc was denied in 2010, and review by the Supreme Court was sought.

The Court, per Scalia, ruled that the placing of the device itself was not authorized by the warrant, since the device was placed outside of the 10-day window and outside of the geographic jurisdiction of the issuing court. Further, entering the parking lot in Maryland and placing the device on the car constituted a trespass and so could not be justified as a warrantless search.

The decision to strike down the search was by a narrow 5–4 vote, with Justice Sotomayor's concurrence making it clear that she was not entirely satisfied with Scalia's reliance on an originalist argument to determine the outcome.[15] In her separate concurrence, and in the separately concurring opinion filed by Justice Alito and joined by Justices Ginsburg, Breyer, and Kagan, the sentiment that the Court was avoiding the issue at hand was clear. Additionally, the matter of whether Scalia could reinvigorate the decision criteria laid out in *Olmstead* and discredit the "reasonable expectation of privacy" standard established in *Katz* is also in question, as much of the concurring opinions address this issue in a direct way. Should this issue be raised again and in a way that does not allow the Court to avoid the issue as easily as it could in the *U.S. v. Jones* case, as it most certainly will, this author does not expect that the Court will follow Scalia's lead on this issue.[16]

WHAT ARE THE COMPETING LEGAL STANDARDS?

It is instructive to look at the actual language of the Fourth Amendment itself.

> The right of the people to be secure in their persons, houses, papers, and effects, against unreasonable searches and seizures, shall not be violated, and no warrants shall issue, but upon probable cause, supported by oath or affirmation, and particularly describing the place to be searched, and the persons or things to be seized.[17]

At the time of the ratification of this amendment, the government had two main ways of directly gaining information about a person. They could force their way into the physical space occupied by the person—that is, into the home, or they could seize the person and use force or threat of force to compel the person to provide information. The Fourth Amendment, according to originalists, was only intended to protect physical spaces. The Fifth Amendment was intended to protect the person.

The standard used to decide the case in *Olmstead v. United States* was based upon the idea that the Fourth Amendment was intended mainly to protect physical spaces. Since there was no actual trespass on the real property of any of the defendants, the Court was able to rule that there had been no violation. This was the type of reasoning used by Justice Scalia to rule in favor of Antoine Jones—that is, that there had been a physical trespass when the tracking device itself was placed on the undercarriage of the vehicle.[18] By contrast, the standard expressed by Justice Harlan in *Katz v. United States* was based upon the idea that the Fourth Amendment protects citizens whenever their expectations of privacy are "reasonable." What the word "reasonable" means is variable and introduces a certain amount of uncertainty. Both justices who tend to favor the state in criminal cases[19] and justices who are more protective of the rights of defendants[20] have been able to use the concept of reasonableness to justify their decisions.

The competing standards of review, hereinafter referred to as the "trespass standard" and the "reasonable expectation of privacy standard," seemingly coexist uneasily. Originalists, such as Justice Scalia, write about physical trespass to property in their decisions,[21] while justices who take the perspective that the Constitution must adapt to changing times are more likely to use the language of reasonableness of a citizens' expectation of privacy.[22] Both standards are open to criticism. The trespass standard can be rightfully criticized because it is incapable of addressing privacy problems posed by the use of information-age computer technology.[23] Indeed, some commentators argue that there was no trespass standard before the Katz case was decided and that the standard itself is unclear.[24] The reasonable expectation of privacy standard can be criticized because its use creates uncertainty as to what the police can and cannot do.[25]

Two Supreme cases that are commonly cited by lower courts seeking federal guidance on the issue of the use of electronic tracking devices, *United States v. Knotts*[26] and *United States v. Karo*,[27] offer some insight into how the doctrine of reasonableness can be applied in cases involving the use of information technology. Both cases were decided in the early 1980s, both involved the transportation of chemicals used in making drugs, and both involved the use of a beeper signal that could be used to determine the location of the property in or on which the beeper was placed. In *Knotts*, the Court upheld the use of the beeper on the grounds that the beeper only served to enhance the ability of the police to track the container in which the chloroform was being transported while it was on public thoroughfares. By contrast, the facts in *Karo* involved tracking a container of ether while it was transported to and stored in James Karo's house. The Court likened this to a search of a home and held that a warrant was required to use this technology.

SOURCES OF DATA

In the modern world, a great deal of information can be garnered without physical intrusion or coercion, via interception of radio signals or tapping signals traveling over a copper wire or fiber optic cable. Humans communicate with each other in a great variety of ways, and actual conversations, face-to-face, are becoming less common. The most obvious source of information comes from the use of cellular phones. All modern units have a Global Positioning System (GPS) chip as well as the memory storage capacity to retain records of numbers called and received, contents of text messages, pictures, videos, and the history of web surfing with the devices. In addition, the location of a cellular phone can be established in a less precise way by the records of its communication with the service provider's towers. Cellular phone service providers store a lot of data about their customers, with what is saved and stored being hidden, to a certain extent, from the public.[28] The police have the ability to ping a phone to determine its location in real-time or to determine its location though access to records of its use from the carrier. This last tool is often referred to as CSLI, or cell site location information. Obviously, smartphones do more things and store more data, and many of the applications available create the potential for third parties to learn a great deal about the user.

Tablet computers have similar capacities, and some even allow for face-to-face conversations via programs such as Skype as well as traditional phone calls. These units have more storage than cell phones, and if such a unit falls into the hands of the police, they would have all sorts of information about the user.

The police also make use of automated camera systems to create databases of the locations of automobiles. The city of New York has a large number of surveillance cameras together with a central, computerized command center that is capable of automatically noting suspicious behavior and alerting human police officers.[29] Other cities are following suit and using large networks of cameras and software capable of analyzing the images for threats.[30] The American Civil Liberties Union (ACLU) has reported that large law enforcement agencies in a number of jurisdictions are using automated scanners that are capable of recognizing license plates to create databases of where cars are at given points in time.[31] While these data-gathering operations are based on observations made in public spaces, thus posing no Fourth Amendment issue, the existence of such efforts does affect the resolution of legal challenges to the use of data devices in that this affects the analysis of what is a reasonable expectation of privacy. One argument in favor of allowing tracking devices to be used is that the exact same location data can be obtained by the low-tech expedient of having a police officer follow a car whenever it is on the road. In essence, there is no reasonable expectation of privacy regarding your whereabouts when you are in public spaces.

The police have obvious and clear uses for data that shows where a person might be at a given time, and cities are investing millions of dollars in these data-gathering efforts.[32] For example, where a suspect is at large and considered dangerous, and he or she is carrying a cellular phone, the police could use that suspect's cell phone to track the person's location as happened in the case of *Pennsylvania v. Rushing*.[33] The media have also documented instances wherein dangerous criminals have been tracked in this way.[34] In another example, a suspect who offers an alibi to the police that suggests that he or she was not in the area where the crime was committed could be verified, or discredited, by the use of records of cell phone location available on the phone itself or held in archival records stored by the service provider. Lastly, in a more benign example, if a crime or even a terror attack occurs in a given location, the police might want to establish a list of potential witnesses by the use of service provider records that list customers who were in the vicinity at the time of the attack.

Data can be retrieved from a cellular phone or other handheld device itself, especially in the case of a smartphone, or from records maintained by the service provider. Common sense would suggest that data stored on the device itself would be subject to greater Fourth Amendment protection since the device is the personal property of the user. This would mean that any question about the admissibility of data from the device itself could be rightfully contested by the owner. Thus, under even the trespass standard, the argument can be made that the device is the property of the person asserting the right and that the police must have a warrant, or at least probable cause, to access the data. By contrast, data stored by the service provider is owned by the service provider, a third party. Historically, at least if one applies the trespass standard, the suspect would have no basis to move for suppression of evidence if no right of the suspect has been the subject of a trespass. This leaves only the reasonable expectation of privacy standard to be applied to cases in which the suspect attempts to exclude evidence obtained from the service provider, and that is a farther reach for a defense attorney.

While there has not been a clear case resolving the issue of what standard of review to apply to cases wherein the police want to introduce evidence obtained from a cell phone, the uneasy coexistence of the two legal standards can be examined by reading the case of *United States v. Jones*.[35] That case presented a situation wherein the police successfully sought a warrant to place a GPS device on a suspect's car. The device was then used to track the whereabouts of the vehicle for 28 days, generating a huge amount of data. The decision by the Supreme Court did not directly address the issue of whether this kind of attention to a particular person's location was in violation of the Fourth Amendment, so trial courts seemingly are free to issue these kinds of warrants. The trial court in the *Jones* case held that the data gathered while the vehicle was in a private garage was inadmissible but that data gathered while the vehicle was in public places was admissible. Ultimately, the court ruled that the placement of the device itself was a violation of the Fourth Amendment and declined to rule on the issue of whether the use of a tracking device could be a violation of the Fourth Amendment. This leaves trial courts without guidance on how to decide whether to issue a warrant for placement of a tracking device and on what terms.

Increasingly, cars come equipped with factory-installed navigation systems. These systems are capable of storing significant amounts of personal information, including private addresses, location data, speed of travel, and dates of travel. Additionally, many newer cars come equipped with what is essentially a tablet computer in the dashboard, with touch-screen controls. These systems potentially allow the occupants of the car to make and accept phone calls, surf the web, access entertainment; other uses will develop as this equipment becomes more common. While this is still fairly new technology and its development may be thwarted by concerns about distracted driving, the potential as a source of information is significant. Moreover, just like with cell phones, the existence of data services like OnStar means that the data are in the possession of a third party. Thus, a suspect's whereabouts, or other information, may be accessed without the necessity of seeking permission from the suspect or a search warrant.[36]

In addition to these obvious devices, modern cars have computer chips as part of a variety of mechanical systems. Most of these chips are designed to increase fuel efficiency, provide the operator with safety information, and facilitate maintenance. These chips do contain information that is potentially useful to the prosecution in a criminal trial. This information includes speed, braking, images from backing cameras, tire pressure, and mechanical condition of a component. The National Highway Traffic Safety Administration has proposed a rule that would require all automobile makers to include an event data recorder (EDR) that would allow for the retrieval of such data in the case of a collision.[37] While such a proposal likely stays under the radar of most citizens, privacy proponents who follow such events are very concerned about the potential for abuse of EDRs and are, at the time of this writing, lobbying for changes to the rule that would allow for the vehicle owner to have control over EDR data. Even without such a rule, it has been reported that 96 percent of new cars have EDRs already.[38]

CURRENT PRACTICES AND OWNERSHIP OF DATA

It is instructive to look at the types of fact scenarios that have given rise to motions to suppress evidence retrieved from the providers of cell phone service. Most often, the dispute over admissibility begins with a government request to a provider of electronic communication services under the auspices of the Stored Communications Act.[39] Section 2703(d) allows for courts to grant orders to release data based on less than probable cause. The language from the statute is "specific and articulable facts." For example, in the case of *United States v. Graham*,[40] the police investigating a series of burglaries were able to obtain several orders for the release of a total of 221 days of cell site location information (CSLI) data for two defendants, based on the Stored Communications Act. In the case of *In the Matter of the Application of the United States of America for an Order Authorizing the Installation and Use of a Pen Register and Trap and Trace Device*,[41] the District Court for the Southern District of Texas denied an application for the Drug Enforcement Agency (DEA) to use a piece of equipment called a stingray, which can intercept cell phone traffic and can analyze that traffic to identify the phone being used by a particular user. As with most of these cases, the request was justified by the government under a federal statute with a legal standard less than probable cause—in this case, the USA Patriot Act's provision that allows for the use of a pen register.

Law enforcement agents are not just interested in cell phones. In the case of *United States v. Ringmaiden*,[42] the Internal Revenue Service tracked the location of the defendant's computer using a cell phone site simulator to mimic a Verizon cell tower as a way to determine the location of the computer's aircard. Since the opinion in that case is focused on the issue of what powers the defendant had to discover information about the technology used to track him, and the government conceded for purposes of analysis that the tracking was a search within the meaning of the Fourth Amendment because it had obtained a search warrant to conduct the search, the case is not useful in terms of determining the legal standard to be applied to analysis of the Fourth Amendment issue of whether the police can use such technology without a search warrant. Still, the case does provide interesting information about the tools available to law enforcement.

In state courts, the facts often begin with a request for data from a cell phone provider, usually based upon a statute. For example, in the case of *Pennsylvania v. Rushing*,[43] the police obtained an order to ping the suspect's phone pursuant to a Pennsylvania wiretap statute[44] after the surviving victims of a horrific multiple murder called police, and told them what had happened, and gave them the suspect's cell phone number. Sometimes, the police ask for data directly from the provider without seeking a court approval. In the case of *Maryland v. Thomas W. Earls*,[45] the police were investigating a string of burglaries. After being told that the defendant intended to harm cooperating witnesses, the police asked, and T-Mobile voluntarily released location data on the defendant.

In contrast to the situation in cases like *U.S. v. Jones*, it is unclear who actually owns the data gathered and stored by the providers of cell phone service. Absent contractual provisions between the customer and the provider that specify that the customer owns the data, data obtained by the service provider voluntarily providing it to the police are not subject to motions to exclude evidence. If the data are obtained via subpoena, it is not at all clear that the customer has standing in court to contest the provision of the data or its use in a criminal trial in which the customer is the defendant. One of the few cases to directly address this type of issue involved the use of Twitter. In the case of *New York v. Harris*,[46] Twitter Inc. sought to quash a subpoena by the New York County District Attorney's Office requesting all sorts of data stored by Twitter. The defendant in the case had unsuccessfully attempted to quash the subpoena, but the trial court had ruled that he lacked standing. In ruling that the subpoena did not violate the Fourth Amendment, the judge was careful to enunciate both the trespass and reasonable expectation of privacy standards and to rule that under neither standard could the company prevail. The determining factor seemed to be the fact that tweets are broadcast to a large number of people, not to just one individual. How this judge might have ruled on a request to access information about a cellular phone account is not clear, as he was careful in stating the limits of his decision to cite the United States Courts of Appeals case of *United States v. Warshak*,[47] which held that there is a reasonable expectation of privacy in e-mails.

It is also significant that Twitter Inc. purposely altered its terms of service contract after receiving the subpoena that was the subject of the case of *New York v. Harris*. The opinion cites to this change of terms of service was made after the trial court had denied the defendant's motion to quash the subpoena and even quotes the new language: "You retain your right to any content you submit, post or display on or through the service."[48] It is not clear from the opinion whether the defendant's initial motion to quash the subpoena would have been granted under the new terms of service. Some courts have held that third parties may allow the police access to social media information without implicating the Fourth Amendment rights of defendants.[49]

It is very clear that the police do want and frequently request data from cellular phone service providers.[50] The common practice is to ask the service provider nicely for the data. If that fails, then the police will ask the prosecutor's office to ask a court to issue a subpoena directly to the service provider.[51] There is no search, within the meaning of the Fourth Amendment, of anything owned by the customer, absent an express agreement between the service provider and the customer that specifies that the account data are the property of the customer. Indeed, since the data are technically owned by the provider, that provider can voluntarily turn over data in response to a request by the police without any judicial supervision, requirement of probable cause, or even "specific and articulable facts."[52]

In essence, when a service provider turns over data to the police, whether in response to a subpoena or in response to a simple request, there is arguably no Fourth Amendment issue, since the search was conducted by a private party. The situation is very much like the facts in the case of *United States v. Jacobsen*[53] in which the Court held that cocaine found in the course of a search of a container by Federal Express was admissible. The defendant in that case had been receiving shipments of cocaine via Federal Express. One of the packages was damaged in the course of handling. The employees of the company, per standard policy, opened the package to see what was in it for purposes of processing an insurance claim for damage to the contents. When they found white powder, they closed up the package and called the federal Drug Enforcement Agency. The DEA agents reopened the container, identified the powder as cocaine, and obtained a warrant to search the addressee's premises. After that search, they arrested the defendant. The Court ruled that the private search was of no constitutional significance because it had been conducted by a private party. Once the drugs had been found by the party, they could report what they found to the police without any violation of the reasonable expectation of privacy of the defendant. In essence, the private search eliminated any expectations of privacy held by the addresses of the package since the police search went no further than the private search. By analogy, the service provider already knows where you were, whom you texted, what web sites you visited, and other data. Providing those data to the police is simply reporting the results of a private search. It is very much like the silver platter doctrine that existed in the interval between the decision in *Weeks v. United States*[54] and the decision in *Mapp v. Ohio*[55] that made the exclusionary rule applicable to the states. During this time, material obtained in violation of the Fourth Amendment rights of a suspect by state and local police, if admissible in their jurisdiction, could be reported to the federal government without fear of the evidence being excluded, despite the fact that the same evidence would not be admissible if seized directly by federal agents.

THE IMPORTANCE OF THE STANDARD OF REVIEW

The potential for a court to simply rule that the turning over of data by a service provider to the police is not a search within the meaning of the Fourth Amendment points out the importance of determining the nature of the constitutional analysis employed to decide such cases. If common law concepts of trespass are to be used to analyze cases in which digital data from archival storage by a service provider or a third party retained to store the data, as Scalia and some other originalists would have it, then the data service customer has no rights at all. If the *Katz* reasonable expectation of privacy standard, as laid out in Justice Harlan's concurring opinion, is applied, then public opinion on the issue of the nature of data usage by individuals becomes very important.[56] Unfortunately, the data on the issue of what Americans expect in terms of privacy is somewhat muddled. Recent poll data gathered in the aftermath of the revelation that the National Security Agency was monitoring cell phone usage was mixed,[57] with two polls finding that the majority of Americans were unhappy with this surveillance[58] and another finding that Americans were willing to give up privacy rights with regard to the use of cell phones in the interests of thwarting terrorist attacks.[59] A poll taken in 2012 showed that many owners of smartphones had considered the potential for invasions of privacy in deciding not to download an application or in deciding to remove a preinstalled application.[60] This also means that lots of people are aware that third parties can monitor their cell phone usage, including determining their location, and that they continue in their behavior despite the knowledge that they can be tracked. In essence, one could argue that there is not a clear national consensus that cellular phone usage is private, or one could argue that most Americans don't like the intrusions into their private lives but tolerate it in the name of national security.

The issue is not so much one of legal analysis as it is one of politics. The origins of the battle over judicial interpretation of the Bill of Rights dates back to the later years of the Warren Court. Conservatives were very keen to change the makeup of the federal judiciary, and criminal procedure cases were very much at the core of conservative objections to many of the Warren Court decisions.[61] President Nixon vowed to get tough on law-and-order issues, declaring a war on heroin soon after he became president, and he made putting conservative judges on federal courts a big issue. By 1980, Reagan made changing the judiciary into a Republican institution a key part of his agenda.[62] It was also during the Reagan era that the very conservative Federalist Society was created, holding its first national conference in 1982.[63]

The nature of change in jurisprudence in a Common Law jurisdiction is that it happens incrementally, with major cases being few and far between. The Supreme Court is reluctant to overturn its own precedent and is naturally shy of political controversy since its own legitimacy often depends on being perceived as politically neutral.[64] Additionally, the actual cases that make major changes are often about relatively technical legal matters. Thus, the difference between determining whether a search in which digital information is obtained based on a trespass standard as opposed to a reasonable expectation of privacy standard is large, even though the statement of the two standards in the abstract might not elicit much of a response from a layperson. This means that discussions of the proper way to analyze a fact pattern may seem far too abstract for the layperson, but once the legal standards are set, the outcomes are easily predicted.

The federal and state cases that have addressed these issues have been divided on the proper standard to use and on the specific issue of whether the use of cellular phones to track a suspect is a violation of the Fourth Amendment.[65] In the federal practice, there have been a number of cases in which challenges have been made to police use of pen registers or the use of pinging to locate someone. None have made it to the Supreme Court level, and very few have made it even to the United States Courts of Appeals. In the federal system, law enforcement officials commonly make their requests for data pursuant to the Stored Communications Act. Nobody at the federal level seems to want to take on the challenge of sorting out the Fourth Amendment implications of this type of law enforcement behavior when the standard of review is uncertain and when cases may be decided by reference to a federal statute.

The Third Circuit's opinion in the case of *In the Matter of the Application of the United States of America for an Order Directing a Provider of Electronic communication Service to Disclose Records to the Government*[66] is often cited as the first case to address the issue of the validity of a request by the police to use cell phone data to locate a suspect. In that case, a federal magistrate in western Pennsylvania had denied a request made under the authority of the Stored Communications Act for the cell site location information (CSLI) of a particular suspect. After the district court affirmed the denial of the request, appeal was made to the United States Court of Appeals. The Third Circuit ruled that a federal magistrate had the power to require a showing of probable cause but that the statute was unclear as to what standard of proof is required since there are competing bases on which a request for data may be based, and the competing bases have apparently different standards of proof. The opinion did state that CSLI is not the equivalent of a tracking device for purposes of Fourth Amendment analysis, which means that the reasonable expectation of privacy standard as applied in the case of *United States v. Karo*[67] does not apply. The opinion reads as though the judge was not comfortable with deciding the matter, even when based on statutory interpretation of the Stored Communications Act.

At the time this chapter was being written, a case was pending in the Fourth Circuit[68] wherein the government acquired CSLI on two Maryland suspects, Aaron Graham and Eric Jordan, over the course of 221 days pursuant to an order issued under the standards of the Stored Communications Act. Under that act, probable cause is not required to obtain an order to gain access to CSLI data from a service provider. Thousands of locations were obtained over the course of the 221 days from Sprint, who complied with the orders without contesting them. The District Court for the District of Maryland denied the motion to suppress the records, and the defendant was convicted of a series of armed robberies.[69] Their appeal is being supported by the American Civil Liberties Union and the Electronic Freedom Foundation, who argue that the lineup of justices in the case of *United States v. Jones* is such that there are five votes for the proposition that the extended tracking of a suspect amounts to a search, which must be based on a warrant.[70] While it is not at all clear that the judges hearing the case will agree with the ACLU's interpretation of the holding in *United States v. Jones*, the case is one that has the potential to force the Supreme Court to revisit the issue raised in that case, which is whether the use of electronic means to track the location of a suspect violates a reasonable expectation of privacy. If this happens, the Court will likely also be forced to clarify the standard of review that is to be used in such cases.

STATE LAWS THAT PROVIDE GREATER PROTECTIONS OF PRIVACY

Under a legal doctrine that was popularized in the 1980s[71] and has been applied to a handful of cases since then,[72] states may grant greater protections to citizens under their own constitutions and statutes than that granted under the federal Constitution. For example, it is well known that by the time that *Mapp v. Ohio* was decided, a majority of states had an exclusionary rule in place.[73] If the legal basis for a ruling that a police activity is forbidden is a state statute or constitution, even if the Fourth Amendment would not have required the exclusion of the evidence, then the federal courts may not overrule that decision.

Two states have directly ruled on the issue of whether their own state constitutions provide greater privacy protection with regard to the use of cell phone data to locate a suspect than that granted under the Fourth Amendment. The first was in the case of *Pennsylvania v. Rushing*.[74] That case involved the use of pinging by the police to find the real-time location of a man who had just committed three brutal murders and, based on testimony from survivors of the attack, was believed to be intent on committing further acts of mayhem. While the court upheld the conviction for murder, the court did rule that Pennsylvania law requires that the police obtain a warrant or demonstrate exigent circumstances and probable cause to use that location tool to find a suspect. The opinion in that case is meticulously crafted, containing useful technical information about the ways that the police can use a cellular phone to track a suspect and a detailed list of lower federal court cases in which police access to cell phone data had been challenged. The opinion also contains a very careful analysis of the facts in the case and will, no doubt, be cited repeatedly by other state courts seeking insight into such cases, given that the federal courts are divided on this issue.

The second case, arising in New Jersey, was *State v. Thomas W. Earls*,[75] decided on July 18, 2013. In that case, T-Mobile voluntarily complied with several requests for tracking information on a cell phone of a suspect believed to be guilty of receiving stolen property. As in the case of *Pennsylvania v. Rushing*, the New Jersey Supreme Court held that a warrant is required to access this kind of information. Additionally, as in Pennsylvania, there was precedent for the principle that the New Jersey Constitution provides greater protection to individuals than does the Fourth Amendment.[76] The opinion is quite explicit that the test to be applied under New Jersey law is the reasonable expectation of privacy standard: "New Jersey case law continues to be guided by whether the government has violated an individual's reasonable expectation of privacy."[77] Further, the Court stated, "Yet people do not buy cell phones to serve as tracking devices or reasonably expect them to be used by the government in that way. We therefore find that individuals have a reasonable expectation of privacy in the location of their cell phones under the State Constitution."[78] The New Jersey Supreme Court also recognized that the practice of obtaining cell phone records from service providers is a commonly used and effective law enforcement tool and ruled that the holding in *Earls* shall not be applied retroactively.[79] Additionally, the opinion clearly invited the appellate court, on remand, to find that exigent circumstances existed that would allow a warrantless search.

While the issue of how to address allegations that the Fourth Amendment has been violated by the police when they access cell phone data remains unresolved, at least two states have provided clear guidance to the police in their states.[80] Given that most criminal prosecutions occur in state practice, and not in the federal realm, the uncertainty is not so damaging to law enforcement efforts as one might think. Even the Stored Communications Act section 2703(d), which is commonly used by federal agents to seek non-content-based data and does not mandate a showing of probable cause, does state that such requests do not have to be granted if they would violate state law. This means that in Montana, Maine, New Jersey, and Pennsylvania, a showing of probable cause is required even for non-content data. Additionally, the state courts that have addressed this issue have made it clear that they will entertain warrantless searches so long as the state can demonstrate a serious public safety interest, such as exigent circumstances involving a dangerous criminal at large or the danger of evidence being destroyed.

THE EFFECTS OF EXISTING FEDERAL STATUTES

While the detailed analysis of the content and interpretations of the Stored Communications Act is beyond the scope of this chapter, the fact that so many requests for data are coming pursuant to section 2703(d) of that act, which does not require a showing of probable cause, does mean that a full analysis of the implications of official behavior requires some consideration of the meaning of that statute. The need to sort out the varying requirements of constitution and statutory law is perhaps more important in the area of electronic privacy than any other area of Fourth Amendment jurisprudence.[81]

First, one should be aware that different legal standards are mandated for requests for information that would reveal the contents of communication. A higher standard of proof is required for requests for data under section 2703(a) or (b), since those provisions cover requests for the actual contents of text messages or other communications. Section 2703(a) mandates a warrant for the contents of communications from a provider of electronic communications service. Section 2703(b) applies only to providers of electronic computing services and allows for avenues of accessing information short of a warrant requirement, in some circumstances. The important issue for this chapter is that the contents of communication are treated with more respect for the privacy of the user of electronic communications services than CSLI, for example. Section 2703(d) only applies to requests for data that does not disclose the contents of the electronic communication.

If a court attempts to apply the reasonable expectation of privacy standard to determine whether a request for data violates the rights of the suspect, the fact that the main federal statute applicable to this area of law mandates a warrant when the content of the communication will be revealed is telling. In essence, the Congress has said that, in its view, the content of such communication is worthy of recognition as the subject of a reasonable expectation of privacy. Given that courts are uncertain as to the standard of review to be followed, judges are attempting to find a result that satisfies both the trespass standard and the reasonable expectation of privacy standard. These judges will have reference to the case construing these statutes as well as, potentially, the records of the debates in Congress when these statutes were enacted or amended.

CONCLUSION

The constitutionality of the common practice of the police using data from cell phones or other mobile devices to track the location of a suspect remains in doubt. The federal judiciary has not provided law enforcement agencies, users of the technology, or service providers with clear guidelines as to what is allowed by the Fourth Amendment. This chapter has discussed the competing legal standards, the trespass standard and the reasonable expectation of privacy standard, but cannot make a reliable prediction of the direction that future litigation on this issue will take. Two states already have made clear statements of what the law is on this issue in their state, but the federal law remains uncertain. Cellular service providers will likely continue to provide information about their customers in response to requests from law enforcement, especially when the suspect is suspected of being involved in terrorist activities. Law enforcement will continue to seek access to records. Courts, faced with the requirement to determine the validity of those challenges, will be forced to find creative ways to justify their decisions. Judges will be required to consider both the trespass standard and the reasonable expectation of privacy standard and then craft decisions that they think are correct and that are sufficiently justified to survive challenges on appeal.

The potential for law enforcement to use location data to solve a large number of crimes is clear. When a suspect's identity is known, and the danger posed by the suspect is plausible, courts will most likely grant forgiveness, based on probable cause and exigent circumstances, for failure to obtain a warrant, even in jurisdictions that mandate a warrant. The danger seen by so many civil libertarians is the use of such data for more routine police work. Will police departments seek records of when and where a text was sent for the purposes of issuing a citation to a driver for texting while driving? Will the courts treat services such as OnStar as service providers subject to the Stored Communications Act, or will that service exist in a gray area wherein the provider of the service sits in the status of any third party? Will it matter for purposes of analysis whether a navigation device is handheld or factory installed? If all cars are mandated to include event data recorders under federal law, it might.

Perhaps the biggest question will be the standard of review the Supreme Court chooses when it is ultimately faced with a Fourth Amendment challenge to the use of location data from a cell phone, car data system, tablet computer, laptop computer, or other handheld device. If the trespass standard is chosen, then, arguably, cell phone service providers are simply third parties who may freely provide information about subscribers without worrying about the Fourth Amendment. They would have to worry about federal and state statutes but not about whether the Constitution prohibits the release of the data. If the reasonable expectation of privacy standard is chosen, then the service providers are not as able to provide data without worrying about violating the Fourth Amendment, but their customers will at least have the ability to contest the validity of the search in court.

In the end, smart criminals will always find a way, and the cops will adapt in a never-ending competition for data supremacy. Osama Bin Laden learned not to use cell phones. Criminals have learned to use prepaid phones and switch them often. When we get to the point where the government can track every new car on the road, the bad guys will drive classic cars. The real trick is to allow the police and the criminals to fight their battles over access to data without compromising the basic integrity of each person's sphere of privacy. This author would opine that without some areas wherein the government simply cannot intrude, there can be no truly free society and, arguably, no real democracy.

NOTES

1. Ellen Nakajima, "Cell Phone Carriers Report Surge in Surveillance Requests from Law Enforcement," *Washington Post Online*, July 9, 2012, accessed July 29, 2013, http://www.washingtonpost.com/world/national-security/cellphone-carriers-report-surge-in-surveillance-requests/2012/07/09/gJQAVk4PYW_story.html.
2. 277 U.S. 438 (1928).
3. Ibid., at 464.
4. *Katz v. United States*, 389 U.S. 347 (1967).
5. *Olmstead v. United States*, note 2, supra.
6. *Goldman v. United States*, 316 U.S. 129 (1942).
7. See, e.g., Rebecca E. Zeitlow, "The Judicial Restraint of the Warren Court (and Why It Matters)," *Ohio St. LJ* 69 (2008): 255, at p. 257, note 7.
8. See, e.g., Philip Bobbitt, *Constitutional Fate: Theory of the Constitution* (New York & Oxford: Oxford University Press, 1984). See, on the issue of conservative activism, Thomas M. Keck, *The Most Activist Supreme Court in History: The Road to Modern Judicial Conservatism* (Chicago: University of Chicago Press, 2004).
9. *Katz v. United States*, note 4, supra.
10. *Katz v United States*, note 4, supra, at 353.
11. *Katz v. United States*, note 4, supra, at 361.
12. Justices Scalia, Thomas, Roberts, and Alito are considered originalists.
13. *United States v. Jones*, 565 U.S. ___, 132 S.Ct. 945 (2012).
14. *United States v. Maynard*, 615 F. 3d 544 (2010).
15. *United States v. Jones*, note 13, supra, Sotomayor concurring, accessed at http://www.law.cornell.edu/supremecourt/text/10-1259#writing-10-1259_CONCUR_4.
16. In a joint amicus curiae brief to the United States Courts of Appeals for the Fourth Circuit in the case of *United States v. Graham*, the Electronic Frontier Foundation, the American Civil Liberties Union, and the National Association of Criminal Defense Lawyers, analyzing *United States v. Jones*, assert on page 16 that the there are five votes on the Court for the principle that the long-term gathering of electronic locational data on a suspect is a search within the meaning of the Fourth Amendment. See that brief at http://www.aclu.org/files/assets/2013.07.02_-_doc_60_-_corrected_aclu_et_al._amicus_brief.pdf (last accessed, July 29, 2013).

17. U.S. Const. amend. IV.

18. Justice Scalia also used the trespass standard to decide in favor of the defendant in the case of Florida v. Jardines, 569 U.S. ___ (2013), accessed August 2, 2013, http://www.supremecourt.gov/opinions/12pdf/11-564_5426.pdf.

19. See, e.g., *California v. Greenwood*, 486 U.S. 35, 37, 40–41 (1988), holding that a search of garbage left at the curb was not a search; and, *Walter v. United States*, 447 U.S. 649 (1980) and United States v. Jacobsen 466 U.S. 109 (1984), both of which involved a search that had been conducted by law enforcement after a private party had conducted a search and then brought the contraband to the attention of law enforcement. In both *Walter* and *Jacobsen*, the Court held that the private search had effectively eliminated a reasonable expectation of privacy on the part of defendants.

20. See, e.g., *Katz v. United States*, note 4 supra; Justice Sotomayor's concurrence in *United States v. Jones*, note 13, supra; and Justice Kagan's concurrence in *Florida v. Jardines*, note 18, supra.

21. See, .e.g., *United States v. Jones*, note 13 supra, and *Florida v. Jardines*, note 18 supra.

22. See, e.g., Sotomayor, concurring in *United States v. Jones*, note 13 supra, and Kagan, concurring in *Florida v. Jardines*, note 18 supra.

23. Justice Sotomayor wrote in her concurring opinion in *United States v. Jones*, note 13, supra, "In cases of electronic or other novel modes of surveillance that do not depend upon a physical invasion on property, the majority opinion's trespassory test may provide little guidance."

24. See, Orin S. Kerr, "The Curious History of Fourth Amendment Searches," *Supreme Court Review* (September 30, 2012); GWU Legal Studies Research Paper No. 2012-107; GWU Law School Public Law Research Paper No. 2012-107, accessed July 29, 2013, http://ssrn.com/abstract=2154611; Orin Kerr, "What Is the State of the *Jones* Trespass Test after *Florida v. Jardines*," The Volokh Conpsiracy, March 27, 2013, accessed July 29, 2013, http://www.volokh.com/2013/03/27/what-is-the-state-of-the-jones-trespass-test-after-florida-v-jardines/; and Bradley Pollina, "*Florida v. Jardines*: Why the Supreme Court did not Say Trespass," *Wake Forest L. Rev. Online* 3(May 2013): 19, accessed July 29, 2013 http://wakeforestlawreview.com/florida-v-jardines-why-the-supreme-court-did-not-say-%E2%80%9Ctrespass%E2%80%9D.

25. See, e.g., Christopher Slobogin and Joseph E. Schumacher, "Reasonable Expectations of Privacy and Autonomy in Fourth Amendment Cases: An Empirical Look at 'Understandings Recognized and Permitted by Society,'" *Duke L J* 42 (1993): 727, in which the authors used surveys and hypothetical fact patterns to measure the consistency of Supreme Court case outcomes with popular attitudes. They found a major disparity. See also, Jim Harper, "Reforming Fourth Amendment Privacy Doctrine," *Am U L Rev* 57 (2008): 1381, at 1386, Referring to the Harlan concurring language in *Katz*, "More importantly for judicial administration, it converted a factual question—had the defendant barred others from access to the information?—into a murky two-part analysis with a quasi-subjective part and a quasi-objective part. It is

an analysis that courts have mangled ever since. And for good reason: It is almost impossible to administer."

26. *United States v. Knotts*, 460 U.S. 276 (1983).

27. *United States v. Karo*, 468 U.S. 705 (1984).

28. Allie Bohm, "How Long Is Your Cell Phone Company Holding on to Your Data?" American Civil Liberties Union Blog of Rights, September 29, 2100, accessed August 6, 2013, http://www.aclu.org/blog/technology-and-liberty/how-long-your-cell-phone-company-hanging-your-data; American Civil Liberties Union, "Cell Phone Location Tracking Public Records Request," March 25, 2013, accessed July 29, 2013, http://www.aclu.org/protecting-civil-liberties-digital-age/cell-phone-location-tracking-public-records-request.

29. Here and Now, "NYC's Web of Cameras Can Catch Unattended Bags," Wednesday, April 24, 2013, accessed August 5, 2013, http://hereandnow.wbur.org/2013/04/24/nyc-surveillance-cameras.

30. Steve Henn, "In More Cities, A Camera On Every Corner, Park and Sidewalk," National Public Radio, Morning Edition, June 20, 2013, accessed August 2, 2013, http://www.npr.org/blogs/alltechconsidered/2013/06/20/191603369/The-Business-Of-Surveillance-Cameras.

31. American Civil Liberties Union, "You Are Being Tracked: How License Plate Readers Are Being Used to Record Americans' Movements," July 2013, accessed July 25, 2013, http://www.aclu.org/files/assets/071613-aclu-alprreport-opt-v05.pdf.

32. American Civil Liberties Union, note 28, supra.

33. *Pennsylvania v. Rushing*, 2013 PA Super 162 (2013), accessed July 26, 2013, http://www.pacourts.us/assets/opinions/Superior/out/J-S25029-12o.pdf.

34. Ellen Nakajima, "Verizon Says It Turned Over Data Without Court Orders," Washington Post online, October 16, 2007, accessed July 29, 2013, http://www.washingtonpost.com/wp-dyn/content/article/2007/10/15/AR2007101501857.html.

35. *United States v. Jones*, note 13, supra.

36. The potential for hacking these systems was explored in a story reported on NPR's Morning Edition on July 30, 2013. See Steve Henn, "With Smarter Cars, the Doors Are Open to Hacking Dangers," July 30, 2013, accessed July 30, 2013, http://www.npr.org/blogs/alltechconsidered/2013/07/30/206800198/Smarter-Cars-Open-New-Doors-To-Smarter-Thieves.

37. National Highway Traffic Safety Administration, "Federal Motor Vehicle Safety Standards: Event Data Recorders," Docket No. NHTSA-2012-0177, Federal Register 777 (240, December 12, 2012): 74144.

38. *USA Today*, "Editorial: 'Black Boxes' Are in 96 Percent of New Cars," January 6, 2013, accessed July 30, 2013, http://www.usatoday.com/story/opinion/2013/01/06/black-boxes-cars-edr/1566098/.

39. 18 U. S. C §2701 et seq.

40. *United States v. Graham*, Criminal No.: RDB-11-0094 (District of Maryland, 2012), retrieved July 27, 2013 from the Electronic Frontier Foundation website, https://www.eff.org/sites/default/files/MDCSLIOpinion.pdf. At the time of writing, the case was under appeal to the United States Court of Appeals, 4th Circuit.

41. *In the Matter of the Application of the United States of America for an Order Authorizing the Installation and Use of a Pen Register and Trap and Trace Device*, ____ F. Supp 2d ____, 2012 WL 2120492 (WD Tex 2012).

42. 844. F. Supp. 2d 982; 2012 U.S. Dist. LEXIS 1506 (D Ariz 2012).

43. *Pennsylvania v. Rushing*, note 33, supra.

44. Ibid., p. 23, fn 3.

45. *State v. Thomas W. Earls*, A-53-11, (068765) (NJ 2013), retrieved July 27, 2013, from http://www.judiciary.state.nj.us/opinions/supreme/A5311StatevThomas-WEarls.pdf.

46. *New York v. Harris*, Criminal Court of the City of New York, County of New York, Docket No: 2011NY080152 (June 30, 2012), retrieved July 25, 2013 from the Electronic Frontier Foundation, https://www.eff.org/node/71147.

47. *United States v. Warshak*, 631 F. 3d 266 (6th Cir 2010).

48. *New York v. Harris*, note 46, supra.

49. See, e.g., *United States v. Meregildo*, 883 F. Supp. 2d 523 (Dist. Court, SD New York 2012), in which the court held that admission of Facebook material listed as private by defendant but provided by Facebook friends of the defendant was not a violation of the Fourth Amendment.

50. Ellen Nakajima, supra, note 1; American Civil Liberties Union, supra, note 31; Declan McCullagh, "Cops to Congress: We Need Logs of American's Text Messages," December 3, 2012, accessed July 28, 2013, http://news.cnet.com/8301-13578_3-57556704-38/cops-to-congress-we-need-logs-of-americans-text-messages/.

51. David Bresnahan, "Gov't Tracking Cell Phones without Court Order," newswithviews.com, January 4, 2006, accessed July 29, 2013, http://www.newswithviews.com/BreakingNews/breaking40.htm; American Civil Liberties Union, note 31, supra; and Ellen Nakajima, note, 34, supra.

52. 18 U.S.C. § 2703(d).

53. *United States v. Jacobsen*, 466 U.S. 109 (1984).

54. *Weeks v. United States*, 232 U.S. 383 (1914).

55. *Mapp v. Ohio*, 367 U.S. 643 (1961).

56. See, Slobogin and Schumacher, note 25, supra, in which it was found that public expectations of privacy, as measured by the researchers, was not in line with the opinions of the Supreme Court.

57. Doug Mataconis, "Initial Polls Seemingly in Conflict on Public Opinion of NSA Surveillance Programs," Outside the Beltway, June 11, 2013, accessed July 29, 2013, http://www.outsidethebeltway.com/initial-polls-seemingly-in-conflict-on-public-opinion-of-nsa-surveillance-programs/. Similarly, divisions are found with regard to the use of surveillance cameras. See, e.g., National Public Radio, "Big Op-Ed: Shifting Opinions on Surveillance Cameras," April 22, 2013, accessed July 29, 2013, http://www.npr.org/2013/04/22/178436355/big-op-ed-shifting-opinions-on-surveillance-cameras.

58. CBS News, "Most Disapprove of Gov't Phone Snooping of Ordinary Americans," June 11, 2013, accessed July 29, 2013, http://www.cbsnews.com/8301-250_162-57588748/most-disapprove-of-govt-phone-snooping-of-ordinary-americans/. See, also, Rasmussen Reports, "59% Oppose Government's Secret Collecting of Phone

Records," June 9, 2013, accessed July 29, 2013, http://www.rasmussenreports.com/public_content/politics/general_politics/june_2013/59_oppose_government_s_secret_collecting_of_phone_records.

59. Washington Post, "Majority Say NSA Tracking of Phone Records 'Acceptable'—Washington Post-Pew Research Center Poll," June 10, 2013, accessed July 29, 2013, http://www.washingtonpost.com/page/2010-2019/WashingtonPost/2013/06/10/National-Politics/Polling/release_242.xml.

60. Jan Lauren Boyles, Aaron Smith, and Mary Madden, "Privacy and Data Management on Mobile Devices: More than Half of App Users have Uninstalled or Avoided an App Due to Concerns about Personal Information," Pew Research Center's Internet and American Life Project, September 5, 2012, accessed July 29, 2013, http://pewinternet.org/~/media//Files/Reports/2012/PIP_MobilePrivacyManagement.pdf.

61. For a description of the rhetoric used by Presidents Nixon and Reagan to justify their efforts to nominate only staunch law and order conservatives to the Supreme Court, see Michael C. Gizzi and R. Craig Curtis, "What Is a Landmark Case? Ranking Search and Seizure Cases Using *Shepard's Citations,*" *Criminal Law Bulletin* 49 (2013): 236, p. 243, notes 31 and 32.

62. Sheldon Goldman, "Reorganizing the Judiciary: The First Term Appointments," *Judicature* 68 (9–10, 1985): 313.

63. Jeffrey Toobin, *The Nine: Inside the Secret World of the Supreme Court* (New York: Anchor Books, 2007), at 15–16.

64. See, e.g., Robert Shapiro, "Objection! Americans' Opinion of Supreme Court Can't Keep Dropping," Christian Science Monitor online, August 5, 2013, accessed August 6, 2013, http://www.csmonitor.com/Commentary/Opinion/2013/0805/Objection!-Americans-opinion-of-Supreme-Court-can-t-keep-dropping; See, also, the description of the Supreme Court deliberations over the disputed presidential election of 2000, in Jeffrey Toobin, note 63, supra, at 165–208.

65. See the opinion in *State v. Thomas W. Earls,* note 45, supra, at 22–23, detailing the variety of outcomes in federal trial court decisions on this issue, and the opinion in *Pennsylvania v. Rushing,* note 33, supra, at 37–38, detailing the hybrid theory employed by a minority of federal courts.

66. *In the Matter of the Application of the United States of America for an Order Directing a Provider of Electronic Communication Service to Disclose Records to the Government,* 620 F. 3d 304 (3rd Cir 2010).

67. *United States v. Karo,* note 27, supra.

68. Bennet Stein, "Fighting a Striking Case of Warrantless Cell Phone Tracking," American Civil Liberties Union, July 1, 2013, accessed July 25 2013, http://www.aclu.org/blog/technology-and-liberty-national-security/fighting-striking-case-warrantless-cell-phone-tracking. The case is *United States v. Graham,* note 40, supra.

69. *United States v. Graham,* note 40, supra.

70. Joint amicus curiae brief, note 16, supra.

71. See, e g., Stewart G. Pollock, "Adequate and Independent State Grounds as a Means of Balancing the Relationship between State and Federal Courts," *Texas L Rev* 63 (March/April, 1985): 977; Two law journals, *University of Puget Sound Law Review*

and the *Texas Law Review* published special issues devoted to this phenomenon in 1985.

72. *Commonwealth v. Edmunds*, 586 A.2d 887, 897–899 (Pa. 1991); *State v. Chrisman*, 100 Wn2d 814 (Washington Supreme Court, 1984).

73. *Mapp v. Ohio*, note 55, supra, at 651.

74. Note 33, supra.

75. *State v. Thomas W. Earls*, note 45, supra.

76. Id., at 3, citing *State v. Reid*, 194 NJ 386, 399 (2008).

77. *State v. Thomas W. Earls*, note 45 supra, at 3.

78. Ibid.

79. Ibid., at 4.

80. Maine (16 MRSA c. 3, sub c. 10, § 642), and Montana (House Bill No 603, http://leg.mt.gov/bills/2013/billhtml/HB0603.htm (last accessed, July 30, 2013) have passed statutes mandating that police obtain a warrant before seeking to track a suspect using his or her cell phone.

81. Timothy Casey, "Electronic Surveillance and the Right to Be Secure," *U C Davis L Rev* 41(2008): 977, provided an excellent summary of the law as of 2008 in terms of what is allowed and not under the Fourth Amendment and the various applicable federal statutes.

9

Drones and Police Practices

John C. Blakeman

Unmanned aerial vehicles, commonly called drones, are a prominent part of the U.S. government's military strategy in the war on terrorism. The military use of drones is appealing for several reasons. Drones are remotely piloted and thus present little risk to the personnel operating them, and they are far cheaper and easier to deploy than manned aircraft. In addition, drones can remain over targets for extended periods of time—far longer than conventional, manned aircraft—and are therefore ideal for the ongoing aerial surveillance of a target or lying in wait to destroy an enemy with a remotely launched missile. Most politicians and a majority of the public support the military use of drones, and as a poll by Gallup showed in 2013, upward of 65 percent of Americans approve of their use to attack terrorists outside of the United States.[1]

While the American public and many politicians are supportive of the overseas use of drones, they view the use of drones within the United States differently. A poll conducted by Monmouth University in 2012 indicated that a large majority (80 percent) support the domestic use of drones for search and rescue missions, and smaller majorities (67 percent) support using them to track down fugitives or control illegal immigration on the border (64 percent). However, that broad support wanes when the public perceives that police will use drones in ways that affect their privacy or aid in day-to-day police activities. Thus, 67 percent of the public opposes the use of drones for a routine police activity such as enforcement of speed limits, and 64 percent are very or somewhat concerned about the use by law enforcement of drones equipped with high-speed cameras that can enhance the ability of police officers to see and record events.[2]

The polling data on drones suggests that the public is concerned about how domestic law enforcement agencies might use drones within the United

States, and that concern is reflected by federal, state, and local lawmakers too. Hearings in the United States Congress in 2013 and 2014, held by the respective judiciary committees in the House of Representatives and Senate, spotlighted the increasing use of drones by law enforcement agencies. Many members of Congress raised concerns about the threats to privacy stemming from the use of drones at home. The same concerns are evident in state politics too, as by mid-2014, over half of the states have either passed or are debating laws regulating the use of drones. Local governments have likewise passed ordinances and resolutions that regulate the use of drones by their corresponding law enforcement agencies.[3]

Accordingly, this chapter focuses on how and why drones are used by law enforcement in the United States and the procedures that police follow when using them. The chapter addresses some of the main public policy and privacy issues raised by their use of drones, with attention given to the policy debates over why drones should be used in law enforcement and the restrictions on their use imposed by the U.S. Constitution's Fourth Amendment search and seizure clause. Applicable cases decided by the United States Supreme Court are likewise discussed, since the Court's decisions will offer guidance on some of the constitutional requirements that police must follow when using drones. Finally, how legislators have responded to put in place procedures for drone use at the federal, state, and even local levels is addressed.

UNMANNED AIRCRAFT SYSTEMS, UNMANNED AERIAL VEHICLES, AND DRONES

Aircraft that are unmanned and instead remotely piloted are referred to as Unmanned Aircraft Systems (UAS), Unmanned Aerial Vehicles (UAV), or drones. For the rest of this chapter, the term "drones" will be used to cover all of the various iterations of unmanned aircraft in use by the military and law enforcement. Drones were primarily developed for military use in the latter half of the 20th century, and they were first used extensively by the Israeli military in the 1982 Lebanon conflict and by the U.S. military in the 1991 Gulf War and in the mid-1990s in the Balkans Conflict.[4] Drones are loosely classified as large or small according to their size and maximum flight altitude.

A large drone is typically the size of a small executive jet and can fly up to 60,000 feet; it can stay aloft for several days and can be used for surveillance, communications relays, and data gathering. A small drone will generally weigh less than 55 pounds, fly below 400 feet, and stay in the air for only a few hours or less; small drones will be used for military reconnaissance, surveillance, and inspections. Large and small drones can be weaponized, too, to carry laser-guided missiles or other armaments. Nano drones, also known as micro drones, are currently in development too, and they can be as small as an insect.[5] The Federal Aviation Administration (FAA) regulates the nation's airspace, and in 2012, Congress passed the FAA Modernization and Reform Act that in part ordered the agency to develop federal regulations for integrating drones into the nation's airspace by 2015.[6]

Manufacturers of drones expect the industry to grow, with the small drone market segment experiencing the greatest growth early on. The law enforcement and commercial markets are gradually emerging.[7] Industry analysts expect local and state government agencies to drive much of the initial growth but also anticipate the development of a private market as businesses figure out how to use drones for their own purposes. Of course, the regulatory issues that all of these new government and private-sector drones pose are complex.

SELECT USAGES OF DRONES BY LAW ENFORCEMENT AGENCIES

The nonmilitary use of drones is expected to grow rapidly, and the FAA forecasts that there will be approximately 10,000 active civilian drones by 2020.[8] The FAA currently grants Certificates of Waiver or Authorization (COA) to public agencies to operate drones and has granted certificates to federal agencies, public universities and research laboratories, and even local fire departments. A Freedom of Information Act request to the FAA by the Electronic Frontier Foundation showed that between 2006 and 2012, the agency issued approximately 750 COAs, although many certificates are of short duration and expire.[9] Nonetheless, the FAA has emerged as the primary federal agency that oversees the domestic use of drones.

Drones are used by federal, state, and local agencies. At the federal level, the Department of Homeland Security (DHS) uses drones the most, although other agencies such as NASA regularly use drones for space and weather-related research. Within DHS, Customs and Border Protection (CBP) uses drones to support federal and state agencies engaged in law enforcement functions. CBP drones have been used to assist the Federal Bureau of Investigation (FBI), the Department of Defense (DoD), Immigration and Customs Enforcement (ICE), and the U.S. Secret Service. At the state and local levels, CBP drones have been used to assist the Texas Rangers, the Minnesota Drug Task Force, and the Pima County, Arizona, sheriff's department. Drones have been loaned for a wide range of purposes too, from border surveillance and drug interdiction to the monitoring of crime scenes. A FOIA lawsuit filed by the Electronic Frontier Foundation forced DHS to disclose that it had loaned its drones to other agencies 700 times between 2010 and 2012.[10]

In one notable example that received national media coverage, DHS used a Predator drone to provide live video feed to the Nelson County, North Dakota, sheriff and the Grand Forks, North Dakota, SWAT team who were engaged in an armed standoff with a farmer and others who refused to allow county officials on their land to recover stray cattle for a local rancher. The video feed was provided for an extended period of time and allowed police to track the individuals' movements on the property, thus facilitating their arrest without a violent gun battle. The farmer, Randy Brossart, was arrested and convicted of several felonies.[11] Brossart sought to dismiss the charges against him, in part because of the use of the drone to survey his property and monitor him without a search warrant. The North Dakota district judge in charge of the case, Judge Joel D. Medd, refused Brossart's request and succinctly noted that the use of the drone had no bearing on the charges against the defendant.[12]

INCENTIVES FOR LAW ENFORCEMENT TO ADOPT DRONES

The use of DHS drone video surveillance by law enforcement in North Dakota illustrates an important incentive for public agencies to use drones for a wide range of public safety purposes. The ability to use live video feed from an unmanned drone that can remain over a location for long periods of time can help police assess how volatile a standoff situation is. Further, live video feed from the air can assist police in locating a subject for arrest, and it can also help with other public safety concerns such as the location of a weapon, an explosive device, or another threat. One commentator has noted that state and local police departments are "eager to equip themselves with drones because they are cheaper and more efficient" than manned aircraft such as helicopters and fixed wing aircraft.[13]

Recent hearings held by the Senate Judiciary Committee on the domestic use of drones offer much detail on some of the incentives for law enforcement to adopt drones.[14]

Benjamin Miller, the director of the Unmanned Aircraft Program for the Sheriff's Department of Mesa County, Colorado, provided important insights in his testimony to the committee. Miller noted that for a geographically expansive county such as Mesa, with 3,300 square miles, operating air units for the sheriff's department is very expensive. The cost to operate a manned helicopter or fixed-wing aircraft can range from $250 to thousands of dollars per hour, exclusive of the initial cost to purchase the aircraft. In contrast, Mesa County can operate an unmanned drone for various purposes for approximately $25 per hour, with a low purchase cost. By way of example,

Miller noted that the county conducted an annual aerial survey of its landfill, which only cost $200 with a drone, compared to the $10,000 spent for a manned survey flight.[15]

Not only are the operating costs of drones far lower than those for manned aircraft, but also, drones are more readily available and adaptable to a wide range of missions. Mesa County primarily uses two small drones, one of which is a backpack-sized helicopter that weighs only 2 pounds and can fly for 15 minutes, and the other is a small fixed-wing UAV that weighs only 8 pounds and can stay aloft for 6 hours. Either drone can fit in the trunk of a police car and be quickly deployed at short notice. They have been used in Mesa County for a wide range of missions, such as search and rescue operations, aerial surveys of forest and structure fires, surveys of crime scenes, and surveys of the county landfill.

The information from the Senate hearings is corroborated by a 2013 report by the Department of Criminal Justice Services (CJS) for the Commonwealth of Virginia.[16] The Virginia General Assembly, in 2012, passed a moratorium on the use of unmanned aircraft by law enforcement until 2015 and asked state agencies to develop protocols for their use. The resulting report by the CJS offered one of the most detailed looks at how states approach the issue of police procedures and drones and at when drones can and should be used. The report indicated that drones can be used in hazardous environments where toxic smoke, fire, hazardous waste, or poor weather would pose a threat to manned flight missions, and they can provide "superior situational awareness while minimizing the danger" for operators. Drones can also be launched from a safe location in close proximity to the place of need, and many types can be launched in close quarters and in a very short amount of time.[17]

In terms of the protocols and operational procedures for drone use, the Virginia report urges agencies to involve the public early on and strongly suggests that law enforcement agencies "engage the community early in the planning process, including ... civil liberties advocates." Moreover, transparency is critical to reassure the community that the agency using drones "is in full compliance with the U.S. Constitution, and federal, state, and local law governing search and seizure." Local media should be involved in a review and comment process for the procedures governing the use of drones as well.[18]

The Virginia report also addressed data collection concerns surrounding the use of drones to gather information through surveillance and applicable Fourth Amendment issues that police agencies must consider. The report noted that Virginia law regulates the use of personal information gathered through criminal investigations; for example, the state law specifies how long that information can be kept by police after it was obtained. However, some police investigations are exempt from those data collection and handling requirements, and the report suggested that some of the information gathered through drone surveillance could be kept indefinitely. Finally, the report acknowledged that "applying the 4th Amendment to [drone] surveillance is new territory for both law enforcement and the courts. The constitutionality of this technology as used by law enforcement will depend on many factors, including how and where the surveillance takes place."[19]

Virginia's report on the protocols for using drones is one of the most detailed policy statements from a state government about how drones should be used by law enforcement. However, other than highly technical specifications about pilot training, flight regulations, and the like, its suggested protocols concerning surveillance and the gathering of evidence are broadly based and offer only limited guidance for police procedures. Perhaps most importantly, the report raised two important concerns. The first is that the procedures regulating drone usage present new policy and constitutional issues for law enforcement and courts to sort out, and the debates over drones and the Constitution are new. The second concern is that police departments should be transparent in their use of drones and should involve a range of stakeholders in the policymaking process, including civil liberties groups and the media.

LAW ENFORCEMENT PROCEDURES FOR DRONES

Federal agencies use drones for a wide range of law enforcement purposes. The Department of Homeland Security maintains the most public use of drones through Customs and Border Protection. CBP uses both remotely piloted drones and drones programmed for autonomous flight (without any pilot supervision from the ground) to monitor the United States' borders with Mexico and Canada. Drones are based primarily on the U.S.-Mexico border, but one drone has been used out of the CBP office in Grand Forks, North Dakota.[20] Other federal agencies use drones for law enforcement purposes, with far less public and political oversight. For example, in June 2012, the Department of Justice reported to Congress that the Bureau of Alcohol, Tobacco, and Firearms had six drone helicopters in its air fleet.[21] The DoJ also noted that the Federal Bureau of Investigation (FBI) had received FAA approval for a few drone missions within the United States but provided no further information about those missions. As the *Washington Post* reported at the time, then-FBI Director Robert Mueller revealed to the Senate Judiciary Committee that "the FBI uses drones 'in a very, very minimal way and very seldom.' He gave no other details, except to say that the agency has 'very few' drones and 'that our footprint is very small.'" Aside from Customs and Border

Protection drones, other federal agencies that use drones for law enforcement purposes have provided very little public information about their usage.[22]

However, more information about the state and local law enforcement use of drones is available. The report by Virginia's Criminal Justice Service discussed above indicated that drones will be used for a wide range of purposes, from accident investigation, disaster management, crowd control, and hostage situations, to support for arrest warrants, VIP security, and explosive ordinance disposal.[23] Similar proposals are seen at the local level too. For example, the police department of Ogden, Utah, proposed using an unmanned blimp for surveillance, and the Houston, Texas, police department faced a public opinion backlash when it sought to use drones with the ability to read license plates for traffic-law enforcement.[24] The police department for the city of Seattle, Washington, drafted a policy and procedures manual that stipulated that drone use will be consistent with the open-view (fields) doctrine of the Fourth Amendment and will only be used to provide support to things such as criminal investigations, missing person searches, hazardous material or natural disasters, hot pursuits, or missions authorized by the department's Homeland Security Bureau.[25] The Seattle manual defines operating procedures to govern the use of drones, but very little attention is given to procedures designed to protect privacy; instead, most procedures focus on operator training and safety regulations.

The debates over the appropriate policies to govern the use of drones are still in their formative stages. As of this writing (in summer 2014), Congress has taken no action to strictly regulate the use of drones by federal agencies other than flight and registration requirements imposed by the FAA, although as discussed below, states and municipalities have recently taken steps to restrict the use of drones. Policymakers tend to approach the issue from two distinct perspectives. For example, some will agree with recommendations by civil liberties groups, such as the American Civil Liberties Union, that law enforcement drones should be regulated by strict rules designed to protect the civil liberties and privacy of Americans, especially their First and Fourth Amendment rights.[26] Others argue that strict limits on the use of drones will not protect privacy and civil liberties and instead will protect criminal wrongdoing and hinder policing. Instead of placing limits on how drones should be used, policymakers should instead require transparency on their use to encourage public accountability.[27] Currently, the Police Foundation, an advocacy group that focuses on practical applications of policing, is using federal grant money to develop a guidebook for law enforcement to navigate "the knowledge gap . . . on the constitutional and legal requirements" for drone use.[28] That the federal Department of Justice is making grant monies available to develop policy manuals for the use of drones indicates the policy debate itself is still very much in its infancy, and perhaps the development of guidelines lags significantly behind the development of the technology itself.

The policy consensus that develops from the debates over how drones should be used will be affected most of all by the constitutional case law surrounding the Fourth Amendment. Legal scholars and policymakers recognize that the Fourth Amendment "is central to the privacy issues" concerning the police use of drones.[29] Although some public safety usages for drones, such as fire suppression, missing person searches, public property surveys, and other uses not associated with criminal investigations, will not normally implicate Fourth Amendment concerns, given the technological sophistication of drones and their ability to record and enhance the naked-eye observations of police through sophisticated cameras and audio equipment, any information gleaned from drone use in a noncriminal setting will raise constitutional questions if it is subsequently used in a criminal investigation. The surveillance capabilities of drones and their ability to stay aloft for far longer than manned flight missions raise fundamental concerns about privacy under the Fourth Amendment too.

DRONES AND THE FOURTH AMENDMENT

The Fourth Amendment's search and seizure clause in the United States Constitution regulates all conditions under which the government may conduct a search and seizure.[30] Our understanding of the amendment's requirements comes primarily from court decisions, and the United States Supreme Court's case law establishes the fundamental principles of law that police must follow.[31] Although the Court's case law is confusing, with one constitutional law scholar referring to it as a "vast jumble of judicial pronouncements that is . . . complex and contradictory," the Court's decisions are still the starting point for Fourth Amendment analysis.[32] Broadly understood, the Court's jurisprudence will focus on two main issues: when a warrantless search is reasonable and therefore constitutional and when a suspect has a reasonable expectation of privacy that requires police to get a warrant prior to a search.

To be sure, the Court is slow to decide disputes concerning new technologies and the government's power under the search and seizure clause. Constitutional law scholars note that the development of new technologies to enhance the human senses—such as high-resolution cameras, listening or tracking devices, or surveillance drones that can stay aloft far longer than manned aircraft—also expand the power of the government when used in law enforcement. For instance, police can remain in a public area and use new technologies to conduct a search in a private space normally covered by the Fourth Amendment and for which police would typically have to have a search warrant.[33] Not only might drones allow police to conduct surveillance of private property from a public space with new technologies, but also, such surveillance may enable the government to track individuals and gather "unprecedented amounts of information about individuals."[34]

Historically, the Court is aware that new technologies disrupt the balance of power between citizens and the police, yet it is often slow to define the constitutional guidelines governing that technology. The Court has yet to address the constitutional limits on the use of drones too. Even so, there are two broad categories of cases that will govern how the search and seizure clause will regulate drones. The first category consists of Court decisions concerning aerial surveillance by police, and the second category of cases focus on the Constitution and use of new technologies by police. Neither category of cases offers a definitive view of constitutional restrictions on the use of drones but only gives guidance for current police procedures.

THE SUPREME COURT AND AERIAL SURVEILLANCE

The Court has decided several cases concerning the constitutional limits on aerial surveillance by law enforcement. Three of the representative cases are discussed below, and at the outset, it is noteworthy that the Justices are divided on the constitutional limits on police use of aircraft. The use of drones poses important distinctions too. For instance, with manned aircraft, police officers are limited to flights of relatively short duration and are also limited in the spaces that they can access. In contrast, many drones are capable of remaining aloft for far longer than manned flight, and when coupled with sophisticated imaging and listening technologies, drones can gather information through surveillance for a much longer time period than manned flights.[35] In addition, small drones can access spaces for surveillance that manned flights cannot access for safety or size reasons, and they can also conduct surveillance from an unobtrusive location. Surveillance drones can be fitted with facial-recognition technology, heat sensors, and even sniffers that can track wireless Internet signals.[36] In a very real sense, the use of drones for aerial surveillance allows law enforcement to overcome the limits of the human senses and human endurance.

In two aerial surveillance cases from 1986, the Supreme Court ruled that the Fourth Amendment does not require a search warrant for aerial searches conducted from publicly navigable airspace. In *California v. Ciraolo*,[37] a five-justice majority ruled that the search and seizure clause was not violated when a police officer, acting on an anonymous telephone tip, hired a private airplane and flew at 1,000 feet above a suspect's house and identified marijuana plants in the suspect's yard. A search warrant was obtained based on the officer's observations.

Chief Justice Warren Burger wrote the majority opinion, reasoned that the officer was in publicly navigable airspace, and observed the drugs in a "physically nonintrusive manner."[38] Even though the marijuana plants were within the curtilage[39] of the suspect's home, any member of the public flying in the same airspace as the police could have observed the plants; therefore, the suspect's expectation of privacy of his garden is unreasonable. As Burger noted:

> The Fourth Amendment protection of the home has never been extended to require law enforcement officers to shield their eyes when passing a home on public thoroughfares. Nor does the mere fact that an individual has taken measures to restrict some views of his activities preclude an officer's observations from a public vantage point where he has a right to be and which renders the activities clearly visible.[40]

Publicly navigable airspace is a public vantage point, and the suspect's expectation of privacy that his garden activities were shielded from an aerial search is unreasonable.

The Court was divided, however, and four justices dissented. The dissent, written by Justice Lewis Powell, argued that the aerial surveillance was indiscriminate, and police conducted a low level flight "solely for the purpose of discovering evidence of a crime within a private enclave into which they were constitutionally forbidden to intrude at ground level without a [search] warrant."[41] Powell's dissent included a trenchant warning too: "Rapidly advancing technology now permits police to conduct surveillance in the home itself, an area where privacy interests are most cherished in our society, without physical trespass."[42]

A companion case, decided with *Ciraolo*, applied the same reasoning to a search conducted under the Clean Air Act. In *Dow Chemical v. United States*, the Court was asked to consider whether the industrial curtilage of a large chemical plant prevented aerial surveillance by the Environmental Protection Agency. After the EPA's request for a follow-up inspection of a Dow plant under the Clean Air Act was denied, the agency hired a commercial photographer to take aerial photographs of the plant from several different altitudes in publicly navigable airspace. The Court ruled that the Dow plant was not "curtilage," but "more comparable to an open field."[43] Thus, it is open to view from the air, and the EPA's warrantless search did not violate the Fourth Amendment. Interestingly, the majority opinion, again written by Chief Justice Burger, mentioned the use of technology in the search. Burger noted that "the surveillance of private property using highly sophisticated surveillance equipment not generally available to the public" might be constitutionally prohibited without a search warrant.[44] However, in the *Dow* case, the photographs were not detailed enough to raise constitutional concerns even though "they undoubtedly give the EPA more detailed information than naked-eye views."[45] This aspect of the Court's opinion could well have ramifications for the use of drones. Even though the Court has upheld the warrantless search of property by law enforcement flying in publicly navigable airspace, the Court suggests that technology used in those aerial searches that enhances the human senses may well be unconstitutional.

Justice Powell again dissented and was joined by three other justices. Powell argued that the EPA's aerial search "penetrated into a private commercial enclave, an area that society has recognized that privacy interests legitimately may be claimed. The photographs captured highly confidential information that Dow had taken reasonable and objective steps to preserve as private."[46] For Powell, Dow had taken justifiable steps to keep private its industrial plant, and its privacy interest stemmed from its need to protect industrial secrets.

One final aerial surveillance case illustrates the Court's reasoning. In *Florida v. Riley*, police conducted a warrantless search from a helicopter flying at 400 feet over a private greenhouse suspected of containing marijuana.[47] The surveillance revealed plants, and a search warrant was subsequently obtained to examine the greenhouse and surrounding property. Again the Court was divided on whether aerial surveillance was allowed under the search and seizure clause. Although five justices ruled that the aerial search without a warrant was constitutional, only four justices agreed to a plurality opinion, and one justice concurred in the result. Thus, the Court was even more divided than in *Ciraolo* and *Dow*, and with only a plurality opinion, there was no majority consensus about the rationale justifying the Court's decision. The four-justice plurality applied the *Ciraolo* precedent and reasoned that since the FAA allows helicopters to fly as low as 400 feet in publicly navigable airspace, the owner of the greenhouse had no expectation of privacy. Thus, the search was permissible without a warrant. Justice Sandra Day O'Connor agreed with the plurality's outcome but disagreed with its reliance on FAA guidelines to determine the limits of the Fourth Amendment. As with *Ciraolo* and *Dow*, four justices dissented. Justice William Brennan's dissent, for example, takes umbrage at police surveillance conducted at the low altitude of only 400 feet, and, perhaps more presciently, highlights the constitutional implications for the future:

> Imagine a helicopter capable of hovering just above an enclosed courtyard or patio without generating any noise, wind, or dust at all—and for

good measure, without posing any threat of injury. Suppose the police employed this miraculous tool to discover not only what crops people were growing ... but also what books they were reading and who their dinner guests were.[48]

For Brennan, the Court's sanctioning of the police surveillance in *Riley* effectively validates warrantless aerial surveillance as long as the police aircraft is in a publicly navigable airspace. In addition, Brennan noted that aerial surveillance technology will evolve to allow police to snoop on suspects unobserved, and while he does not refer directly to the use of drones, his dissent certainly alludes to the emerging ability of police to enhance their naked-eye observations with emerging technologies.

The aerial surveillance cases showed the Court's support for the use of police aircraft to conduct warrantless searches for evidence of criminal wrongdoing as long as the aircraft operates within airspace open the public. Yet that the Court was divided in all of the cases, with only a plurality opinion in *Riley*, will make it challenging for the justices to simply extend the jurisprudence of aerial surveillance to cover the use of drones. To be sure, the cases do not offer a ready answer to constitutionality of drones for aerial surveillance.

THE SUPREME COURT AND EMERGING TECHNOLOGIES

Although the Court's decisions on aerial surveillance validate police procedures that use aircraft without a search warrant, its decisions on emerging technologies may indicate otherwise. The Court has long grappled with how new technologies affect the balance of power in the Fourth Amendment between the government and citizens.[49] Drones present the Court with similar dilemmas since they can be outfitted with a myriad of technologies to dramatically enhance police surveillance capabilities. With high-resolution cameras, for example, or devices to track wireless signals or cell phones, police can now record and preserve images or tracking signals that greatly enhance their ability to observe and follow suspects. To be sure, the technology itself not only raises privacy concerns but also, the government's ability to preserve information gleaned through surveillance presents significant questions about civil liberties.

Two recent cases show that the Court is willing to define constitutional limits on the power of police to use certain technologies to conduct surveillance. In *Kyllo v. U.S.* from 2001, the issue was whether law enforcement could use a thermal imaging device to detect an inordinate amount of heat coming from the defendant's house, which the police suspected was generated by lamps used to grow marijuana indoors. Based on the thermal images, police procured a search warrant and discovered the drugs.[50] In a more recent case from 2012, *U.S. v. Jones*, the issue was whether it was a search under the Fourth Amendment when police installed a GPS (Global Positioning System) device on a suspected drug trafficker's car without a search warrant in order to track his movements for almost one full month.[51]

In both cases, the Court ruled that the police use of the thermal device and GPS without a search warrant violated the Fourth Amendment. In *Kyllo*, Justice Scalia wrote the majority opinion and noted that although the police observed the suspect's home from a public street where they had a right to be, they were "engaged in more than naked-eye surveillance of a home." Thus, "obtaining by sense-enhancing technology any information regarding the interior of the home that could not otherwise be obtained without physical intrusion . . . constitutes a search—at least where (as here) the technology in question is not in general public use."[52] The principle established by the Court is that technology that is not in general public use cannot be used without a search warrant by police. However, the Court was divided, with four justices dissenting and willing to uphold the constitutionality of the use of the thermal device. Justice John Paul Stevens wrote the main dissent and argued that the imaging device does not see through the wall of the home in question, but only measures heat emanating off of the wall. Thus, "all that the infrared camera did . . . was passively measure heat," and there were no Fourth Amendment concerns since the device did not penetrate the walls of the house.[53] Stevens seems to draw a line between the types of technologies used by police. Devices that simply measure aspects of a building's exterior, such as heat, are presumptively constitutional, but those that peer into a dwelling's interior raise significant constitutional questions.

The most recent case, *U.S. v. Jones*, found all of the justices in agreement that attaching a GPS tracking device to a suspect's car was a search within the meaning of the Fourth Amendment. However, the Court did not address whether the search was constitutionally allowable since that argument was not addressed by lower courts. Although Justice Antonin Scalia's opinion for the Court limited the decision to the sole question of whether a search had occurred, he recognized that the government's use of technology to track suspects is problematic, and the Court may have to deal with those concerns in future cases. Justice Samuel Alito penned a concurring opinion, joined by three other justices, in which he noted that the "emergence of many new devices that permit the monitoring of a person's movements" make it "relatively easy and cheap" for the government to engage in long-term monitoring of suspects. Moreover, the use of these devices will "continue to shape the average person's expectations about the privacy of his or her daily movements."[54] Although Alito does not reach the question of whether the GPS search was unconstitutional, his concurring opinion showed an awareness that the Court will be faced with questions concerning technology and the Fourth Amendment, and the public's expectations of privacy concerning government searches—upon which much of the Court's reasoning in Fourth Amendment cases is based—will evolve as the technology evolves.

The Court's cases in aerial surveillance and police uses of emerging technologies do not provide crystal-clear guidance on how police procedures should govern the use of drones. Regarding aerial surveillance, the Court's decisions establish that police do not generally need a search warrant to conduct manned police flights in publicly navigable airspace, at altitudes determined by the FAA, to survey outdoor properties for illicit drugs. With the use of emerging technologies, the Court's cases establish that using a thermal camera that is not publicly available to measure the heat on a building's exterior is a violation of the Fourth Amendment, and the use of GPS to conduct long-term electronic monitoring of a suspect's whereabouts is a search under the Constitution, although the constitutionality of that search was not addressed.

The Court's reasoning in the aerial surveillance cases is premised on manned flights using naked-eye observations; thus, it is an open question as to whether the justices will extend those cases to cover unmanned drone flights that can remain aloft for longer periods and that use sense-enhancing technologies. And again, the sense-enhancing technologies used by police, from thermal cameras to GPS devices (and others not yet developed) will perhaps pose separate questions for the justices. Indeed, the Court may decide to separate the two issues anyway, thus determining whether unmanned drone flights and the use of sense enhancing technology used with drones may become two separate constitutional lines of inquiry and debate. If that occurs, then the constitutional law governing police procedures and the use of drones becomes even murkier than it currently is.

POLICE PROCEDURES, CONGRESS, AND STATE LEGISLATURES

Justice Alito noted in his opinion in the GPS case that one of the best ways to resolve conflicts over the Fourth Amendment and new technologies is through the political process. As he put it, "a legislative body is well suited to gauge changing public attitudes, to draw detailed lines, and to balance privacy and public safety in a comprehensive way."[55] There is an ongoing debate in Congress and state legislatures about the constitutional limits on the use of drones by law enforcement, and members of Congress and state legislatures are acutely aware of the Fourth Amendment concerns that surround the police use of drones.[56] Many bills to regulate drones have been introduced into Congress in the past few years, yet none have passed. One bill proposed by Senator Rand Paul would have required a search warrant for almost all instances when police use drones for surveillance.[57] Another bill was introduced in the House of Representatives to prohibit the use of drones for aerial surveillance of farmland,[58] and yet another would require all agencies using drones to provide a public statement on information collected by the drones and how that information will be used.[59] In early 2014, the National Association of Criminal Defense Lawyers (NACDL) reported that there were at least 10 separate bills pending in the House or Senate to regulate a wide range of drone usages, from prohibiting the use of drones to kill citizens within the United States, to putting in place stringent federal privacy protections for information gathered by drone surveillance.[60]

The congressional debates over police use of drones have not resulted in any legislation. Of far more significance is the debate that is occurring within state legislatures and local governments. The PEW Charitable Trust's project on state politics and the National Conference of State Legislators (NCSL) have recently tracked local and state regulations on drones. Pew noted that while most state legislators do not object to the military use of drones abroad, "they cringe at the possibility that domestic police forces will violate people's privacy by using them in regular policing."[61] Some examples provided by the Pew study illustrate the trend. In early 2013, the city council of Charlottesville, Virginia, passed a nonbinding resolution calling for a ban on the use of information gathered by drones in federal and state courts. At the same time, Florida's legislature held committee hearings on a pending bill that would require search warrants for drone surveillance, and when a sheriff's department captain from Orlando specifically asked for an exception to the law that would allow for crowd surveillance by drones at large events like professional sports games or political rallies, the committee's response was a resounding no.[62] The Virginia legislature passed a two-year moratorium on the police use of drones in early 2013 as well. There are significant policy issues concerning privacy and the nonpolice, or civilian, use of drones too.[63]

Regarding police procedures and drones, state legislatures have begun to actively regulate their police forces. The NCSL tracks state legislation on drones and noted that in 2013, most state legislatures (41) considered bills regulating their use, and 16 of them passed those bills into law. In the first half of 2014, 35 states considered drone legislation, with 4 states passing laws by early summer.[64] The NCSL's analysis of state drone legislation indicates that state regulations predominately affect police procedures, with a few bills also regulating drone usage by the general public. A few states, such as North

Dakota and Maryland, have appropriated monies to create FAA-designated unmanned aircraft test sites in order to facilitate the testing of drones and to create the foundation for the commercial development and manufacturing of drones. Regardless of the wide range of state measures debated or passed, the NCSL shows that approximately two-thirds of the laws passed or currently debated regulate police procedures in two main ways: the first is by forcing law enforcement to adhere to federal and state constitutional guidelines when using drones, and the second is by regulating how police can use drones to collect date, retain that data, and use that data as evidence in criminal proceedings.

It is important to note that even if the U.S. Supreme Court rules that the police use of drones does not require a search warrant, states are still free to put in place stricter requirements on their police agencies and per their constitutions and criminal justice systems. Utah and Indiana, for instance, now require police in almost all circumstances to procure search warrants prior to drone surveillance. Montana law prohibits the use of information obtained by a drone from being used to procure a search warrant, and other states prohibit the use of drone evidence obtained outside of statutory or court-granted authority (such as a search warrant). Although the federal government has exclusive power to regulate most aspects of unmanned flight in the nation's air space, and that power is extensive enough to cover many things from air safety to overall flight operations, states retain the power to put in place restrictions on how their law enforcement agencies use drones for public safety purposes. Thus, states have authority under their own constitutions to require police to adhere to certain procedures such as procuring a search warrant prior to using drones for surveillance, and they have put restrictions on what police can survey and how the information gathered from drone surveillance will be used and stored for future analysis. Federalism provisions in the United States Constitution are not violated by state laws either, as long as those laws do not interfere with the national government's exclusive control over how national airspace is used. Finally, provided state laws follow the minimum requirements of the Fourth Amendment as interpreted by the Supreme Court, states can put in place even stricter requirements on their own police agencies to further protect the constitutional rights of their citizens.

ONWARD AND UPWARD

It should be no surprise that state and local police have adopted the use of drones, given the range of incentives to do so. Unmanned aircraft are far cheaper to operate than manned aircraft, cost much less to purchase and maintain, are easy to employ quickly for public safety missions, and offer an array of advanced technological features such as high-resolution cameras or wireless signal sniffers that allow police to collect information that cannot be perceived by human senses. Coupled with the rapid development of the drone industry itself, the adoption of drones by police forces has outpaced the development of procedures designed to protect the constitutional and privacy rights of American citizens. As one current scholar of drones and the law puts it, "the only certain aspect of the debate about unmanned aircraft and privacy is that it will be contentious."[65]

The Supreme Court's jurisprudence on the Fourth Amendment, aerial surveillance, and new technologies offers some guidance to law enforcement for defining constitutional procedures to govern their use of drones. However, that guidance will most certainly be of limited use until the justices decide a case concerning the use of drones by police within the United States. As noted above, the Court's decisions on emerging technology and the Fourth Amendment are often years behind the initial use of the technology in question; thus, it is unlikely that the Court will address the police use of drones in the short term. Meanwhile, the use of drones will most likely continue to increase, and police departments sense that the incentives to use them outweigh the legal and constitutional drawbacks.

Yet it is clear that federal and state lawmakers perceive not only the threats that drones pose to the rights and liberties of the people but also that the procedures governing the use of drones are in flux. Although Congress has been unable to pass a bill regulating the police use of drones, many state legislatures have done just that, with a majority of states now having passed laws or are currently debating bills that defined the procedures that police will follow when using drones for law enforcement purposes. Thus, the procedures governing the use of drones are defined by state and local governments, and this will remain the case until the Supreme Court starts to deal with the constitutional issues raised by the law enforcement use of unmanned aircraft and until Congress begins to regulate them at the federal level.

NOTES

1. Gallup Poll, March 25, 2013, In U.S., 65 Percent Support Drone Attacks on Terrorists Abroad, http://www.gallup.com/poll/161474/support-drone-attacks-terrorists-abroad.aspx. Last accessed May 30, 2014.

2. See in general "U.S. Supports Some Domestic Drone Use," Monmouth University Poll, Monmouth University, West Long Branch, New Jersey. Released June 12, 2012. Available at www.monmouth.edu/polling. Last accessed May 30, 2014.

3. For a comprehensive discussion of state and local regulations, see the resources at the National Center for State Legislatures (www.ncsl.org) and the PEW Foundation's project on the states (www.pewstates.org).

4. John Villasenor, ""Observations from Above: Unmanned Aircraft Systems and Privacy," *Harvard Journal of Law and Public Policy*, Vol. 36, No. 2 (2013), 464.

5. For extensive background on the history, development, and types of drones, see the following: Richard M. Thompson, II, "Drones in Domestic Surveillance Operations: Fourth Amendment Implications and Legislative Responses," Congressional Research Service, CRS Report R42701 (April 3, 2013), 1–3; and Gerald L. Dillingham, et al., "Unmanned Aircraft Systems: Measuring Progress and Addressing Potential Privacy Concerns Would Facilitate Integration into the National Airspace System," United States Government Accountability Office, GAO-12-981 (September 2012), 2–7.

6. FAA Modernization and Reform Act of 2012, Pub. L. No. 112–95.

7. See GAO Report, supra note 5, 11.

8. Bart Elias, "Pilotless Drones: Background and Considerations for Congress Regarding Unmanned Aircraft Operations in the National Airspace System," Congressional Research Service, CRS Report R42718 (September 10, 2012), 4.

9. Jennifer Lynch, "FAA Releases List of Drone Certificates—Many Questions Left Unanswered," (April 19, 2012). Electronic Frontier Foundation. https://www.eff.org/deeplinks/2012/04. Last accessed April 30, 2014.

10. See Jennifer Lynch, "Customs & Border Protection Loaned Predator Drones to Other Agencies 700 Times in Three Years According to Newly Discovered Records," January 14, 2014. Electronic Frontier Foundation. https://www.eff.org/deeplinks/2014/01. Last accessed May 5, 2014.

11. For media coverage, see Jason Koebler, "Court Upholds Domestic Drone Use in Arrest of American Citizen," *U.S. News and World Report*, August 2, 2012. http://www.usnews.com/news/articles/2012/08/02. Last accessed on April 10, 2014. See also Stephen J. Lee, "North Dakota farmer to spend months in jail for terrorizing," *Bakken Today*, January 15, 2014. http://www.bakkentoday.com/event/article/id/36024/. Last accessed June 3, 2014.

12. The case is *North Dakota v. Randy Brossart*, District Court for Nelson County, North Dakota. Case Number 32-2011-CR-00049, 00071. August 1, 2012. A copy of the Judge Medd's memorandum order is available at The National Association for Criminal Defense Attorneys: https://www.nacdl.org/uploadedFiles/files/news_and_the_champion/DDIC/Brossart%20Order.pdf. Last accessed June 3, 2014.

13. Hillary B. Farber, "Eyes in the Sky: Constitutional and Regulatory Approaches to Domestic Drone Deployment," *Syracuse Law Review*, vol. 64 (2014), 3.

14. "The Future of Drones in America: Law Enforcement and Privacy Considerations," Hearings before the Committee on the Judiciary, United States Senate, 113rd Congress, 1st Session, (March 20, 2013). Serial No. J-113-10. U.S. Government Printing Office, 2013.

15. Testimony of Benjamin Miller, Unmanned Aircraft Program Manager, Mesa County Sheriff's Department, ibid. 6.

16. "Protocols for the Use of Unmanned Aircraft Systems (UAS) by Law-Enforcement Agencies," Report of the Department of Criminal Justice Services, House Document No. 12, Commonwealth of Virginia, Richmond, Virginia. October 2013.

17. Ibid., 5.
18. Ibid.
19. Ibid., 9.
20. See Chad C. Haddal and Jeremiah Gertler, "Homeland Security: Unmanned Aerial Vehicles and Border Surveillance," Congressional Research Service (July 8, 2010). CRS Report RS21698.
21. Craig Whitlock, "FBI has Received Aviation Clearance for at Least Four Domestic Drone Operations," *Washington Post*, June 20, 2013. http://www.washingtonpost.com/world/national-security/fbi-has-received-aviation-clearance-for-at-least-four-domestic-drone-operations/2013/06/20/a040edb6-d9df-11e2-8ed8-7adf8e-ba6e9a_story.html. Last accessed June 3, 2014.
22. As part of its FOIA lawsuit against the FAA, the EFF sought information on the FBI's use of drones. To date, the FBI has refused to comply. https://www.eff.org/deeplinks/2013/06/why-wont-fbi-tell-public-about-its-drone-program. Last accessed June 4, 2014.
23. Supra note 14, 2.
24. Farber, supra note 15, 11.
25. The EFF procured Seattle's draft manual through a FOIA request. See https://www.eff.org/document/seattle-police-department, last accessed June 7, 2014. Publication of the manual online created a public outcry against the city, and the police department was forced to cancel its drone program. See http://www.usnews.com/news/us/articles/2013/02/07/seattle-mayor-ends-police-drone-efforts. Last accessed June 7, 2014.
26. See testimony by Christopher R. Calebrese, Legislative Counsel, American Civil Liberties Union, to the Judiciary Committee of the House of Representatives. "Eyes in the Sky: The Domestic Use of Unmanned Aerial Systems," Hearing Before the Subcommittee on Crime, Terrorism, Homeland Security, and Investigations of the Committee on the Judiciary, U.S. House of Representatives, 113th Congress, 1st Session, May 17, 2013. U.S. Government Printing Office, Serial No. 113-40. Pages 26–39.
27. See the testimony of law professor Gregory McNeal, ibid., 29.
28. See http://www.policefoundation.org/content/unmanned-aerial-vehicles-policing, Last accessed April 10, 2014. The federal grant is from the Office of Community Oriented Policing Services of the Department of Justice. See "Development of Guidelines for Drone Usage as a Surveillance Tool by Local Law Enforcement Agencies," http://www.cops.usdoj.gov/pdf/2013AwardDocs/CPD/2013-CPD-AppGuide.pdf. Last accessed June 7, 2014.
29. Villasenour, supra note 4, 475.
30. The Fourth Amendment reads: "The right of the people to be secure in their persons, houses, papers, and effects, against unreasonable searches and seizures, shall not be violated, and no warrants shall issue, but upon probable cause, supported by oath or affirmation, and particularly describing the place to be searched, and the persons or things to be seized."
31. See in general Henry J. Abraham and Barbara A. Perry, *Freedom and the Court* (Lawrence, KS: Kansas University Press, 2003).

32. Akhil Reed Amar, *The Constitution and Criminal Procedure: First Principles* (New Haven, Connecticut: Yale University Press, 1997), 1; and pages 1–45.

33. Orin S. Kerr, "An Equilibrium Adjustment Theory of the Fourth Amendment," *Harvard Law Review*, vol. 125 (December, 2011), 497.

34. Farber, supra note 13, 6.

35. Farber, supra note 13, 13–14.

36. Ibid., 17.

37. *California v. Ciraolo* 476 U.S. 207 (1986).

38. Ibid., 214.

39. Curtilage is the area recognized by the common law as covering the "intimate activity" associated with the "sanctity of a man's home and privacies of life." Ibid., 213. The curtilage offers enhanced privacy protection.

40. Ibid., 213.

41. Ibid., 225.

42. Ibid., 227.

43. *Dow Chemical Co. v. United States*, 476 U.S. 227, 239 (1986).

44. Ibid.

45. Ibid.

46. Ibid., 247.

47. 488 U.S. 445 (1989).

48. Ibid., 463.

49. For an historical discussion of the Court and technology, see in general Kerr, supra note 33, *passim*.

50. 533 U.S. 27 (2001).

51. 132 S.Ct. 945 (2012). The government had procured a search warrant to install the GPS device, but the warrant had expired before law enforcement could install it. Lower federal courts treated the search as a "warrantless" search.

52. 533 U.S. 27, 35.

53. Ibid., 44.

54. 132 S.Ct. 945, 964.

55. Ibid., 965.

56. See the House and Senate Judiciary Committee Hearings, supra notes 14 and 25. For state political developments, see the National Center for State Legislatures, supra note 3.

57. Preserving Freedom from Unwanted Surveillance Act of 2012, S. 3287, 112th Congress (2012).

58. Farmer's Privacy Act of 2012, H.R. 5961, 112th Congress (2012).

59. Drone Aircraft Privacy Act of 2012, H.R. 6199, 112th Congress (2012).

60. See www.nacdl.org/usmap/news/27169. Last accessed May 5, 2014.

61. Maggie Clark, "States Seek Legal Limits on Domestic Drones," February 22, 2013. Available at www.pewstates.org/projects/stateline/headlines/states-seek-legal-limits-on-domestic-drones. Last accessed April 2, 2014.

62. Ibid., page 3.

63. See in general Margot E. Kaminski, "Drone Federalism: Civilian Drones and the Things They Carry," *California Law Review*, vol. 4 (May 2013).

64. See "2013 Unmanned Aircraft Systems (UAS) Legislation," and "2014 State Unmanned Aircraft Systems (UAS) Legislation," both available at www.ncsl.org/research/civil-and-criminal justice. Last accessed April 30, 2014.

65. Villasenor, supra note 4, 516.

10

So Long, Stakeout? GPS Tracking and the Fourth Amendment

Maureen Lowry-Fritz and Artemus Ward

On February 5, 2011, Aaron Graham and Eric Jordan used a gun to rob a McDonald's restaurant. After the thieves departed, eyewitnesses provided their descriptions to police, and they were found and arrested 10 minutes later. Officers recovered their cell phones and sought to use them to determine the defendants' past locations and movements, as there had been a series of armed robberies in the area in the prior months. Pursuant to the Stored Communications Act (SCA) of 1986, the police asked a judge order the defendants' cell phone providers to turn over data including the locations and movements of the defendant's in the prior months. Both the judge and the cell phone companies complied. Graham and Jordan filed suit to suppress this information, claiming that the data obtained by police was too broad and constituted 24-hour dragnet surveillance. In *U.S. v. Graham* (2012), a federal district court ruled against the defendants, and the case is currently before the Fourth Circuit Court of Appeals as of this writing.[1]

The case raises a number of important questions about the rights of defendants under the Fourth Amendment in light of the ubiquity of location-based data. Under the SCA, a judge's order is not a warrant for Fourth Amendment purposes and can therefore be issued under a lesser standard of review. Is the SCA constitutional? If so, how much evidence can be obtained by law enforcement? If not, must a warrant be obtained to retrieve location-based data? Furthermore, is there a constitutional or statutory difference between obtaining historic location information—as was obtained in *Graham*—and real-time tracking? The Department of Justice contends that warrants are necessary for real-time tracking but not for past location information, because with the latter, there is no reasonable expectation of privacy over the data.

In 2013, a number of states passed laws requiring law enforcement to obtain warrants before tracking individuals either in real-time or historically from location data. These developments illustrate that the recent explosion of location-based data will almost certainly dominate Fourth Amendment law in the future.

In this chapter, we discuss the state of technology and the Fourth Amendment with a specific focus on the use of Global Positioning System (GPS) technology. We begin with a brief overview of the key U.S. Supreme Court precedents on the topic, including the Court's recent decision in *U.S. v. Jones* (2012)—its first on GPS tracking.[2] We also explain what GPS technology is and how it is being used by law enforcement. Finally, we analyze the possible legal approaches that could be taken in the post-*Jones* landscape and suggest what we think is most likely going forward.

TECHNOLOGY AND THE FOURTH AMENDMENT: FROM *KATZ* TO *JONES*

Over time, technological advances have forced the Supreme Court to continually reconsider how the Fourth Amendment applies to new law enforcement techniques. Dating back to common law, judges determined whether searches were valid based on the physical intrusion of property or place. In *Olmstead v. U.S.* (1928), the Court applied this common law trespass standard to uphold the use of electronic wiretaps that were placed on telephone wires outside the defendants' homes. Writing for the majority, Chief Justice William Howard Taft said, "There was no searching. There was no seizure. The evidence was secured by the use of the sense of hearing and that only. There was no entry of the houses or offices of the defendants."[3]

But technology prompted the justices to abandon the old standard and adopt a new one more protective of individual rights. Specifically, in *Katz v. U.S.* (1967)—another wiretapping case, but in a phone booth outside the home—the Court said that the Fourth Amendment is triggered when an individual has a reasonable expectation of privacy. This new test considered the extent of an individual's public exposure and whether the information sought by law enforcement was either intimate—which was presumably protected—or mundane—which presumably was not.[4] But the Court has recently split on the appropriate standard. Some justices continue to rely on the reasonable expectation of privacy standard, while others have returned to the common law trespass regime, which they feel is less subjective and more faithful to the intent of the framers.

In the cases discussed below, the justices considered everything from electronic wiretapping to thermal imaging and tracking devices. The Court made plain that individuals have more constitutional protections in the home than they do once they venture outside. The justices struck down warrantless wiretapping and disallowed police use of tracking or thermal imaging devices to detect and monitor persons or objects in the home. Yet outside the home, they allowed a person's movements to be tracked by an electronic device as long as it was voluntarily accepted by the individual, but not if it was surreptitiously planted by law enforcement. We briefly discuss each of these developments before turning to our analysis of possible paths forward for Fourth Amendment law and technology.

The Reasonable Expectation of Privacy

Katz v. U.S. (1967) marked the dawning of the modern era of surveillance jurisprudence by broadening Fourth Amendment rights. The decision marked a significant expansion of the Court's previous position, which had approached searches and seizures from the common law perspective, emphasizing property and trespass law (e.g., *Olmstead v. U.S.*, 1928; *Goldman v. U.S.*, 1942; *Silverman v. U.S.*, 1960). Forty years prior to *Katz*, in his *Olmstead* dissent, Justice Louis Brandeis cautioned that physical intrusion may not continue to serve as the determining factor of privacy invasion in a world where, "[s]ubtler and more far-reaching means of invading privacy" are becoming available.[5] The *Katz* majority responded to developments in technology—as Brandeis had forewarned—holding that "the reach of [the Fourth Amendment] cannot turn upon the presence or absence of a physical intrusion into any given enclosure . . ."[6] *Katz* marked a transition to a new perspective that the Fourth Amendment protected "people, not places." Specifically, the Court ruled that the placement of an electronic listening and recording device outside a public phone booth constituted a search, which violated the privacy on which Katz reasonably relied and expected. By finding that an individual has a "reasonable expectation of privacy" while conversing in a closed telephone booth, the Court held that citizens' privacy rights are protected anywhere and everywhere that conditions give rise to the expectation.

Public Exposure and the Reasonable Expectation of Privacy Standard

Roughly 15 years after *Katz*, the Court faced a markedly different set of circumstances in *U.S. v. Knotts* (1983). Unlike the telephone booth wiretapping scenario presented in *Katz*, *Knotts* involved law enforcement agents placing a beeper into a container that was then sold to the defendant.

The beeper (which the defendant subsequently placed in his car) enhanced the police officers' ability to monitor the defendant's movement along the highway. The Court sided with law enforcement, holding that this type of surveillance technology revealed no additional information than what the police could have obtained through a mere visual surveillance of the defendant's vehicle from public vantage points. Writing for a unanimous Court, Chief Justice William Rehnquist noted, "Nothing in the Fourth Amendment . . . prohibited the police from augmenting the sensory facilities bestowed upon them at birth with such enhancement as science and technology afforded them . . ."[7] He reasoned that an individual travelling by vehicle on public thoroughfares possesses no reasonable expectation of privacy in his transitions from one place to another, given that such activity could easily be monitored by law enforcement agents even without the beeper.

One year later, the Court considered a case that presented a nuance to the circumstances presented in *Knotts*. In *U.S. v. Karo* (1984), agents from the Drug Enforcement Agency (DEA) used a combination of beeper and visual surveillance equipment that—similar to *Knotts*—was placed in a can that was subsequently moved among the homes of the codefendants. Using the surveillance data, the officers tracked the defendants to a home and obtained information that they could not have obtained through mere observation from outside the house. While accepting the *Knotts* rationale regarding the constitutionality of monitoring the beeper on its journey through public thoroughfares, the Court drew the line at the point at which the beeper moved into the defendant's home. The majority reasoned that different constitutional concerns arise when agents use technology to obtain information regarding defendants' activities within their private residences that they would not have uncovered without entering the residence. As such, the Court ruled that the beeper monitoring constituted a Fourth Amendment search at the moment the beeper entered the home—even without an actual physical intrusion into the home by the agents.[8] Drawing on the precedents established in *Katz* and *Knotts*, the Court held that a Fourth Amendment violation occurs when, "the Government surreptitiously employs an electronic device to obtain information that it could not have obtained by observation from outsides the curtilage of the home."[9]

The "Intimacy Doctrine"

As a new millennium commenced, technological advances continued to change the way the judiciary dealt with surveillance-related matters. In *Kyllo v. U.S.* (2001), the Court was confronted with the intimacy doctrine, which holds that searches revealing sensitive personal information are more likely to trigger a reasonable expectation of privacy than those that uncover ordinary information.[10] The Court faced a scenario where agents, by directing a thermal imager at the defendant's home, detected heat lamps that they believed were being used for a residential marijuana-growing operation. The Court considered whether the surveillance strategy violated the defendant's Fourth Amendment rights, even though no physical intrusion into the home occurred. Applying the *Katz* "reasonable expectation of privacy" standard, the Court recognized that physical intrusion of property and premises alone does not completely determine Fourth Amendment protections. The Court held that the thermal imager aimed at the defendant's home constituted a search even without physical intrusion on the premises, since the technology uncovered intimate information that would otherwise be inaccessible absent a physical intrusion. The Court's holding, however, was not without limitations. Writing for the majority, Justice Antonin Scalia stated, "Where... the Government uses a device that is *not in general public use*, to explore details of the home that would previously have been unknowable without physical intrusion, the surveillance is a 'search' and is presumptively unreasonable without a warrant"[11] (emphasis added). Scalia's reasoning implied that the use of technology to uncover sensitive personal information might not constitute a Fourth Amendment search and seizure if the technology in question was of a type that is commonly used among the general public.

GPS Technology

The term "telematics" refers to the utilization of telecommunication technology for sending, receiving, and collecting information.[12] Global Positioning Satellite (GPS) technology is a form of telematics that uses satellites and receivers to determine precise and accurate positions. GPS is a precise navigation and positioning tool, originally developed by the Department of Defense (DOD) in 1973 to enable military personnel to accurately determine their locations worldwide. In 1983, civilian users were permitted access to GPS, and today, GPS can be utilized in military, commercial, and scientific venues by anyone in possession of a GPS device such as SatNa, a mobile phone, or a handheld GPS device. The GPS used in the United States is called Navstar, and it is still funded and operated by DOD. Other nations have their own GPS systems at varying stages of testing or development.[13] The implications for the Fourth Amendment are plain: GPS technology allows law enforcement agents to monitor the activities of suspects without reliance on mobile or physical surveillance.[14] Accumulating continuous and real-time location data, GPS tracking provides police offers with descriptive and detailed accounts of behavior that is simply not feasible through visual and physical surveillance.[15]

GPS is a system comprised of roughly 30 satellites orbiting the Earth. The satellites orbit the earth twice daily, transmitting location and time data to receiving equipment on the ground. Atomic clocks positioned on each of the satellites provide an accurate reference of time, which is critical to precise functioning. Each of the GPS satellites is synchronized, so that they all transmit their signals simultaneously in a single instant. The signals move at the speed of light and arrive at the receivers at varying times, due to the different distances between the satellites and receivers. Precise positions are generated through a process called "trilateration." As mentioned above, each satellite sends data regarding time and its position at regular intervals. A GPS receiver then calculates its own distance from each satellite, based on the length of time it took for the data to arrive. Once the receiver acquires information regarding the distance to at least three satellites, it can identify its own precise location. When the receiver calculates the distance to four GPS satellites, it can determine its position in three dimensions.

Today, many citizens utilize GPS technology for a wide variety of purposes. Alison M. Smith, legislative attorney for the Congressional Research Service, notes the many ways that members of the general public have come to use GPS technology in their everyday lives.[16] Drivers of cars, boats, and planes rely on GPS devices for directional and navigational purposes as they travel. Some car manufacturers install GPS devices in their vehicles as a standard option, which enables a central command center to track a vehicle's location at all times for the purposes of locating a lost or stolen vehicle or providing assistance in the case of an accident. Pet owners purchase collars equipped with GPS devices to find their lost pets. The development of the smartphone has led to a massive explosion in recent years of average citizens routinely accessing GPS data for navigational, mapping, and directional planning. Even parents use GPS technology to track their teenage children. As Adam Koppel observes, GPS is now largely woven into the fabric of Americans' everyday functioning.[17]

In addition to civilian use, rapidly advancing GPS technology has become a fundamental feature of government. For example, it has transformed the business of gathering evidence and tracking suspected criminals. Until recently, police officers and other law enforcement officials were forced to rely exclusively on their own sensory capabilities to follow suspects and gather evidence. This could involve the physical tracking of suspects' movements and activities by foot and vehicle, over the course of weeks and sometimes months. It might also entail methodical patrolling of large territories or countless hours of sitting in unmarked cars, waiting for a suspect to enter or exit a specified location. Following suspected criminals in this manner required a great deal of time, workers, and resources. Furthermore, since the process relied solely on human observational powers, it was fraught with potential inaccuracies and errors.

Introduction of computerized GPS tracking devices and systems, however, has tremendously enhanced government capacity to target, monitor, and ultimately apprehend suspected criminals. Law enforcement agents can either access factory-installed equipment or place a GPS tracking device on a suspect's car, boat, or motorcycle, to monitor the individual's activity. Similarly, as most cellular phones include GPS technology, law enforcement agents can secure information from cellular telephone service providers so that they can track suspects' activities and locations in real time.[18]

The GPS device is encoded to transmit an electronic signal through a cellular tower to a receiving unit roughly every five seconds, yielding the specific latitude and longitude of the suspect's vehicle.[19] Officers access this data through a smartphone or tablet in nearly real-time speed from virtually any remote location. This combination of cellular and satellite technology provides up-to-the-minute information that enables officers to determine where a suspect travels, where he or she stops, and how long that person remains at a particular location—evidence that may ultimately result in an arrest.

A number of recent cases illustrate the various ways in which law enforcement authorities employ GPS technology to assist them in their day-to-day operations. In *United States v. Garcia* (2007), Wisconsin police placed a memory tracking unit (a type of GPS device) under the bumper of a vehicle driven by a known methamphetamine manufacturer.[20] The tracking device allowed the police to determine the travel history of the suspect since the installation of the device. They learned that he had been repeatedly visiting a large tract of land. After securing consent from the landowner, the police searched the property and discovered methamphetamine manufacturing equipment and ultimately the suspect himself. One year later, law enforcement agents in Virginia became aware of a rash of attacks on women in Alexandria and Fairfax County. After nearly a dozen cases in six months, the police department identified a convicted rapist as a likely suspect. Officers placed a GPS tracking device on his vehicle and monitored his activities, ultimately seizing him as he was about to commit another assault.[21] In *People v. Weaver* (2009), New York police affixed a GPS device to the vehicle of Scott Weaver, a suspected burglar. Following two months of uninterrupted tracking, the police determined that Weaver had driven to a recently burglarized store.[22] This evidence corroborated witness testimony that Weaver had been seen casing the establishment, looking for its vulnerable points.

Even more recently, federal investigators and other law enforcement agencies have been utilizing a newer and more powerful type of technology referred to as the stingray. This type of surveillance equipment, used for the past two decades as a counterterrorism measure, is a compact and sophisticated device capable of tracking cellular phone signals through cars, houses, and even insulated buildings.[23] The stingray acts as a fake cell tower that enables law enforcement agents to identify the whereabouts of a suspect by sucking up phone data such as texts, e-mails, and location information. When a suspected criminal uses his cell phone, the stingray tricks the phone into sending its signal back to the police and collects data from potentially thousands of additional cell phone users in the surrounding areas. This enables law enforcement agents to harvest and catalogue massive amounts of numbers, information, and even communication content.[24]

Given the extensive use of GPS technology by law enforcement agents, concerns over constitutional rights have necessarily arisen. The Supreme Court waded into the issue for the first time in *U.S. v. Jones* (2012). The case also allowed the Court to revisit its legal approach to the Fourth Amendment. Yet, as we discuss below, rather than clarifying the law, the justices only muddied the waters.

Return of the Property-Trespass Standard

Jones involved a scenario where government agents attached a GPS device to a vehicle and used it to monitor the vehicle's movements. Evidence of the defendant's activities was admitted in the trial court, and ultimately, the defendant was convicted. The *Jones* appeal presented the Court with a real-life scenario involving the "dragnet-type law enforcement practices" that it had envisioned 30 years earlier in *Knotts*.[25] The Court considered the question of whether 24-hour surveillance of the defendant—made possible due to the attachment of a GPS tracking device to the defendant's vehicle—constituted a Fourth Amendment search and seizure.

In a unanimous ruling, the Court determined that the surveillance in question did, indeed, constitute a search under the Fourth Amendment. Justice Scalia's narrowly written majority opinion focused exclusively on an objective common law trespass test, rather than the more subjective "reasonable expectations" standard. Referring to privacy rights that existed when the Fourth Amendment was adopted, Scalia posited that "[w]hat we apply is an 18th-century guarantee against unreasonable searches, which we believe must provide at a minimum the degree of protection it afforded when it was adopted."[26] For Scalia, the act of physically attaching a GPS device to a vehicle amounted to trespassing on private property and was therefore enough to constitute a Fourth Amendment search.

Scalia's emphasis on common law trespass rights contrasted with the reasonable expectation perspectives set forth in the separate concurring opinions authored by Justices Samuel Alito and Sonia Sotomayor. Alito argued that an act of trespass was neither necessary nor sufficient to give rise to a constitutional violation. He asserted that the majority opinion erroneously emphasized the attachment of the GPS rather than the use of the GPS, and he warned that the majority's approach would likely yield incongruous results. For example, he said that under Scalia's approach, attaching a GPS device for a nominal amount of surveillance would be unconstitutional, but long-term vehicular or aerial tracking would be constitutional. Alito acknowledged that while the subjectivity of the *Katz* "reasonable expectation of privacy" standard renders it incapable of precisely determining the point at which the surveillance of Jones's vehicle triggered Fourth Amendment protections, it should nevertheless be employed in order to avert illogical and contradictory holdings.[27]

Sotomayor also analyzed *Jones* through a reasonable expectation perspective, contending that the majority was misguided in grounding its opinion on 18th-century trespass law. Quoting from *Kyllo*, Sotomayor said that a Fourth Amendment search occurs when law enforcement violates a "subjective expectation of privacy that society recognizes as reasonable."[28] She framed the threshold question as, "whether people reasonably expect that their movements will be recorded and aggregated in a manner that enables the Government to ascertain, more or less at will, their political and religious beliefs, sexual habits, and so on."[29]

Both Alito and Sotomayor observed that Scalia's narrow emphasis on physical intrusion of property overlooks the reality that modern day technology makes long-term surveillance without intrusion upon premises highly likely. It is precisely such a scenario that this chapter addresses. Specifically, the following section speaks to the question of whether warrantless GPS surveillance—without a physical intrusion on property—triggers Fourth Amendment protections.

NAVIGATING PRIVACY IN A POST-*JONES* WORLD: TYPOLOGY OF POSSIBLE LEGAL APPROACHES

As GPS equipment advances and becomes a factory-installed standard feature in vehicles, cellular telephones, and mobile devices, the aggregate effect of such technology may render the individual's life completely visible and accessible to others. It seems possible that—but for the intervention of social, legal, governmental, or other forces—there will be no nook or cranny on earth where one might evade surveillance or scrutiny. A natural concern results: In this era in which "privacy-destroying technologies" render individuals' lives readily transparent, what becomes of one's reasonable expectation of privacy?[30] This section suggests five possible legal approaches to the question of whether—absent a physical trespass of property—warrantless GPS surveillance constitutes a Fourth Amendment search and seizure. Table 10.1 outlines these approaches.

One judicial approach to the current state of affairs may be a complete abdication of Fourth Amendment protections. As Scott McNealy, CEO of Sym Microsystems, stated, "You have zero privacy. Get over it."[31] Judges may eventually determine that GPS-based tracking via vehicles, cell phones, or mobile devices is sufficiently prevalent so as to no longer implicate the Fourth

TABLE 10.1 Typology of Fourth Amendment Legal Standards for Warrantless GPS Surveillance

Name	Standard / Test
Katz Approach	Two-prong standard: (1) Whether a physical trespass of property occurred. (2) Whether the defendant's "reasonable expectation of privacy" is violated.
Quantitative/Mosaic	Whether the quantity of information retrieved through the aggregate of government surveillance activity violated the defendant's expectation of privacy.
Sequential	Whether government conduct constitutes a search at each incremental step in the surveillance process. Three questions: (1) Did a search occur? (2) Was the search unconstitutional? (3) Is a remedy appropriate?
Public Space	"Minimal Exposure": Whether the defendant's actions revealing private information were even minimally exposed to the public. "Likelihood of Public Observation": Whether the defendant's actions revealing private information were likely to be observed by members of the public.
Third Party	Whether the defendant willingly revealed or disclosed information to a third party.

Amendment and to consequently justify exemption from warrant requirements. Another legal rationale for this approach may be that law enforcement agents can already access extensive personal information using less sophisticated technology, and GPS merely improves the quality of the evidence.[32] While such arguments warrant mentioning, such a complete renouncement of Fourth Amendment protections is unlikely. What follows is a discussion of more feasible approaches to warrantless GPS cases.

Katz Standard: Property Intrusion and Reasonable Expectation

In light of the *Jones* majority and concurring opinions, it is possible that the outcome of future GPS cases may be based on the *Katz* dual standard of trespass and reasonable expectation of privacy. The Court's unanimous decision in *Jones*, however, should not mislead one into thinking that a clear precedent has been set. In reality, disagreement among the justices in the *Jones* case left a number of important questions unanswered. Justice Scalia emphasized a common law trespass rationale, while Justices Alito and Sotomayor focused on the "reasonable expectation of privacy" and quantity of information rationales. Whether one or more of these standards will guide future cases remains to be seen.

Indeed, the *Jones* majority opinion reintroduced the *Olmstead* property-based framework, which would enable the Court to objectively analyze future cases involving physical intrusion onto premises. Simply put, given an act of trespass upon property, the common law property intrusion standard would invariably trigger Fourth Amendment protections. However, the *Jones* majority opinion does not address the possible scenario of government surveillance that does not involve the physical trespass of property. Recall the *Karo* and *Kyllo* cases, in which the government obtained evidence against the defendants without committing physical trespass of property. In facing such circumstances, the Court may invoke the reasonable expectations of privacy test set forth in *Katz* and addressed by Alito in his *Jones* concurrence.

In that instance, the question would be whether the warrantless surveillance without physical intrusion violated the defendant's reasonable expectation of privacy. This component of the *Katz* test is subjective in nature, and its application lacks specific guidelines. On the one hand, some may suggest that since no person reasonably expects to be monitored by law enforcement or the government on a 24/7 basis for several consecutive weeks, such use of GPS surveillance technology will likely be viewed as a violation of one's privacy. On the other hand, others may suggest that as technology becomes increasingly advanced and decreasingly expensive, it will become more common, thereby reducing the likelihood that utilization of this technology will be regarded as a violation of one's privacy. Given the subjective nature of the *Katz* reasonable expectation of privacy standard as well as the divergent viewpoints on the role of continuous surveillance afforded by GPS technology, future judicial decisions will likely hinge on the unique facts of the case.

The facts will likely pertain to the duration of the surveillance activities. As Renee McDonald Hutchins posits, "It is entirely consistent with existing precedent to understand the level of proof required from one who challenges covert tracking as bearing an inverse relationship to the length of time such surveillance is conducted."[33] For example, a scenario in which the defendant is monitored for a 24-hour period may not trigger Fourth Amendment protections. However, a case in which the government monitors a defendant for a three-to-four-week period may be viewed as involving a Fourth Amendment search. The question then becomes, what if the duration of warrantless surveillance falls somewhere between two days and four weeks? The threshold inquiry may be whether the law enforcement agency could have conducted continuance surveillance for that period without the use of the technology in question. Could a team of law enforcement agents, rotating shifts and using unmarked cars, aerial tracking, and other non-property-invasive strategies have accomplished an equivalent amount of surveillance and uncovered the same evidence? If so, a case can be made that the defendant's reasonable expectation of privacy has not been violated. If not, then a case can be made that the monitoring amounted to a Fourth Amendment search and seizure.

Quantitative or Mosaic Standard: How Much Information Is Obtained

As mentioned in the previous section, another potential approach to future warrantless, nonintrusion GPS surveillance cases is the view that advances in technology will require a broader, more expansive application of the Fourth Amendment. In this modern age of powerful technology, limitless storage capacity, and an insatiable drive to accumulate data, information has become dynamic and exhaustive. This has significant potential implications for the role of the Fourth Amendment. As Melissa Arbus notes, "The rapid development of advanced surveillance technologies has the potential to fall outside the purview of the Fourth Amendment."[34] Individuals are subjected to surveillance in countless transactions over the course of days, months, and years. Information is recorded by private companies as well as by federal, state, and local agencies. This data is accumulated and stored in permanent, malleable, and transportable electronic databases and harvested over and over as time passes.

In this post-*Jones* world, the Court may soon be asked to develop a test that provides privacy protections in light of the 21st century "era of ubiquitous technology."[35] The approach set forth in this section acknowledges the likelihood that the ever-increasing power, availability, and accessibility of technology will necessitate a broader application of Fourth Amendment protections. To put it in stronger terms, the Court's failure to provide protections from warrantless GPS searches may render the Fourth Amendment impotent and obsolete.

Therefore, one possible approach to future GPS cases would be for the Court to regard warrantless GPS surveillance as a Fourth Amendment search and seizure, worthy of privacy protections. Hutchins observes that "if the Fourth Amendment is to enjoy continued vitality in practice and not just theory, we must find ways to reclaim its relevance within the existing constitutional framework."[36] She posits that even in light of pervasive GPS surveillance technology, the Fourth Amendment can still effectively constrain observation, and as a result, protect privacy rights. This position recalls the historical purpose of the Fourth Amendment to protect individuals from a potentially tyrannical government and overzealous law enforcement institutions and urges its continued relevance in a modern and technology-laden world.

Such an approach recognizes that the quantity of information retrieved through the use of warrantless GPS is a relevant factor in determining whether or not the surveillance constitutes a Fourth Amendment search and seizure. Thus, the Court may deem unconstitutional law enforcement use of sense-augmenting technologies to gather a vast amount of information—even if that information is largely unimportant and mundane. In *Dow Chemical Company v. U.S.*, the Supreme Court held that while the government's aerial photographs of a physical plant did not trigger a Fourth Amendment search and seizure, a different outcome might have occurred if greater amounts of information had been uncovered. The majority observed, "It may well be, as the Government concedes, that surveillance of private property by using highly sophisticated surveillance equipment . . . such as satellite technology, might be constitutionally proscribed absent a warrant. But the photographs here are not so revealing of intimate details as to raise constitutional concerns."[37]

David Gray and Danielle Citron recommend a technology-centered perspective on protecting Fourth Amendment rights in an approach they refer to as "quantitative privacy."[38] Gray and Citron base their position on a proposition set forth in the *Jones* concurring opinions, specifically that "citizens have Fourth Amendment interests in substantial quantities of information about their public or shared activities, even if they lack a reasonable expectation of privacy in each of the constitutive particulars."[39] By blending privacy law and Fourth Amendment jurisprudence, Gray and Citron argue that the *Jones* concurrences suggest that "government access to technologies capable of facilitating broad programs of *continuous and indiscriminate monitoring* should be subject to the same Fourth Amendment limitations applied to physical searches"[40] (emphasis added). Gray and Citron regard Justice Sotomayor's concurrence in *Jones* as an endorsement of a "doctrine of quantitative privacy."[41] Justice Sotomayor observed that in today's digital age, individuals cannot avoid relinquishing large amounts of personal information to third parties such as search engines, social network sites, cell phone and Internet service providers, and so forth. These third parties then accumulate, use, and share immense quantities of personal data without the individual's knowledge. It would be unreasonable, Gray and Citron contend, to assert that the Fourth Amendment grants government agents unfettered authority to employ technology capable of this degree of monitoring. Rather, the scholars assert that the use of warrantless GPS-enabled tracking technology, which provides law enforcement agents with second-by-second location data, should constitute a Fourth Amendment search and seizure.

Sequential Standard

The sequential standard is an alternative to the mosaic approach that considers whether a law enforcement agent's conduct constitutes a search at each incremental and discrete step in the surveillance or search process—as opposed to assessing government conduct in the aggregate. Much like a series of snapshots, each step of government action is analyzed in isolation. If none of the sequential steps of policy action constitutes a search, then the Fourth Amendment is not triggered, and no violation has occurred. Orin Kerr advocates this approach, describing it as, "the foundation of existing search and seizure analysis."[42] The sequential approach involves three questions: (1) What is a search? (2) When is a search unreasonable and, as a result, unconstitutional? (3) When does an unconstitutional search justify a remedy?

To determine whether a specific instance of police action constitutes a search, one may look to case law for guidance. Kerr observes that Supreme Court decisions regarding the reasonable expectation of privacy clearly distinguish between inside and outside surveillance. More specifically, conduct is regarded as violating an individual's reasonable expectation of privacy when a law enforcement agent trespasses into or onto a private and enclosed area, thereby revealing the interior content of that space. That moment of exposure is regarded as a search by Fourth Amendment standards.[43] On the other hand, a search does not occur when circumstances involve surveillance of the exterior of a property,[44] something that is already visible to the public[45] or to public thoroughfares.[46]

If it is determined that a search did indeed occur, the next question is whether the search was constitutionally reasonable or unreasonable. Traditionally, the Court deemed a search reasonable when law enforcement had obtained a warrant prior to the search. In more recent years, however, the Court has developed a different method that purports to balance the government interests with the citizen's interests. Kerr describes the Court's contemporary approach to this inquiry balancing the government's interests in utilizing an investigatory strategy against the privacy interests at stake.[47] The Court attempted to find this equilibrium in *United States v. Place* (1983), stating, "We must balance the nature and quality of intrusion on the individual's Fourth Amendment interests against the importance of the governmental interests alleged to justify the intrusion."[48] This type of analysis requires an assessment of the "totality of the circumstances," to determine the extent to which the search interferes with an individual's privacy as well as the extent to which the search is necessary for the advancement of the government's legitimate interests.[49]

If it is determined that the Fourth Amendment has been triggered, the Court must finally consider whether remedies apply in the case. A remedy might be appropriate, for example, if an unconstitutional search causes a piece of evidence to be unearthed. According to Kerr, the determination of causality requires consideration of whether the unconstitutional conduct was the "but for" grounds or the "proximate cause" of the sighting of the evidence at issue.[50] Kerr observes that the existence of cause necessitates the recognition of an isolated event and an assessment of the degree to which that event contributed to the outcome. For this reason, he contends that this causal determination process is well-suited to the sequential approach to surveillance activities.

Public Space Standard: From Minimal Exposure to Likelihood of Public Observation

A fourth possible approach to future warrantless GPS surveillance cases might derive from what has been referred to as the public space standard. This approach has two general iterations. The first asserts that when a defendant's actions are even minimally exposed to the public, Fourth Amendment protections will not be afforded. We can term this the "minimal exposure" standard. The second iteration claims that Fourth Amendment protections are triggered when the defendant acts within areas of public space that allow for anonymity. We can call this the likelihood of public observation standard. We discuss each of these in turn.

The minimal exposure standard stems from *U.S. v. White* (1971). In this case, a government informant secretly recorded a number of conversations with the defendant in a residence, a restaurant, and a vehicle. The agents heard and captured evidence in which the defendant admitted his personal involvement in a number of narcotics transactions. The Supreme Court held that the evidence was admissible in court because Fourth Amendment protections do not apply to information exposed to the public.[51] For many years, the judiciary continued to accept any nominal amount of public exposure as dispositive, refusing to consider the likelihood of public observation as a factor when determining the scope of Fourth Amendment protections.

Roughly 30 years later, the Court altered its perspective when it rendered its decisions in *Bond v. U.S.* (2000) and *Kyllo*. These two cases prompted the Court to contend with a previously unexplored notion referred to as privacy in public. In *Bond*, a border-patrol agent squeezed a passenger's luggage in a random search for drug contraband. The Court held that the act of squeezing constituted a Fourth Amendment search and seizure and was deemed unconstitutional.[52] In *Kyllo*, the Court held that agents' use of a thermal imager directed at defendant's house constituted a search, even without a physical trespass on the premises.[53] According to Elizabeth Canter, *Bond* and *Kyllo* provide a more realistic approach, suggesting that courts should consider the likelihood of public observation when determining the scope of Fourth Amendment protections.[54] These decisions expanded the threshold inquiry, thereby potentially subjecting more surveillance activity to Fourth Amendment proscriptions. Canter explains this shift in jurisprudence, observing that "there is a dynamic relationship between the Fourth Amendment right and the Fourth Amendment remedy that facilitated the metamorphosis evidenced in *Bond* and *Kyllo* . . . the anemia of Fourth Amendment remedies enabled and may have provided some hydraulic pressure toward an expanded Fourth Amendment right."[55]

One possible route toward this expansion of Fourth Amendment protections is a transition away from the *Katz* principle that the Fourth Amendment protects "people, not places." Rather, Marc Jonathan Blitz argues that "courts can often best protect privacy in public life by focusing on *places* rather than the people who act in them (emphasis added).[56] Blitz posits that courts might best protect individuals' expectations of privacy in public by recognizing and defending those aspects of society and public space that permit anonymity and other privacy-related advantages to sufficiently thrive. Individuals navigate through public spaces, subjected to monitoring by electronic tollway systems, intelligent transport systems, and roadway technology that collects data regarding weather, traffic, and travel conditions. Even if this technology does not capture and/or reveal intimate thoughts or activities, the aggregation of data over time just might. According to Blitz, the fact that people willingly submit to these varieties of tracking technology should not leave them without any constitutional safeguards or protections. As such, courts should protect the aspects of public space that permit anonymity and privacy of actions occurring within its realm to the same degree that they have protected residences. As Blitz observes, "People also need privacy and anonymity in many aspects of public life."[57]

Third-Party Doctrine

What is referred to as the third-party doctrine is a fifth possible approach to future warrantless surveillance cases. The doctrine is fairly straightforward: By revealing to a third party, the individual relinquishes all of his or her Fourth Amendment protections over the information surrendered. To put it another way, an individual cannot claim to hold a reasonable expectation of privacy regarding information he or she has disclosed to a third party. It is often referred to in terms of assumption of risk. For example, in *Smith v. Maryland* (1979), the Supreme Court held that the petitioner "assumed the risk" of disclosure of his financial documents, and therefore it would be unreasonable for him to expect such information to remain private.[58] While this doctrine may seem to be easily applicable to arms-length transactions, the principle becomes less clear when related to data accumulated through GPS surveillance. Does an individual assume the risk of information disclosure when he or she travels to a bank, restaurant, or across state lines with a factory-installed GPS system or OnStar in the car or a smartphone in his or her pocket?

How applicable is the third-party doctrine to a scenario in which one subjects him- or herself to GPS surveillance by that person's own car or phone? Might the doctrine under such circumstances render the Fourth Amendment meaningless? Stephen Henderson contends that the third-party doctrine must be strictly interpreted if the Fourth Amendment is to provide individuals with any constitutional protections in this modern age. He asserts that "knowing exposure" of information to a third party is not the appropriate test.[59] Rather, courts should make the threshold question whether the individual affirmatively and deliberately transmitted the content with the intent that a third party *utilized* the information. Henderson explains that since the framers of the Constitution were bound only by those intrusions that were available in their day, the Supreme Court's initial interpretations would logically be grounded in common law trespass law. However, he contends that their directive that individuals be free from unreasonable searches and seizures is sufficiently malleable to permit courts to provide continued protections in a world of obtrusive technology. In this sense, framing the directive in terms of the individual's affirmative conveyance for the purpose of third-party use will enable the judiciary to preserve society's reasonable expectation of privacy.

CONCLUSION

The use of GPS tracking, cell phone location information, and social media check-ins will no doubt continue to give rise to innumerable questions about the privacy rights of individuals. For example, on June 27, 2013 a divided New York Court of Appeals held in *Cunningham v. New York State Dept. of Labor* that the state can use GPS tracking without obtaining a warrant to monitor its employees during working hours.[60] Courts will no doubt continue to grapple with such questions in the coming years. The Supreme Court's precedents in this area suggest that no clear standard is likely to emerge anytime soon. Still, we have attempted to provide a number of legal options that we think are most likely to hold sway going forward.

Future litigation will likely involve the question left unanswered in *Jones*. Specifically, is warrantless GPS surveillance, absent a physical intrusion on property, a violation of Fourth Amendment protections? As Justices Alito and Sotomayor noted, the *Jones* majority opinion fails to acknowledge the reality of today's—and more important, tomorrow's—modern-day technology. We already live in a society where advanced technology is commonplace and long-term surveillance of citizens is conducted without any hint of physical trespass on property. The question remains whether law enforcement is constitutionally permitted to use the data accumulated through these means.

How will the Court decide under such circumstances? Will it base the constitutionality of the search upon the amount of information accumulated throughout the surveillance? Will it deem a search unconstitutional if the individual's actions were subject to public observations? Will surveillance be permitted when the individual willingly gives information to a third party? Or will the Court hold that we must expect a diminishing level of constitutional protection for privacy interests, given that technology continues to augment the capacity of surveillance equipment? Such a decision would invariably reduce Fourth Amendment protections to obsolescence. While each of these inquiries has been alluded to in past judicial decisions, it is unlikely that the Court will hang the fate of the Fourth Amendment on such a definitive or bright-line test.

While *Jones* provided the Court with the opportunity to apply the *Katz* test to GPS-related facts, the majority opted to base its opinion on the overriding presence of physical intrusion of property. Simply put, the facts in *Jones* obviated the need to address the larger question of warrantless GPS surveillance conducted without trespass. However, in light of rapidly developing advances in technology, it is unlikely that the Court will be able to avoid this larger question much longer.

When faced with the question of warrantless GPS surveillance conducted without a physical trespass of property, it is likely that the Court will decide future cases by applying the reasonable expectation of privacy precedent established in *Katz* and its progeny to the specific facts of the case at hand. Simply put, the facts will likely determine the outcome. In previous cases, the Court has ruled on the basis of very specific factual nuances. In future cases, the Court will again be forced to contend with specific contextual matters: the type of technology at issue (i.e., GPS, stingray, memory tracking); the means through which it is initiated (i.e., factory installed, third-party arrangements, contractual agreement with service provider); the individual's willingness to share the information; the duration of the surveillance (days versus months); the likelihood of public observation (in home versus in public); the amount of information accumulated by the government (large versus small); the type of information accumulated (mundane versus intimate); and so forth. These considerations will likely result in a pragmatic approach that adjusts to both factual circumstances and advances in technology. Thus, it is likely that in this post-stakeout world of real-time virtual surveillance, the Court will rely on both precedent and contemporary context to determine the parameters of Fourth Amendment protections in warrantless GPS scenarios.

NOTES

1. *United States v. Graham*, 846 F.Supp. 384 (D. Md. 2012).
2. *United States v. Jones*, 565 U.S. ____ (2012).
3. *Olmstead v. United States*, 277 U.S. 438 (1928).
4. *Katz v. United States*, 389 U.S. 347 (1967).
5. *Olmstead*, 277 U.S. at 473–4. Brandeis dissent.
6. *Katz*, 389 U.S. at 353.
7. *United States v. Knotts*, 460 U.S. 276 (1983), p. 283.
8. *United States v. Karo*, 468 U.S. 705 (1984).
9. *Karo*, 468 U.S. at 715.
10. *Kyllo v. United States*, 533 U.S. 27 (2001).
11. *Kyllo*, 533 U.S. at 29.
12. Elizabeth E. Joh, "Discretionless Policing: Technology and the Fourth Amendment," *California Law Review* 95 (2007): 199–234.
13. pocketgpsworld.com
14. Steven Penney, "Reasonable Expectations of Privacy and Novel Search Technologies: An Economic Approach," *The Journal of Criminal Law and Criminology* 97 (2007): 477–529.

15. April A. Otterberg, "GPS Tracking Technology: The Case for Revisiting Knotts and Shifting the Supreme Court's Theory of the Public Space under the Fourth Amendment," *Boston College Law Review* 46 (2005): 661–704.

16. Alison Smith, *Law Enforcement Use of Global Positioning (GPS) Devices to Monitor Motor Vehicles: Fourth Amendment Considerations*, CRS Report R41663 (Washington, DC: Library of Congress, Research Service, February 28, 2011).

17. Adam Kopel, "Warranting a Warrant: Fourth Amendment Concerns Raised by Law Enforcement's Warrantless Use of GPS and Cellular Tracking," *University of Miami Law Review* 64 (2010): 1061, 1064.

18. Michael Isikoff, "The Snitch in Your Pocket," *Newsweek*, February 19, 2010.

19. Rob Cerullo, "GPS Tracking Devices and the Constitution," *The Police Chief: The Professional Voice of Law Enforcement*, July 2013.

20. *United States v. Garcia*, 474 F.3d 994 (7th Cir. 2007).

21. Ben Hubbard, "Police Turn to Secret Weapon: GPS Device," *Washington Post*, August 13, 2008, http://articles.washingtonpost.com/2008-08-13/news/36792290_1_gps-device-crime-fighting-tool-investigative-tool.

22. *People v. Weaver*, 909 N.E.2d 1195 (N.Y. 2009).

23. Ellen Nakashima, "Little-Known Surveillance Tool Raises Concerns by Judges, Privacy Activists," *Washington Post*, March 27, 2013., Clarence Walker, "New Hi-Tech Police Surveillance: The 'StingRay' Cell Phone Spying Device," *Global Research*, April 13, 2013.

24. Andrea Peterson, "Meet Stingrays, The Surveillance Tech The Government Doesn't Want To Talk About," *ThinkProgress*, May 17, 2013.

25. *Jones*, 565 U.S. ____

26. Ibid.

27. Ibid., Alito concurrence, Sotomayor concurrence.

28. *Kyllo*, 533 U.S. at 33.

29. Ibid., at 4.

30. A. Michael Froomkin, "The Death of Privacy?," *Stanford Law Review* 52 (2000): 1461–1543.

31. Ibid., 3.

32. John S. Granz, "It's Already Public: Why Federal Officers Should Not Need Warrants to Use GPS Vehicle Tracking Devices," *The Journal of Criminal Law and Criminology* 95 (2005): 1325–62.

33. Renee McDonald Hutchins, "Tied Up in *Knotts*? GPS Technology and the Fourth Amendment," *UCLA Law Review* 55 (2007): 455.

34. Melissa Arbus, "A Legal U-Turn: The Rehnquist Court Changes Direction and Steers Back to the Privacy Norms of the Warren Era," *Virginia Law Review* 89 (2003): 1776.

35. Susan W. Brenner, "The Fourth Amendment in an Era of Ubiquitous Technology," *Mississippi Law Journal* 75 (2012): 4.

36. Hutchins, "Tied Up in Knotts? GPS Technology and the Fourth Amendment," 412–413.

37. *Dow Chemical Company v. U.S.*, 476 U.S. 227 (1986).

38. David Gray and Danielle Citron, "A Technology-Centered Approach to Quantitative Privacy," *Social Science Research Network* (August 14, 2012), 30, http://ssrn.com/abstract=2129439.

39. Ibid., 1.

40. Ibid., 5.

41. Ibid., 13.

42. Orin Kerr, "The Mosaic Theory of the Fourth Amendment," *Michigan Law Review* 110 (forthcoming), 316.

43. *Karo*, 468 U.S. at 712.

44. *New York v. Class*, 475 U.S. 106 (1986).

45. *Katz*, 389 U.S. 347.

46. *Kyllo*, 533 U.S. 27.

47. Kerr, "The Mosaic Theory of the Fourth Amendment."

48. *United States v. Place*, 462 U.S. 696 (1983).

49. *Samson v. California*, 547 U.S. 843 (2006).

50. Kerr, "The Mosaic Theory of the Fourth Amendment."

51. *U.S. v. White*, 401 U.S. 745 (1971).

52. *Bond v. U.S.*, 529 U.S. 334 (2000).

53. *Kyllo*, 533 U.S. 27.

54. Elizabeth Cantor, "A Fourth Amendment Metamorphosis: How Fourth Amendment Remedies and Regulations Facilitated the Expansion of the Threshold Inquiry," *Virginia Law Review* 95 (2009): 155–203.

55. Ibid., 2.

56. Marc Jonathan Blitz, "Video Surveillance and the Constitution of Public Space: Fitting the Fourth Amendment to a World That Tracks Image and Identity," *Texas Law Review* 82 (2012): 1364.

57. Ibid., 1481.

58. *Smith v. Maryland*, 442 U.S. 735 (1979).

59. Stephen Henderson, "Nothing New Under the Sun? A Technologically Rational Doctrine of Fourth Amendment Search," *Mercer Law Review* 56 (2005): 511.

60. *Cunningham v. New York State Dept. of Labor*, 2013.

11

Drones, Domestic Surveillance, and Privacy: Legal and Statutory Implications

David L. Weiden

The use of drones, or unmanned aerial vehicles, has been the subject of considerable controversy in legislative and public arenas. Drone aircraft equipped with weapons are being used with greater frequency by the military in combat and intelligence-gathering operations. There is no question that unmanned drone aircraft provide a significant military advantage in wartime missions and help prevent casualties. However, it is the use of drones for domestic surveillance and monitoring of citizens that has engendered significant discussion and disagreement. The technology that makes drones so effective in combat missions can also be used within the borders of the United States to gather information on citizens and to assist with law enforcement. The *Washington Post* recently reported that federal and state law enforcement agencies are increasingly borrowing drone aircraft from other agencies and using these unmanned aerial vehicles at a much greater rate than previously known.[1] Nearly 700 drone operations were conducted by the Customs and Border Protection department for other federal agencies in the period from 2010 to 2012.[2]

The specter of drone aircraft being unleashed domestically without the prior acquisition of search warrants by the police has raised constitutional questions regarding the right to privacy for citizens. Lawmakers across the nation are grappling with the tradeoff between privacy and security interests, and there is no question that judges will be required to confront legal questions regarding the boundaries of the Fourth Amendment as it applies to domestic drone usage and warrantless surveillance.

This chapter will examine drone technology, the early state and federal legislative responses to drones, major U.S. Supreme Court cases regarding the Fourth Amendment's search jurisprudence, and how these precedents may apply to domestic drone surveillance. The chapter concludes with an analysis of the possible constitutional approaches that may be employed by the Supreme Court on the use of unmanned aerial vehicles for surveillance by law enforcement.

OVERVIEW OF DRONE TECHNOLOGY

Drones, also known as unmanned aerial vehicles, are aircraft that operate without a pilot aboard. Rather, drones typically fly by remote-control operation. These remote systems can be controlled by a ground-based pilot or programmed to fly preset patterns.[3] Obviously, the lack of an onboard pilot makes drones particularly valuable in military or law enforcement operations. Manned vehicles are generally more expensive to purchase and operate, and the risk of harm to the pilot imposes limits on their use. Unmanned aerial vehicles can be equipped with weapons for a variety of combat missions or can be used strictly for noncombat operations, such as to gain intelligence or engage in reconnaissance.

In 2013, the Department of Defense budgeted nearly six billion dollars for drone aircraft, but this figure does not include the expenditures on remotely piloted vehicles by the Department of Homeland Security or other federal and state agencies.[4] One report estimates that annual spending on drone technology will exceed $11 billion per year in the next decade.[5] Proponents of drone technology note that the manufacture of these vehicles will provide a substantial boost to the economy. Barry reports that there are an estimated 100 companies and military contractors that are involved in the development and manufacture of drone vehicles, and these organizations have given substantial campaign contributions to those members of the U.S. Congress who are known to support drone use.[6]

Drone technology is being used extensively by the U.S. military, but it is becoming increasingly popular with state and local law enforcement due to the relatively low cost of surveillance drones.[7] Law enforcement agencies have used drones for border enforcement, to apprehend suspects, to gather information in criminal investigations, for search-and-rescue operations, and for many other applications. Department of Homeland Security grants have enabled even smaller law enforcement agencies to purchase unmanned aerial vehicles.[8] For example, a Homeland Security grant of $258,000 allowed the Texas Montgomery County sheriff's office to purchase a ShadowHawk drone vehicle.[9] However, regarding the overall utility of drone aircraft, one analyst notes that unmanned aerial vehicles have not yet been proven to be cost effective, at least in the case of border security.[10]

Unmanned aerial vehicles can also be used in a wide variety of commercial applications. For example, the Internet retailer Amazon recently announced plans for delivery of packages using micro-drone vehicles.[11] Drone aircraft can range in size, from that of a regular commercial plane to a vehicle no larger than a hummingbird. The advances in creating very small and less expensive unmanned aerial vehicles have greatly increased the number of applications for which they are used. In short, drones are fast becoming an integral tool for private and governmental organizations. However, it is the use of unmanned aerial vehicles for domestic surveillance of U.S. citizens that has generated the most controversy.

Stanley and Crump note that surveillance drones utilize technology that enables these aircraft to gather information in a highly effective manner. These technologies include high-power zoom lenses capable of providing high-resolution photographs, infrared and ultraviolet night vision, and radar that can see through walls and track individuals even when they are inside buildings.[12] Unmanned aerial vehicles can be equipped with automated license plate readers to track automobiles easily. Facial-recognition technology allows drones to track movement patterns of individuals, and emerging technologies will tie facial recognition to governmental biometrical databases.[13] Acoustical surveillance is possible with certain drones, as laser optical microphones can be installed and used to listen to distant conversations.[14] Unmanned aerial vehicles can even be equipped with technology to detect certain odors at a molecular level. Thus, drones can be used to ascertain the presence of drugs, explosives, or even individual colognes and perfumes.[15] Finally, it is important to note that many drone aircraft are highly efficient and can operate continuously for extended periods—days, weeks, or perhaps even longer as the technology improves.[16]

Stanley and Crump also report that the air force is developing a new system known as the Gorgon Stare, which will involve "a large number of cheap, autonomous UAVs working in concert like a swarm of insects."[17] Takahashi notes that advanced technologies such as the Gorgon Stare can be used to gather enormous amounts of data that can be stored and later mined to assist in surveillance and monitoring.[18] In other words, drone vehicles using this technology could be used to constantly monitor and track all citizens, not just those citizens suspected of criminal activity. These drones have been described as providing omnipresent surveillance, a prospect that alarms many observers.[19]

Future technologies already in development include the miniaturization of drones to the size of insects and drones that do not fly at all (snake bots) but, rather, move in foliage or even climb trees to observe individuals and gather information.[20] Other emerging technologies include synthetic-aperture radar, a highly accurate form of radar that can even identify footprints and tire tracks.[21]

Clearly, drone surveillance technology is developing at a rapid pace, and it is highly likely that unmanned aerial vehicles will be able to monitor nearly every aspect of American society within several years, if not sooner, aided by these highly sophisticated, multimodal technologies. Indeed, the scope of technologies used by drones is unprecedented and allows for levels of surveillance that had previously been impossible using standard techniques. These existent and emerging technologies raise serious questions regarding the trade-off between security and privacy.

Drones and Law Enforcement in the United States

Law enforcement agencies are already using drones extensively within the United States. The Department of Homeland Security is currently using unmanned aerial vehicles for border security and to apprehend those bringing drugs and other contraband into the country. The Homeland Security agency has also been testing drones for use in other situations involving criminal investigations.[22] The Department of Defense, Federal Bureau of Investigation, Secret Service, and Environmental Protection Agency have all introduced plans to begin using drone aircraft, and recent reports indicate that some of these agencies have already started employing unmanned aerial vehicles.[23]

Local law enforcement agencies have also begun to use drone technology to assist in criminal investigations and arrests. Thompson reports that over 300 local police departments and municipalities have received authorizations from the FAA to operate drones in domestic airspace.[24] In 2011, a North Dakota sheriff used a Predator drone to locate and arrest several armed men who had been accused of stealing cows. Although there had been a search warrant issued for the missing cattle, the attorney for the defendants argued that the use of the drone vehicle exceeded the permissible scope of the warrant.[25] The North Dakota case is almost certainly a harbinger of the future, as state and local law enforcement agencies discover the advantages of drone technology for surveillance and investigations and increase their utilization of these vehicles, and criminal defendants seek to exclude evidence obtained through the use of drones.

DRONES AND LAWMAKERS: STATE AND FEDERAL LEGISLATIVE RESPONSES

The rapid development of drone technology and usage has created a truly astonishing bipartisan coalition of groups concerned with the privacy implications of unmanned aerial vehicles. Both the American Civil Liberties Union and the Heritage Foundation, among others, have vehemently opposed the use of drones to surveil American citizens.[26] Given that many liberal and conservative groups generally agree on this issue, it is likely that some form of federal legislation governing domestic drones will eventually be passed, along with an increasing number of state laws. In this section, some of the features of proposed federal legislation and existing state laws regulating unmanned aerial vehicles will be discussed to examine how these laws will affect domestic drone surveillance.

At this time, there is a patchwork of laws governing the use of drone aircraft for domestic surveillance, but no overall regulatory framework has yet emerged. There have been a number of bills introduced in the U.S. Congress regulating the use of drones and protecting citizens' privacy rights, but these bills have not yet been enacted into law. In late 2013, the Federal Aviation Administration released a document that outlined how unmanned aerial systems would be integrated into the national airspace. The roadmap states that there will be six test sites where research will be conducted as to the safe and effective integration of drones with other aircraft.[27] As to privacy considerations, the report states that "although the FAA's mission does not include developing or enforcing policies pertaining to privacy or civil liberties, experience with the UAS test sites will present an opportunity to inform the dialogue in the IPC and other interagency forums concerning the use of UAS technologies and the areas of privacy and civil liberties."[28] Thus, the FAA report does not provide any standards for the protection of privacy rights but simply calls for those regulations to be developed.[29]

A large number of state legislatures have introduced bills to regulate the use of unmanned aerial vehicles. The ACLU reports that, in 2013, 45 states introduced drone bills, and 9 states enacted such laws.[30] The majority of these state statutes required law enforcement agencies to obtain a probable cause warrant before a drone can be used to gather evidence in a criminal case.[31] In addition, other states placed limits on how long information acquired through the use of drones can be retained.[32] The sheer number of states considering local restrictions on the use of unmanned aerial vehicles in criminal cases indicates that there is substantial concern regarding privacy rights. However, because federal law generally preempts state law, it is unclear at this point as to what effect state legislation will ultimately have on the use of remotely piloted aircraft, especially those operated by federal authorities.

At the federal level, there has been considerable legislative activity in the Congress and executive agencies. In addition to the recently issued Federal Aviation Administration roadmap, the U.S. Congress also considered the issue of drone aircraft in both hearings and proposed legislation. The Senate Committee on Commerce, Science, and Transportation held a hearing on

January 15, 2014, to address multiple issues regarding the development and regulation of unmanned aerial vehicles.

In addition, drone aircraft have been the subject of several Congressional bills, most notably S. 1639, the Drone Aircraft Privacy and Transparency Act of 2013, introduced by Democratic Senator Edward Markey of Massachusetts.[33] That bill is currently being considered by the Committee on Commerce, Science, and Transportation. Given that the vast majority of bills introduced in Congress fail to be enacted into law, it is difficult to predict the bill's chances of success. However, it is instructive to examine the text of the bill, especially in light of the state laws regulating unmanned aerial vehicles. The section of that bill relating to privacy reads:

(a) In General—A governmental entity (as defined in section 2711 of title 18, United States Code) may not use an unmanned aircraft system or request information or data collected by another person using an unmanned aircraft system for protective activities, or for law enforcement or intelligence purposes, except pursuant to a warrant issued using the procedures described in the Federal Rules of Criminal Procedure (or, in the case of a State court, issued using State warrant procedures) by a court of competent jurisdiction, or as permitted under the Foreign Intelligence Surveillance Act of 1978 (50 U.S.C. 1801 et seq.).

(b) Exception-
  (1) IN GENERAL—Subsection (a) shall not apply in exigent circumstances (as defined in paragraph (2)).
  (2) EXIGENT CIRCUMSTANCES DEFINED- Exigent circumstances exist when—
    (A) a law enforcement entity reasonably believes there is an imminent danger of death or serious physical injury; or
    (B) a law enforcement entity reasonably believes there is a high risk of a terrorist attack by a specific individual or organization and the Secretary of Homeland Security has determined that credible intelligence indicates there is such a risk.[34]

The privacy regulations in the Markey bill are in line with many of the state laws regulating drones. This bill includes an exigent circumstances exception, which allows law enforcement to act without a warrant when there is a reasonable belief of imminent death or serious bodily injury; the bill also allows police to forego a warrant if there is a "high risk of a terrorist attack" as determined by the Secretary of Homeland Security. Both of these exceptions correspond to similar provisions in the state laws, as discussed below.

The only significant departure in the Markey bill from the state statutes is that the bill allows for warrantless surveillance to occur as allowed by the Foreign Intelligence Surveillance Act. The Foreign Intelligence Surveillance Act allows the President and Attorney General to authorize electronic surveillance of a foreign government or a faction of a foreign government for a period of up to one year, so long as the information sought is intended to protect the United States against potential attack or acts of terrorism.[35]

Interestingly, another bill seeking to regulate drones was introduced in 2013 by Republican Representative Austin Scott of Georgia. That bill, which is currently under consideration in the Judiciary Committee, is similar to the Senate bill, except that it does not include language allowing for warrantless surveillance pursuant to the Foreign Intelligence Surveillance Act.[36] Again, it is difficult to predict the chances of passage for both of these bills, but it appears very likely that additional legislation will be proposed in Congress on the topic of domestic drone surveillance in the future.

In individual states, a large number of bills were introduced for regulating drones, in 2013. Although some of these bills failed, others have been held over to next year's legislative session and may be passed during the next legislative session. Drone legislation was also passed in Maine, but the governor vetoed the bill. In addition, legislation regulating drone aircraft was passed in New Jersey, but Governor Chris Christie pocket-vetoed the bill by failing to sign it before the mandatory deadline.[37]

As for the drone statutes that were enacted into law in the states, many of them contain similar provisions, but there are some significant differences that will impact how unmanned aerial vehicles can be used in each of these states. The first state law to be enacted regarding restrictions on unmanned aerial vehicles was in Idaho. That statute requires law enforcement authorities to obtain a warrant in most situations before using a drone in a criminal investigation. The legislation also substantially restricts the use of drones for private purposes.[38] The critical section of the Idaho law reads as follows:

> Absent a warrant, and except for emergency response for safety, search and rescue or controlled substance investigations, no person, entity or state agency shall use an unmanned aircraft system to intentionally conduct surveillance of, gather evidence or collect information about, or photographically or electronically record specifically targeted persons or specifically targeted private property including, but not limited to:
>
> (i) An individual or a dwelling owned by an individual and such dwelling's curtilage, without such individual's written consent;

(ii) A farm, dairy, ranch or other agricultural industry without the written consent of the owner of such farm, dairy, ranch or other agricultural industry.[39]

It is interesting to note that the Idaho legislature specifically excluded drug and controlled-substance investigations from the warrant requirement. The legislature also allowed a safety exception but did not define how the term "safety" was to be interpreted. It is an open question as to how broadly or narrowly Idaho law enforcement agencies will interpret the safety exception.

Oregon also enacted legislation placing restrictions on the use of drone aircraft. It is instructive to compare Idaho's law with the Oregon statute. The Oregon law reads in the relevant part:

A law enforcement agency may operate a drone, acquire information through the operation of a drone, or disclose information acquired through the operation of a drone, if:

(a) A warrant is issued authorizing use of a drone; or
(b) The law enforcement agency has probable cause to believe that a person has committed a crime, is committing a crime or is about to commit a crime, and exigent circumstances exist that make it unreasonable for the law enforcement agency to obtain a warrant authorizing use of a drone.[40]

Thus, Oregon does not exclude controlled-substance investigations from the warrant requirement or create a public safety exception as does Idaho, but the law does allow for law enforcement agencies to forego obtaining a warrant when exigent circumstances exist. Still, the Oregon law is considerably more restrictive on governmental agencies' use of drone vehicles than the Idaho statute.

Tennessee also passed a law in 2013 regulating law enforcement's use of remotely piloted aircraft. The Tennessee statute provides for an exception to the warrant requirement in cases of terrorism but requires that the federal Secretary of Homeland Security verify the risk of a terrorist attack. The relevant section of the Tennessee law is as follows:

Except as provided in subsection (d), no law enforcement agency shall use a drone to gather evidence or other information.

(d) This section shall not prohibit the use of a drone:
(1) To counter a high risk of a terrorist attack by a specific individual or organization if the United States secretary of homeland security determines that credible intelligence indicates that there is such a risk;

(2) If the law enforcement agency first obtains a search warrant signed by a judge authorizing the use of a drone;
(3) If the law enforcement agency possesses reasonable suspicion that, under particular circumstances, swift action is needed to prevent imminent danger to life;
(4) To provide continuous aerial coverage when law enforcement is searching for a fugitive or escapee or is monitoring a hostage situation; or
(5) To provide more expansive aerial coverage when deployed for the purpose of searching for a missing person.[41]

The Tennessee law also limits the exigent circumstances exception found in the Oregon statute by only allowing law enforcement authorities to proceed without a warrant if there is "reasonable suspicion" that immediate action is needed to "prevent imminent danger to life." This provision is considerably more restrictive than the Oregon law, which allows the police to act using drones when there is probable cause to believe a crime has been or is being committed and there is not time to obtain a judicially issued warrant.

The Florida legislature also enacted a drone law in the 2013 legislative session. The Florida statute is similar to the Tennessee legislation, in that it also requires the Secretary of Homeland Security to validate a "high risk of a terrorist attack" to use a drone without prior judicial approval. The Florida law states, in part:

This act does not prohibit the use of a drone:

(a) To counter a high risk of a terrorist attack by a specific individual or organization if the United States Secretary of Homeland Security determines that credible intelligence indicates that there is such a risk.
(b) If the law enforcement agency first obtains a search warrant signed by a judge authorizing the use of a drone.
(c) If the law enforcement agency possesses reasonable suspicion that, under particular circumstances, swift action is needed to prevent imminent danger to life or serious damage to property, to forestall the imminent escape of a suspect or the destruction of evidence, or to achieve purposes including, but not limited to, facilitating the search for a missing person.[42]

The Florida legislation also contains language authorizing law enforcement to use drone aircraft without a warrant when there is "imminent danger to life." However, the Florida law also includes the categories of "serious damage to property" and prevention of "destruction of evidence" to the exigent circumstances exception. Florida appears to be the only state that allows police to forego obtaining a probable cause warrant when using drones if there is the possibility of damage to property interests.

Illinois, too, has passed drone legislation. The Illinois statute is similar to the laws passed in Oregon, Tennessee, and Florida. It reads, in part:

> This Act does not prohibit the use of a drone by a law enforcement agency:
>
> (1) To counter a high risk of a terrorist attack by a specific individual or organization if the United States Secretary of Homeland Security determines that credible intelligence indicates that there is that risk.
>
> (2) If a law enforcement agency first obtains a search warrant based on probable cause issued under Section 108-3 of the Code of Criminal Procedure of 1963. The warrant must be limited to a period of 45 days, renewable by the judge upon a showing of good cause for subsequent periods of 45 days.
>
> (3) If a law enforcement agency possesses reasonable suspicion that, under particular circumstances, swift action is needed to prevent imminent harm to life, or to forestall the imminent escape of a suspect or the destruction of evidence. The use of a drone under this paragraph (3) is limited to a period of 48 hours. Within 24 hours of the initiation of the use of a drone under this paragraph (3), the chief executive officer of the law enforcement agency must report in writing the use of a drone to the local State's Attorney.[43]

The Illinois statute also contains language allowing for the warrantless use of drones when there is a federal determination of a risk of a terrorist attack, and the law includes the exigent circumstances exception in cases of "imminent harm to life," escape of a suspect, or destruction of evidence.

Montana also passed a law regulating the use of drone aircraft by police and law enforcement authorities. The Montana statute is one of the shortest and simplest of all the state laws recently enacted. It reads, in total:

> (1) In any prosecution or proceeding within the state of Montana, information from an unmanned aerial vehicle is not admissible as evidence unless the information was obtained:
>
> (a) pursuant to the authority of a search warrant; or

(b) in accordance with judicially recognized exceptions to the warrant requirement.
(2) Information obtained from the operation of an unmanned aerial vehicle may not be used in an affidavit of probable cause in an effort to obtain a search warrant unless the information was obtained under the circumstances described in subsection (1)(a) or (1)(b) or was obtained through the monitoring of public lands or international borders.
(3) For the purposes of this section, "unmanned aerial vehicle" means an aircraft that is operated without direct human intervention from on or within the aircraft. The term does not include satellites.[44]

Thus, the Montana law excludes some of the exceptions to the warrant requirement found in other state legislation; however, the Montana statute does allow for "judicially recognized exceptions to the warrant requirement." In other words, the Montana law allows for certain exceptions to the warrant requirement found in Montana case law. Thus, this statutory language allows for the possibility of exceptions to the warrant requirement to be created through judicial precedent in Montana state and federal courts.

There are a number of commonalities among the state drone laws discussed above. Each of those statutes requires law enforcement agents to obtain probable cause warrants before using unmanned aerial vehicles in criminal investigations.[45] Most of the laws include provisions granting police the right to forego the procurement of a warrant when certain exigent circumstances exist. Several of the laws also specifically exclude terrorist activities from the warrant requirement. Two states—Tennessee and Illinois—also place limits on how long data obtained through the use of drones can be retained.[46] Tennessee requires that data obtained through drone surveillance on individuals that are not the target of an investigation must be deleted as soon as possible, no later than 24 hours after collection. The Illinois statute requires that such information must be deleted within 30 days. Finally, Montana and Oregon both require law enforcement agencies to obtain warrants even if they are using third-party drone aircraft.[47] That is, those states require that the police obtain a warrant before utilizing drone surveillance, regardless of who owns the drone aircraft used in the investigation.

What emerges from the analysis of the first state laws regulating the use of unmanned aerial vehicles is a consensus that a judicially issued probable cause warrant must be obtained before law enforcement may proceed with the use of a drone, and any evidence obtained in contravention of these procedures may not be used at trial. However, nearly all of the states do provide for certain exceptions to the warrant requirement, thus freeing law enforcement to act quickly in certain situations. It is likely that federal legislation will proceed largely along these lines, and, as noted above, several federal bills that have been introduced have included similar language.

There is, however, one outlier among the states: Texas. The Texas statute differs significantly from every other state's laws, in that it does not impose a warrant requirement upon law enforcement officials before using a drone vehicle in a criminal investigation or for domestic surveillance. What might be called the Texas model of drone regulation represents a very different approach to the use of unmanned aerial vehicles by governmental officials, and this model allows virtually unlimited usage of drone vehicles by Texas law enforcement. Bohm reports that the Texas law imposes very strict regulations on the use of private drones but few meaningful limitations on the use of unmanned aerial vehicles by police.[48] Indeed, the Texas statute merely requires that the Texas Department of Public Safety shall adopt regulations governing the use of drones and that each law enforcement agency must provide a written report every two years detailing the number of times that a drone was used in a criminal investigation. The Texas law states, in part:

> Rules for Use by Law Enforcement. The Department of Public Safety shall adopt rules and guidelines for use of an unmanned aircraft by a law enforcement authority in this state.
>
> Reporting by Law Enforcement Agency. (a) Not earlier than January 1 and not later than January 15 of each odd-numbered year, each state law enforcement agency and each county or municipal law enforcement agency located in a county or municipality, as applicable, with a population greater than 150,000, that used or operated an unmanned aircraft during the preceding 24 months shall issue a written report to the governor, the lieutenant governor, and each member of the legislature and shall:
> (1) retain the report for public viewing; and
> (2) post the report on the law enforcement agency's publicly accessible website, if one exists.
> (b) The report must include:
> (1) the number of times an unmanned aircraft was used, organized by date, time, location, and the types of incidents and types of justification for the use;
> (2) the number of criminal investigations aided by the use of an unmanned aircraft and a description of how the unmanned aircraft aided each investigation;

(3) the number of times an unmanned aircraft was used for a law enforcement operation other than a criminal investigation, the dates and locations of those operations, and a description of how the unmanned aircraft aided each operation;
(4) the type of information collected on an individual, residence, property, or area that was not the subject of a law enforcement operation and the frequency of the collection of this information; and
(5) the total cost of acquiring, maintaining, repairing, and operating or otherwise using each unmanned aircraft for the preceding 24 months.[49]

It is difficult to reconcile the Texas law with the other states' legislation, especially given Texas's traditional distrust of governmental intrusion. One report indicated that the Texas legislation was introduced by a state representative to prevent environmental and animal rights groups from being able to use drone aircraft to gather information for their causes.[50] It is possible that the forthcoming regulations from the Texas Department of Public Safety may incorporate additional privacy restrictions on the use of drones by law enforcement agencies. However, it seems likely that the Texas model of drone regulation will remain the minority standard at both state and federal levels, given the considerable support for privacy protections and limitations on the usage of drones by law enforcement agencies.

THE FOURTH AMENDMENT AND PRIVACY RIGHTS

The Fourth Amendment of the United States Constitution guarantees citizens the right "to be secure in their persons, houses, papers, and effects, against unreasonable searches and seizures" and also mandates that "no Warrants shall issue, but upon probable cause, supported by Oath or affirmation, and particularly describing the place to be searched, and the persons or things to be seized."[51] The amendment provides that Americans shall have a degree of privacy that cannot be violated by law enforcement agencies in the absence of probable cause that a crime has been committed. However, the meaning of these provisions has been subject to a number of rulings by the U.S. Supreme Court, and these judicial interpretations have changed over time as the societal understanding of the concept of privacy has evolved.

The Supreme Court's early approach to the question of an improper police search focused upon the authorities entering the actual physical property of a suspect. In the case of *Olmstead v. United States* (1928), the Court held that direct physical entry by law enforcement was required for a search to have occurred, and therefore the placement of wiretaps outside of the defendant's home did not constitute an unlawful search.[52] In other words, the Court ruled that wiretaps placed outside of a suspect's home that gathered information from telephone conversations inside of the home did not require prior judicial approval because authorities never actually entered the residence.[53] This property-centered approach to the definition of a search remained the applicable legal standard for nearly 40 years.[54]

However, the Supreme Court changed direction in regard to Fourth Amendment search-and-seizure jurisprudence in the landmark case of *Katz v. United States* (1967), where the modern approach to the expectation of privacy was delineated.[55] The Court ruled that the act of wiretapping a public telephone booth constituted a search, because the defendant had an expectation that his phone conversation would be private. In other words, the Court reasoned that the Fourth Amendment applies to individuals rather than being attached to real property only.[56] This decision marked a change in the Court's previous case law, which had emphasized individuals' right to be free from unreasonable searches on their property only.[57] The Court summarized this point by noting, "The Government's activities in electronically listening to and recording the petitioner's words violated the privacy upon which he justifiably relied while using the telephone booth, and thus constituted a 'search and seizure' within the meaning of the Fourth Amendment. The fact that the electronic device employed to achieve that end did not happen to penetrate the wall of the booth can have no constitutional significance."[58]

The ruling in *Katz* was later extended in *Smith v. Maryland* (1979), where the Court held that a person must have an actual, subjective expectation of privacy, and there must be a societal acceptance that this expectation of privacy is reasonable.[59]

Nearly 20 years after the *Katz* case, the Supreme Court would examine the issue of aerial surveillance in the case of *California v. Ciraolo* (1986).[60] In that case, the Court ruled that the police could conduct an aerial search without a warrant, because the homeowner did not have a reasonable expectation of privacy. The defendant in the case was growing marijuana in his backyard, which did contain privacy fences, but the marijuana plants were visible from planes flying overhead. Thus, according to the Court, there was no legitimate expectation of privacy, and the prosecution could use the evidence at trial.[61] The Court opined, "The Fourth Amendment simply does not require the police traveling in the public airways at this altitude to obtain a warrant in order to observe what is visible to the naked eye."[62]

This standard was later extended in *Dow Chemical v. United States* (1986), where the Court held that no warrant is required when law enforcement officers fly over open fields located in private industrial areas.[63] The Court noted, "EPA's taking, without a warrant, of aerial photographs of petitioner's plant complex from an aircraft lawfully in public navigable airspace was not a search prohibited by the Fourth Amendment. The open areas of an industrial plant complex such as petitioner's are not analogous to the 'curtilage' of a dwelling, which is entitled to protection as a place where the occupants have a reasonable and legitimate expectation of privacy that society is prepared to accept."[64] The Court further stated that another relevant factor was that the EPA was using a conventional camera commonly used in mapmaking, not some rare, commercially unavailable technology.[65]

The ramifications of *Ciraolo* for drone surveillance and privacy seem clear. According to the standard expressed in that case, it is likely that there is no expectation of privacy for any information that is obtained through aerial drone surveillance if that information is exposed and visible. That is, if the evidence can be observed by a manned aircraft, then it seems that the legal standard will apply even if the vehicle is operated remotely. However, the legal standard in *Ciraolo* will likely be inapplicable to evidence that is obtained through radar technology that can see inside buildings and structures.

The issue of technology that allows for information inside buildings to be acquired was addressed by the Court in *Kyllo v. United States* (2001).[66] In that case, the police used thermal imaging to analyze the amount of heat within the interior of a private home from the outside, and they had not obtained a warrant prior to using the thermal imaging device. The police had suspected that the defendant was growing marijuana in his home and was using numerous heat lamps for that purpose. The Court used the *Katz* standard to rule that the use of such sense-enhancing technology constituted a search, and thus a probable cause warrant was required.[67] The majority in the case stated that using "sense-enhancing technology [to obtain] any information regarding the interior of the home that could not otherwise have been obtained without physical 'intrusion into a constitutionally protected area,' constitutes a search."[68] The *Kyllo* standard reinforces the presumption that surveillance drones utilizing radar and other sense-enhancing technology to obtain information inside buildings will likely constitute a search, and thus require prior judicial approval and a search warrant.

In the recent case of *United States v. Jones* (2012), the Supreme Court held that the action of the police in attaching a Global Positioning System (GPS) device to an automobile in order to track the movements of the suspect constituted a search, and thus prior authorization by securing a warrant was required.[69] The Court partially returned to the property-based approach in *Olmstead* as to which circumstances constitute a search. Most relevant to the question of drone technology and privacy was Justice Alito's concurring opinion, which stated, in part, "Under this approach, relatively short-term monitoring of a person's movements on public streets accords with expectations of privacy that our society has recognized as reasonable. . . . But the use of longer term GPS monitoring in investigations of most offenses impinges on expectations of privacy. For such offenses, society's expectation has been that law enforcement agents and others would not—and indeed, in the main, simply could not—secretly monitor and catalogue every single movement of an individual's car for a very long period."[70] Justice Alito's concurrence appears to apply directly to the usage of long-term drone surveillance and suggests that such surveillance will likely require a probable cause judicial warrant. However, it is unclear as to whether his approach will be joined by a majority of the Court.

REGULATING DRONES IN THE COURT

Although both federal and state legislatures will continue to enact statutes that place limitations on the use of domestic drone surveillance, it is highly likely that these laws will be challenged, and the state and federal courts will be called upon to address the question of how—or if—the Fourth Amendment may limit the use of unmanned aerial vehicles without a warrant. At this point, there is little judicial or scholarly consensus upon the appropriate jurisprudential approach to the regulation of domestic surveillance drones. Indeed, one commentator contends that the emerging drone multimodal technology is so sophisticated that, even with a judicial warrant, the use of unmanned aircraft exceeds the permissible scope of a search under the Fourth Amendment.[71]

In this section, previous doctrinal frameworks used to interpret Fourth Amendment privacy rights will be discussed in relation to drone use, and a new legal approach will be analyzed as a possible answer to the conundrum of balancing security and privacy interests.

As discussed above, the Supreme Court's earliest approach to the question of a permissible police search is found in the case of *Olmstead v. United States* (1928).[72] In that case, the Court ruled that actual physical entry by police was required for a search to have occurred.[73] The property-centered approach to the Fourth Amendment definition of a search remained the applicable legal standard for generations before being overturned in the *Katz* case.

Regarding the use of drones for domestic surveillance, the *Olmstead* standard appears to be inapplicable for these situations.[74] Obviously, surveillance technology has improved exponentially since 1928, and so the property-centered approach to the Fourth Amendment has been largely superseded. Drone aircraft can be equipped with high-power zoom lenses, radar, and infrared and ultraviolet night vision to gather information without ever physically entering a structure. Because unmanned aerial vehicles can acquire a massive amount of information without physical entry, the *Olmstead* jurisprudential framework is not sufficient for these analyses, and is unlikely to be revived by the Court in future drone cases. In any case, the *Olmstead* physical-entry standard was largely abandoned by the Court in later decades.[75]

As discussed in the previous section, the Court radically transformed its Fourth Amendment jurisprudence in the case of *Katz v. United States* (1967), where the justices moved away from the property-centered doctrine of *Olmstead* to a standard that emphasized a defendant's reasonable expectation of privacy.[76] The *Katz* standard was expanded in the case of *Smith v. Maryland* (1979) to include a societal acceptance that this expectation of privacy is reasonable.[77]

The *Katz* reasonable expectation of privacy test may serve as the jurisprudential framework upon which the Fourth Amendment regulation of domestic drones may rest. The linchpin of this analysis would rest upon the determination of whether an individual had a reasonable expectation of privacy against highly sophisticated drone surveillance technologies. It is possible that the Supreme Court would view the extended operation of an unmanned aerial vehicle, equipped with infrared optical and advanced acoustical technology, as being far beyond the reasonable societal and individual expectations of personal privacy and thus require prior judicial approval before the use of such drone aircraft.

It is also possible that, regarding the societal acceptance of domestic drone surveillance, the Court could be influenced by shifting public opinion. One of the few surveys on this issue found that 65 percent of citizens opposed the use of drones for police work in the United States, which indicates that the American citizenry does not accept domestic drone surveillance as being reasonable or societally acceptable at this time.[78] In addition, another survey found that the American public is supportive of the use of weaponized drones to launch airstrikes against suspected terrorists in foreign countries, but only 25 percent supported such strikes in the United States.[79] Obviously, the use of drones for military strikes is a very different issue than the use of drone aircraft for domestic surveillance, but the survey does provide another indication that the American people are not comfortable with the use of unmanned aerial vehicles within the boundaries of the United States. Overall, it seems likely that public opposition to the usage of drones for governmental surveillance and monitoring within the borders of the nation will increase as knowledge about these technologies increases.

Thus, it is feasible that the Supreme Court could utilize the existing *Katz* standard to analyze domestic drones in regard to the Fourth Amendment.

It is possible that the Court could find that, buttressed by public opinion, society does not find the use of unmanned drone aircraft to exist within a reasonable expectation of privacy.[80] By using the *Katz* reasonable expectation of privacy test in this manner, the Court can provide a doctrinal framework for the regulation of domestic drone aircraft that would not require a radical jurisprudential innovation.

There is one other legal framework that the Supreme Court may use for the regulation of domestic drone aircraft used by law enforcement authorities. As discussed above, in the recent case of *United States v. Jones* (2012) involving the use of GPS tracking by police, the Supreme Court considered a new approach to Fourth Amendment search jurisprudence.[81] Kerr has dubbed this new doctrine the "mosaic theory."[82] Kerr notes that when using traditional Fourth Amendment doctrine, judges examine each action of the police sequentially. That is, judges tend to view each step by law enforcement in isolation to determine if a search has occurred within the meaning of the Fourth Amendment.[83] By way of example, Kerr observes, "If an officer sees suspects preparing for a robbery, stops them, and pats them down for weapons, the court will consider the viewing, the stopping, and the patting down as distinct acts that must be analyzed separately. Each step counts as its own Fourth Amendment event and is evaluated independently of the others."[84] This step-by-step analysis is foundational in traditional Fourth Amendment doctrinal analysis but may not be well suited to the examination of the domestic drone surveillance, given that there may be hundreds of discrete actions taken by a particular drone aircraft.

However, four justices in the *Jones* case embraced the mosaic theory, which rejects traditional sequential analysis and instead views the aggregated actions of the police.[85] That is, rather than analyze each separate act of the police to determine if a search has occurred, mosaic legal theory instead looks to the totality of the law enforcement actions to ascertain if a mosaic of information has been obtained that constitutes an impermissible search beyond the boundaries of the Fourth Amendment. Kerr notes, "Instead of asking if a particular act is a search, the mosaic theory asks whether a series of acts that are not searches in isolation amount to a search when considered as a group. The mosaic theory is therefore premised on aggregation: it considers whether a set of nonsearches aggregated together amount to a search because their collection and subsequent analysis creates a revealing mosaic."[86]

The application of mosaic theory to analysis of domestic drone surveillance may present a useful and relevant legal framework for this issue.[87] Rather than analyze each minor act of data acquisition by a drone to determine if a search has occurred, the mosaic analysis would aggregate all of those actions by the drone aircraft to determine if a warrantless search has occurred in violation of a citizen's privacy rights. It is highly likely that the Court would rule that such an unreasonable search had occurred if this doctrinal analysis were to be adopted in a drone case. In other words, if a drone vehicle were to obtain hundreds of pieces of data on a suspect over a period of days or weeks, it is likely that the aggregation of these data would constitute an illegal search, even if any one of those data points would not constitute a search by itself. Thus, the new legal framework suggested in the *Jones* case represents another promising approach for the Fourth Amendment analysis of domestic drone surveillance by law enforcement.

CONCLUSION

The use of unmanned aerial vehicles for purposes of surveillance of American citizens will undoubtedly continue to be highly controversial. Opinion polls suggest that citizens are uncomfortable with this practice, and both liberal and conservative interest groups have expressed opposition to the use of these unmanned vehicles for domestic monitoring. Furthermore, there has been a striking number of state and federal bills introduced to govern the use of these aircraft by law enforcement, and there will certainly be more attempts in legislative arenas to create new frameworks for regulation.

But, whatever form these state and federal laws may take, the United States Supreme Court will almost certainly be asked to weigh in on the issue. The legal standard that will be used by the Court is open to speculation. This article has suggested that the traditional *Katz* standard should be sufficient for the Court to analyze the legal issues, but the emerging mosaic theory of Fourth Amendment jurisprudence suggested in the *Jones* case presents another promising approach for the doctrinal analysis of drones and the Fourth Amendment.

Although the Court has not yet ruled upon the constitutionality of warrantless domestic drone surveillance, the growing reliance of law enforcement on these technologies virtually guarantees that these practices will soon be challenged in state and federal courts. The direction that the Court takes in these cases may very well determine the future course of privacy issues in our rapidly changing technological and legal environment.

NOTES

1. Craig Whitlock and Craig Timberg, "Border-Patrol Drones Being Borrowed by Other Agencies More Often than Previously Known," *Washington Post*, January 14, 2014.

2. Ibid.

3. Timothy T. Takahashi, "Drones and Privacy," *Columbia Science & Technology Law Review* XIV (2012): 72–114.

4. Tom Barry, "Drones Over the Homeland: How Politics, Money and Lack of Oversight Have Sparked Drone Proliferation, and What We Can Do," Center for International Policy report, April 2013, http://www.ciponline.org/images/uploads/publications/IPR_Drones_over_Homeland_Final.pdf.

5. Phil Mattingly, "FBI Uses Drones in Domestic Surveillance, Mueller Says," *Bloomberg*, June 19, 2013, http://www.bloomberg.com/news/2013-06-19/fbi-uses-drones-in-domestic-surveillance-mueller-says.html.

6. Barry, "Drones Over the Homeland: How Politics, Money and Lack of Oversight Have Sparked Drone Proliferation and What We Can Do."

7. See generally, Abigail R. Hall and Christopher J. Coyne, "The Political Economy of Drones," *Defence and Peace Economics* (2013). doi:10.1080/10242694.2013.833369.

8. Gene Healy, "Drones Pose a Threat to Americans' Privacy," *DC Examiner*, May 21, 2012.

9. Barry, "Drones Over the Homeland: How Politics, Money and Lack of Oversight Have Sparked Drone Proliferation and What We Can Do."

10. Ibid. Barry also notes that drone aircraft at prone to costly breakdowns and require expensive maintenance. Ibid. at 25.

11. Steve Banker, "Amazon and Drones—Here Is Why It Will Work," Forbes.com, December 19, 2013, http://www.forbes.com/sites/stevebanker/2013/12/19/amazon-drones-here-is-why-it-will-work/.

12. Jay Stanley and Catherine Crump, "Protecting Privacy from Aerial Surveillance: Recommendations for Government Use of Drone Aircraft," American Civil Liberties Union Report, December 2011.

13. Stanley and Crump, "Protecting Privacy from Aerial Surveillance: Recommendations for Government Use of Drone Aircraft," 5–6.

14. Takahashi, "Drones and Privacy," 88–89.

15. Ibid., 89–90.

16. David H. Dunn, "Drones: Disembodied Aerial Warfare and the Unarticulated Threat," *International Affairs* 5 (2013): 1237–1246.

17. Ibid., 6.

18. Takahashi, "Drones and Privacy," 91–92.

19. Anna Mulrine, "Not your average drone: New technology the US military is developing," *Christian Science Monitor*, December 6, 2013, http://www.csmonitor.com/World/Security-Watch/2013/1206/Not-your-average-drone-new-technology-the-US-military-is-developing-video.

20. Rachel L. Finn and David Wright, "Unmanned Aircraft Systems: Surveillance, Ethics and Privacy in Civil Applications," *Computer Law & Security Applications* 28 (2012): 184, 187.

21. Tom A. Peter, "Drones on the US border: Are they worth the price?" *Christian Science Monitor*, February 5, 2014, http://www.csmonitor.com/USA/2014/0205/Drones-on-the-US-border-Are-they-worth-the-price.

22. Richard M. Thompson II, "Drones in Domestic Surveillance Operations: Fourth Amendment Implications and Legislative Responses," *Congressional Research Service report*, April 3, 2013.

23. Whitlock and Timberg, "Border-Patrol Drones Being Borrowed by Other Agencies More Often than Previously Known," *Washington Post*, January 14, 2014.

24. Ibid., 3.

25. Takahashi, "Drones and Privacy," 74–77.

26. See, e.g., Stanley and Crump, "Protecting Privacy from Aerial Surveillance: Recommendations for Government Use of Drone Aircraft," and Paul Rosenzweig, Steven P. Bucci, Charles D. Stimson, and James Jay Carafano, "Drones in U.S. Airspace: Principles for Governance," *Heritage Foundation report*, September, 2012.

27. Federal Aviation Administration, "Integration of Civil Unmanned Aircraft Systems (UAS) in the National Airspace System (NAS) Roadmap," 2013.

28. Ibid., 11.

29. See also Chris Schlag, "The New Privacy Battle: How the Expanding Use of Drones Continues to Erode Our Concept of Privacy and Privacy Rights," *Journal of Technology Law & Policy* XIII (2013): 21. Schlag calls for a "specifically developed consumer protection agency or a created body within the FAA dedicated solely to drone technology [that] would be responsible for implementing and overseeing compliance with the law." Ibid. at 21.

30. Allie Bohm, "Status of Domestic Drone Legislation in the States," *American Civil Liberties Union*, January 22, 2014, https://www.aclu.org/blog/technology-and-liberty/status-domestic-drone-legislation-states.

31. Allie Bohm, "The Year of the Drone: An Analysis of State Legislation Passed This Year," American Civil Liberties Union, November 2013, https://www.aclu.org/blog/technology-and-liberty/year-drone-roundup-legislation-passed-year.

32. Ibid.

33. A number of bills regarding drone surveillance were introduced during the current legislative session in Congress, and it is highly likely that even more will be considered in future years. A good site to track both federal and state drone legislation is the National Association of Criminal Defense Lawyers, Domestic Drone Information Center, http://www.nacdl.org/domesticdrones/.

34. S. 1639, 113th Cong. (2013).

35. 50 U.S.C. 1801 et seq. The act has been amended numerous times, and there is substantial controversy as to how much freedom the federal government should be granted to surveil suspected terrorists without a warrant. See generally Orin S. Kerr, "Updating the Foreign Intelligence Surveillance Act," *University of Chicago Law Review* 75, (2008): 225–243.

36. H.R. 972, 113th Cong. (2013); see also S. 1106, 113th Cong. (2013).

37. Andrew Seidman, "Moriarty Bill, Drone Measure among Christie Pocket-Vetoes," *Philly.com*, January 24, 2014, http://articles.philly.com/2014-01-24/news/46519456_1_trenton-gov-christie-law-enforcement.

38. Allie Bohm, "The First State Laws on Drones," American Civil Liberties Union, April 15, 2013, https://www.aclu.org/blog/technology-and-liberty-national-security/first-state-laws-drones.

39. Idaho Senate Bill 1134, Idaho Legislature, 2013.

40. Oregon House Bill 2710, Oregon Legislative Assembly, 2013.

41. Tennessee Public Chapter No. 470, Senate Bill No. 796, Tennessee General Assembly, 2013.

42. Florida Senate Bill 92, Florida Legislature, 2013.

43. Illinois Public Act 098-0569, Illinois General Assembly, 2013.

44. Montana Code Annotated 46-5-109, 2013.

45. Bohm, "The Year of the Drone: An Analysis of State Legislation Passed This Year."

46. Ibid.

47. Ibid.

48. Ibid.

49. Texas House Bill No. 912, Texas Legislature, 2013.

50. Dan Solomon, "Texas's Drone Law is Pretty Much the Opposite of Every Other State's Drone Law," *The Daily Post*, September 17, 2013, http://www.texasmonthly.com/daily-post/texasss-drone-law-pretty-much-opposite-every-other-states-drone-law.

51. U.S. Const. Amend. IV.

52. *Olmstead v. United States*, 277 U.S. 438 (1928).

53. Ibid.

54. See generally Peter Winn, "*Katz* and the Origins of the 'Reasonable Expectation of Privacy' Test," *McGeorge Law Review* 40 (2009): 1–13.

55. *Katz v. United States*, 389 U.S. 347 (1967).

56. Ibid.

57. See, e.g., *Olmstead v. United States*, 277 U.S. 438 (1928).

58. *Katz*, 389 U.S. at 353.

59. *Smith v. Maryland*, 442 U.S. 735 (1979). Regarding the *Katz* case and societal expectations, Winn notes, "Thus, an objectively reasonable expectation of privacy necessarily must reference other norms independent of the idea of privacy itself. The test is not just what the judge says it is; the test must also incorporate a long tradition of what other judges and lawmakers have declared the law to be in the past. This tradition includes as an important aspect those norms underlying society's objective expectations of privacy—among which a central place is held by the law of property." Winn, "*Katz* and the Origins of the 'Reasonable Expectation of Privacy' Test," 8–9.

60. *California v. Ciraolo*, 476 U.S. 207 (1986).

61. Ibid.

62. *Ciraolo*, 476 U.S. at 215.

63. *Dow Chemical v. United States*, 476 U.S. 227 (1986).

64. *Dow Chemical*, 476 U.S. at 228.

65. Ibid.

66. *Kyllo v. United States*, 533 U.S. 27 (2001).

67. Ibid.

68. *Kyllo*, 533 U.S. at 34.

69. *United States v. Jones*, 132 S.Ct. 945 (2012).

70. *Jones*, 132 S.Ct. at 964.

71. Takahashi, "Drones and Privacy," at 111–112. Takahashi argues that the use of archived data is an unreasonable violation of privacy, even if a warrant was obtained beforehand: "It is probable that the archived information will clearly exceed what is permissible for police to obtain under the plain view doctrine. Any law enforcement action that draws upon archived data obtained by legitimate drone flights could trigger a Fourth Amendment challenge." Ibid., at 112.

72. *Olmstead*, 277 U.S. 438 (1928).

73. Ibid.

74. Sean Sullivan, "Domestic Drone Use and the Mosaic Theory," Paper No. 2013-02, University of New Mexico School of Law, Legal Studies Research Paper Series, February, 2013.

75. However, Justice Scalia asserted in *United States v. Jones* (2012) that the property-based standard in *Olmstead* had not been overturned in *Katz*. *Jones*, 132 S.Ct. at 950. Thus, it is not clear that the *Olmstead* doctrine has been nullified.

76. *Katz*, 389 U.S. 347 (1967).

77. *Smith*, 442 U.S. 735 (1979).

78. Rasmussen Reports, "65% Oppose Use of Drones for U.S. Police Work," October 28, 2013, http://www.rasmussenreports.com/public_content/politics/general_politics/october_2013/65_oppose_use_of_drones_for_u_s_police_work.

79. Alyssa Brown and Frank Newport, "In U.S., 65% Support Drone Attacks on Terrorists Abroad," Gallup Politics, March 25, 2013, http://www.gallup.com/poll/161474/support-drone-attacks-terrorists-abroad.aspx.

80. But see Sullivan, "Domestic Drone Use and the Mosaic Theory," who does not believe that the *Katz* standard can be used for legal analysis of domestic drones.

81. *Jones*, 132 S.Ct. 945 (2012).

82. Orin S. Kerr, "The Mosaic Theory of the Fourth Amendment," *Michigan Law Review* 111 (2012): 311–354. See also Sullivan, "Domestic Drone Use and the Mosaic Theory."

83. Kerr, "The Mosaic Theory of the Fourth Amendment," 315–316.

84. Ibid. at 316 (citation omitted).

85. Justice Scalia, the author of the majority opinion, did not address the issue of mosaic theory analysis. *Jones*, 132 S.Ct. 945.

86. Ibid. at 320.

87. Sullivan, "Domestic Drone Use and the Mosaic Theory."

12

21st-Century Developments in Fourth Amendment Privacy Law

Timothy O. Lenz

The Fourth Amendment is arguably the most widely criticized area of criminal law. Legal scholars describe the law of search and seizure as an "embarrassing,"[1] "archaic,"[2] "Rube Goldberg contraption."[3] Supreme Court justices across the ideological spectrum think that search-and-seizure law is long overdue for a major "overhaul"[4] and call for new judicial and statutory approaches. The dissatisfaction, which seems to be a "near-permanent condition,"[5] has made the Fourth Amendment the most dynamic area of criminal law. This chapter examines the dissatisfaction, the changes in the Court's Fourth Amendment doctrines, and the proposed alternative approaches. The main theme of the chapter is that digital technology presents new challenges for finding the right balance between granting the government enough power to effectively fight crime and protect national security while also limiting government power, consistent with the Fourth Amendment's original purpose—which is to guarantee limited government by protecting privacy.

The chapter examines three main developments. The first development is the expansion of the number of exceptions to the rule that the Fourth Amendment requires search warrants. This includes limiting the exclusionary rule's impact on the admissibility of evidence. The second development is the impact of science and technology on law enforcement methods and missions. Scientific advances and digital technology have greatly increased government capability to conduct electronic searches, which are subject to fewer legal limits than physical searches. As a result, technology has eroded Fourth Amendment protections for informational privacy. The third development is the increased integration of domestic law enforcement (policing) and national security. Counterterrorism policy has brought elements of the legal regime that governs national security to the legal regime that governs domestic law enforcement.

LEGAL REGIMES

The chapter devotes the most attention to developments in Supreme Court case law interpreting Fourth Amendment privacy rights. However, search-and-seizure law is best understood as a legal regime. The term "legal regime" refers to the formal rules and policies as well as the "informal practices" that govern an area of law. The legal regime governing search and seizure includes the text of the Fourth Amendment; Supreme Court case law; congressional statutes relating to criminal justice, national security, and the right to privacy; treaties and other international agreements; and the administrative rules, regulations, and policy statements promulgated by the executive branch agencies that are responsible for fighting crime, providing national security, and protecting the borders by implementing statutes and court rulings. Therefore, some attention is paid to Congress and the executive branch.

The Supreme Court uses judicial doctrines—rules, principles, or concepts—to give concrete meaning to ambiguous language such as "unreasonable search and seizure." Beginning in the early 1970s, a conservative working majority on the Court changed search-and-seizure law by applying doctrines that reflect the get-tough-on-crime values and policies that are elements of the crime-control model of justice.[6] But the justices are still searching for ways to ensure that the Fourth Amendment remains relevant to life in a digital age, where technological advances create new ways for the government to search and surveil that are largely outside the current Fourth Amendment legal regime for protecting informational privacy. In fact, the judicial, legislative, and executive branches are searching for a limiting principle that can protect privacy while allowing the government to effectively fight crime and protect national security. This is not a new search. Technology often upsets the balance between government power to search and the limits on those powers. Wiretap technology that allowed the government to listen to telephone conversations upset the balance. Today, computers and digital technology have upset the balance by greatly enhancing the government's ability to conduct electronic search and surveillance.

Political liberals and conservatives and legal liberals and conservatives have for decades disagreed on where the balance should be struck in crime and national security cases. This ideological saliency complicates efforts to fit digital technology into the Fourth Amendment. The institutional dialogues among the Court, Congress, and the executive branch about where to strike the balance between government power and individual rights are especially lively on matters related to digital technology's impact on informational privacy.

CONTINUITY AND CHANGE

The main theme of the chapter is that the development of the law of search and seizure is the story of the constant struggle to apply the Fourth Amendment to contemporary circumstances. One key to understanding political and constitutional conflicts in the United States is the strong tension between continuity and change. The struggle to adapt the Constitution to modern life while remaining true to the original founding values reflects the desire to maintain continuity with the past while changing with the times. This tension exists in other countries, but it is especially strong in the United States, where the commitment to the Constitution and the founding values of politics and government embodied in the Constitution is so strong that it has been described as "biblical worship."[7] At almost any moment in American history, it could be said that "constitutional argument in our time is largely a revived and sophisticated version of a much older debate reaching back to the 'founding' disputes between Federalists and anti-Federalists."[8] Many of today's debates about politics, government, and public policy are differences of opinion about how to maintain continuity with founding values—traditional values such as limited government, individual freedom, and equality—while adapting to a world where economic, political, social, technological, and scientific changes are a fact of life. This certainly applies to political and legal debates about how to reconcile digital technology with the enduring values that are embodied in the Fourth Amendment.

THE MEANING OF THE FOURTH AMENDMENT

The Fourth Amendment declares the right of the people to be secure in "persons, houses, papers, and effects, against unreasonable searches and seizures." It also requires that a search warrant be issued only "upon probable cause . . . and particularly describing the place to be searched, and the persons or things to be seized." These Fourth Amendment rights are conditional rights.

Conditional Rights

The Fourth Amendment is part of the legal web of rights. It is officially linked to the Fifth Amendment, which provides that no person "shall be compelled in any criminal case to be a witness against himself . . . ," because a search can produce incriminating evidence. The Fourth Amendment can affect First Amendment rights, because investigating crimes, national security threats, terrorism, whistleblowing, or "leaking" can have a chilling effect on freedom of expression. These kinds of investigations are especially worrisome when the government is not investigating a particular crime or individual suspect but merely watching or searching for illegal activity. Innocent people and activities are almost certainly going to be caught when the government casts a wide net and then sorts between the intended catch (the targets) and the unintended catch. The 4th Amendment can also affect 14th Amendment rights to not be deprived of due process or equal protection of the laws. Finally, the Fourth Amendment is officially linked to privacy rights by protecting persons, places, and things from unreasonable search and seizure.

None of these constitutional rights is absolute; all are conditional. The Fourth Amendment prohibits only unreasonable search and seizure. The right to be free from unreasonable search or seizure is a place-based right that means different things in a home, a bus depot, an airport, or a motor vehicle on a public roadway. Informational privacy rights vary depending on the form of communication. And the legal regime governing search and seizure is based on the concept of a reasonable expectation of privacy. The fact that the "ultimate touchstone" of the Fourth Amendment is "reasonableness"[9] makes the Fourth Amendment dynamic (it changes with the times) and subjective (its meaning is relative in the sense that it is based on personal expectations). The concept of a reasonable person, a reasonable doubt, and a reasonable expectation of privacy are closely related to public opinion. And public opinion—specifically, reasonable expectations of privacy—plays a surprisingly important role in the Supreme Court's interpretation of the Fourth Amendment. The role that public opinion plays in shaping constitutional law is often overlooked because so much attention is paid to the Supreme Court, Congress, and the executive branch. The Court's reasonable expectation of privacy doctrine has a subjective component and an objective component. The subjective component is the individual's expectation of privacy. The objective component is the contemporary public's (or modern society's) expectation of privacy.

Search-and-seizure law is a dynamic area of law because expectations of privacy change. Expectations diminish as people grow accustomed to surveillance cameras in almost all public and private (commercial) spaces, as consumers grow accustomed to business use of personal data for customer convenience, and as citizens grow accustomed to government use of personal data for public safety and parents and children grow accustomed to the use of spyware as a parenting tool to monitor and control children's activities. As monitoring of almost all activities becomes the new normal, expectations of privacy decrease, and government power increases. This dynamic is reflected in statements indicating that it is a waste of time to worry about Big Brother because technology, not the police or National Security Agency, is making privacy "an antiquated concept."[10]

The Role of the Supreme Court

The official functions of American courts are dispute resolution and law interpretation. Dispute resolution is deciding a case; law interpretation is deciding what a law means. The Supreme Court's unofficial function is making legal policy. Much of the criticism of the Court is directed at this policymaking function. Criminal justice and national security are ideologically salient issues that have consistently divided conservatives and liberals over the last half century. Get-tough-on-crime conservatives tend to be hawkish on national security and defense, and soft-on-crime liberals tend to be more dovish on national security and defense. On criminal justice issues, liberals tend to support the due-process model of justice, which consists of a set of values and policies that emphasize procedural protections for individuals who are accused of crimes. Conservatives tend to support the crime-control model of justice, which consists of a set of values and policies that emphasize effective crime fighting. These political ideologies are also reflected in the legal ideologies of Supreme Court justices whose voting records and jurisprudence in criminal cases mirror political thinking about crime.

The Warren Court (1953–1969) earned its reputation as a policymaking court because its criminal justice rulings reveal a clear preference for broad policy rulings, rather than narrowly deciding cases, and a clear preference for the due process model of justice.[11] The Burger Court (1969–1986) began the rightward movement in criminal law, the Rehnquist Court (1986–2005) continued it, and now court watchers look at the Roberts Court to see where the conservative working majority will take criminal law. The Rehnquist and Roberts Courts are more restraintist in the sense that their judicial deference to police, prosecutors, and prison officials supports executive discretion. The Roberts Court continues to write broad policy rulings, but unlike the Warren Court, its rulings reflect the values and policies associated with the crime control model of justice. Criminal law and national security may seem to be very different areas of law, but the crime control model closely resembles what might be called a national security model of justice. Both models balance individual rights and government power with a thumb on the side of public safety, and both value effective government power—which generally means support for search, seizure, and surveillance.

Court Watching

Court watching makes sense because the Court regularly interprets the meaning of the words of the more than 200-year-old document. The Fourth Amendment declares the right to be free from unreasonable searches and provides procedural and substantive requirements for search warrants. These two provisions of the Fourth Amendment limit government power by declaring the right to be free from unreasonable searches and by specifying procedural and substantive requirements for search warrants. The two provisions also illustrate two problems with reading the Constitution to understand what the words mean. It is impossible to read broad words such as "unreasonable" and vague phrases such as "probable cause" to know what they mean. Such broad language invites legal conflicts over the meaning of the words. But more precise, specific language is not necessarily the solution, because specific language also creates problems. The Fourth Amendment specifically lists four things to be protected—persons, houses, papers, and effects. Does this mean that it protects only these four things? No.

The Court does not read the Fourth Amendment literally to protect only the four things that are specifically mentioned. The Fourth Amendment is read more broadly as a provision of the Constitution that was originally intended to guarantee limited government by protecting the right to privacy. This broad reading has kept the two-centuries-old amendment relevant in a modern world with automobiles, airplanes, drones, smartphones and other electronic communication devices, cameras everywhere, thermal imaging devices, body scans and facial recognition software, and scientific methods of criminal investigation and national security protection that rely on DNA analysis. But the broad reading of the Fourth Amendment also invites conflicting interpretations. It is ironic that the Fourth Amendment's broad language and specific language make it a very dynamic area of constitutional law that is continually being shaped by legal, political, and technological forces.

Fourth Amendment cases, like civil liberties cases generally, require balancing individual rights and government power. Fighting crime and protecting national security are two basic government powers, but how government fights crime and protects national security is political. The politics of criminal justice and national security reflect the broader ideological and partisan conflicts about the nature and scope of government power. These political conflicts are evident in the voting records and the jurisprudence of the Supreme Court justices.

Umpires, Referees, or Players?

According to the sports metaphor that is sometimes used to describe the appropriate role for judges, said judges should be umpires or referees, not players. The metaphor of judges seems logical, but Supreme Court justices are not merely umpires who call balls and strikes. The justices are players because the Court, as an institution, makes the rules of the game. It does not just call balls and strikes—it defines the strike zone. The voting records of individual justices mark them as conservative or liberal, Republican or Democrat. In search-and-seizure cases involving warrants or the admissibility of evidence, liberal justices have smaller strike zones that make more evidence inadmissible, and conservative justices have larger strike zones that make more evidence admissible. In both cases, the way the rules are defined affects how the game is played and who wins and loses.

The Court's informal role is to adapt search-and-seizure law to political and technological changes while striving to remain true to the Fourth Amendment's original purpose. The Court has assumed this role because there seems to be no viable alternative to judicial policymaking. Two possible alternatives are constitutional amendments and federal legislation. Amendments are unlikely to be the way that the Fourth Amendment is adapted to contemporary expectations of government power to search and seize, because the Constitution is so hard to amend. Informal amendment by judicial interpretation is a convention, or accepted practice, that lessens the need for formal amendment. The other alternative to judicial interpretation, federal statutes, is not really viable because Congress does not have a good track record of keeping federal law current with technological and scientific developments. Legislation tends to lag behind technological innovations. Technology is more likely to lead and the law follow than the reverse.

The Technology Driver

Technological and scientific advances have historically presented challenges to the conventional reading of the Fourth Amendment. Motor vehicles gained increased mobility, which meant that evidence could be long gone by the time a police officer got a warrant to search a vehicle. So, the Court ruled that a warrantless search of a vehicle could be justified if necessary to preserve evidence. Motor vehicles also challenged the place-based conception of Fourth Amendment rights. Does a person have the same reasonable expectation of a privacy right to be free from unreasonable search and seizure while traveling on a public road that he or she has while at home? No. The old saying, "A man's home is his castle," is based on the belief that the government's power to search a person's home is greatly limited. The home-as-castle concept is still used as a metaphor for protection from government. The Court adapted search-and-seizure law to motor vehicles and telephones by broadly interpreting the Fourth Amendment as intended to protect the political principle of limited government. The Court's decisions in early wiretap cases show how the law accommodates technology.

Early Wiretap Rulings

Wiretap technology upset the existing balance between government power and limits by greatly enhancing the ability to investigate crimes and get evidence by listening to telephone conversations without having to physically search or seize anything. The Supreme Court's early rulings in wiretap cases show how it fits technological advances into the Fourth Amendment framework. In *Olmstead v. U.S.* (1928), the Court held that the police did not need to get a search warrant to place a listening device outside a public telephone booth and listen to a telephone call by a man suspected of criminal activity. According to the Court, the bugging was not a search or seizure because the police did not physically search or seize anything—they merely used technology to listen to a telephone conversation. Justice Brandeis dissented. He argued that the nation's founders intended to protect spiritual and material things, beliefs and property when they created "the right to be let alone—the most comprehensive of rights, and the right most valued by civilized men . . ." He also believed that it was necessary to protect that right by deeming every "unjustifiable" government intrusion upon privacy, "by whatever the means employed, a violation of the Fourth Amendment."[12] The Court eventually accepted this reading of the Fourth Amendment in *Katz v. U.S.* (1967).

There is still broad consensus on the Court that the Amendment was intended to protect the principle of limited government and the right to privacy, not just the four things that are specifically mentioned. This is somewhat surprising because originalism—the primarily conservative theory that judges should decide cases according to the original meaning of the words in the legal text—might seem to be irrelevant to efforts to apply an amendment adopted in 1791 to a world with 21st-century technologies. But Justice Scalia, one of the most conservative members of the Court and a leading advocate of originalism, uses the Fourth Amendment to explain why originalism is relevant to debates about what the Fourth Amendment says about smartphones and drones and GPS technology.[13] He explains that originalism is not strict construction or literalism. Originalism does not require reading the Fourth Amendment literally to protect only the four things that are specifically mentioned, because the Amendment was originally intended to guarantee the principle of limited government, not just to protect persons, houses, papers, and effects. So, an originalist can read the over-200-year-old Amendment to make it relevant to a world with automobiles, telephones, aircraft, thermal-imaging devices, computers, DNA and genomic testing, and a broad array of smart electronic communication devices in the age of digital information. Originalism does not solve the problem of interpreting the Fourth Amendment, but the conservative justices have used originalism to change judicial doctrines.

Developments in Two Rules of Fourth Amendment Law

The warrant requirement and the exclusionary rule are two rules of Fourth Amendment law. The requirement that the government get a search warrant before it conducts a search is considered a basic element of the American system of justice. By contrast, the exclusionary rule, which prohibits using illegally obtained evidence in court to obtain a conviction, is controversial. Both rules have been substantially changed over the years to reflect conservative thinking about crime. Some legal scholars think that both rules are pale shadows of their former substance because they have been "hollowed out" by conservative justices.[14]

William Rehnquist, who was appointed as associate justice in 1972 and chief justice in 1986, had a professional interest in criminal justice. He wrote 25 majority opinions in Fourth Amendment cases during his tenure as chief justice (from 1986–2005) and had a "substantial impact" on the law.[15] His successor, Chief Justice Roberts, is not particularly interested in criminal justice and has not written many Fourth Amendment opinions. He is comfortable assigning search-and-seizure cases to a justice, such as Scalia, who generally writes crime-control opinions. Scalia has written so many majority opinions on Fourth Amendment cases, and he writes with such "vision, force, and perseverance," that the Roberts Court is "Scalia's Court" when it comes to the Fourth Amendment.[16]

DOES THE FOURTH AMENDMENT REQUIRE SEARCH WARRANTS?

The answer to the question whether the Fourth Amendment requires search warrants is as follows: The general rule is that a warrant is required for a search to be reasonable, and obtaining a warrant requires the police to convince a judge that there is probable cause that the search of a particular person, place, or thing will produce evidence of particular criminal activity. But there are many exceptions to the rule:

- **Consent**. A person can knowingly give up the right to require the police to get a warrant before a search. However, a person's consent to a warrantless search must be knowing and voluntary.

- **Stop-and-frisk.** A police officer may stop and frisk a person without getting a warrant because a stop is not an arrest, and a frisk is not a search.[17]
- **Plain Sight (or Plain Smell).** If police officers have a right to be where they are, they may seize contraband or evidence that is in plain sight or smell without first getting a warrant. This exception applies to a broad range of settings: "open fields"; buildings; garbage left at the curb for trash removal; aerial surveillance of backyards; and drugs, chemicals, or explosives detected by specially trained dogs.
- **Search Incident to an Arrest.** A search incident to an arrest must be for officer safety or to preserve evidence.
- **Hot Pursuit.** An officer who is in hot pursuit of a fleeing suspect does not have to stop the pursuit and get a search warrant before search and seizure.
- **Exigent Circumstances.** Emergencies can justify warrantless searches. The ticking-time-bomb scenario is often used to justify this exception to the rule that searches require warrants.
- **Motor Vehicles.** The mobility of motor vehicles justifies warrantless searches to prevent the evidence from literally being driven away while an officer requests a search warrant. The increased use of e-warrants, which allow law enforcement officers to obtain search warrants electronically, reduces the need for a motor vehicle exception.
- **Special Needs.** The government has a compelling interest in public safety that justifies requiring government employees who perform certain sensitive jobs, such as railway workers and customs officials, to take drug tests without warrants, probable cause, or individualized suspicion.[18]
- **Schools.** The Warren Court extended some constitutional rights to students in public schools. The Rehnquist and Roberts Courts limited the scope of such rights by treating schools as special places with special missions, ruling that school officials who are responsible for providing a safe environment do not have to get warrants to search students or their possessions. Such warrantless searches include random drug tests for students who participate in certain school activities. Mass school shootings have also tilted the balance toward safety—with the notable exception, perhaps, of Second Amendment gun rights.
- **Administrative Inspections.** The Court originally held that the Fourth Amendment did not apply to administrative inspections (i.e., noncriminal searches). Today, the Fourth Amendment applies to administrative settings, but the protections are not as strong. Warrants are not required to search heavily regulated businesses, to enforce game laws, to

conduct business inspections to ensure food safety or enforce immigration laws, or to maintain safety and order in public housing or public transportation. Warrantless inspections of restaurants, salons, food-processing plants, and chemical plants are constitutional if the government demonstrates that doing so serves a substantial government interest in regulation. This includes airport searches and inspections. Routine border searches do not require warrants, but special or nonroutine border searches do require warrants. In fact, the national border zone, which extends inward 100 miles from the border, has its own set of search and seizure rules. As border security, immigration control, and hiring illegal aliens are transformed from regular public policy issues into issues of national security and criminal law, the point at which individual rights are balanced against government power slides toward government power.
- **National Security**? The question mark is appropriate because the Court has not officially recognized a national security exception to the warrant requirement, but gathering national security intelligence can be a "special need," and the legal regime governing national security is different from the legal regime governing ordinary criminal investigations.

There are now so many exceptions to the warrant requirement that it is not accurate to say that the Fourth Amendment requires a warrant in order to be a reasonable search and seizure. The changes in the law reflect both changes in expectations of privacy and conflicting meanings of privacy.

Two Concepts of Privacy

Reading search-and-seizure cases reveals that the justices have different conceptions of privacy. One conception of privacy applies to an act that a person does alone or that has no impact on anyone else. This conception of privacy is a liberty-based right that is based on the distinction between self-regarding actions, which are private, and actions that affect others, which are public. The second conception of privacy is based on the distinction between acts that government has a legitimate interest in regulating, which are therefore public, and acts that the government does not have a legitimate interest in regulating, which are therefore private even if the action involves more than one person.

According to this second conception of privacy, making a telephone call, sending an electronic communication, or conducting an Internet search (or in another context, having an abortion), may be considered private even though it is a shared (and therefore public) act, because the government has no legitimate interest in it. This conception of privacy ensures limited government by protecting acts and things from government power even though they do involve more than one person. These private-yet-public acts and information are at the center of the debate about how to ensure limited government in a digital age, wherein cameras and computers can monitor virtually everything. The Court's dissatisfaction with the current state of search-and-seizure law is based in part on the sense that current doctrines leave too much shared action and information unprotected.

The Property Rights Approach

One proposed remedy is to strengthen the property-rights approach to Fourth Amendment rights. Conservatives tend to support property rights more than privacy rights, while liberals tend to support privacy rights more than property rights, so the conservative justices are following Justice Scalia's lead in reviving the old common law trespassing doctrine as a property-rights approach to search and seizure. The Court had abandoned the trespass (or property-based) reading of the Fourth Amendment and adopted the privacy reading in *Katz* when it held that the Fourth Amendment was intended to protect "people and not places."

The privacy reading has old roots. In *Boyd v. U.S.* (1886), the Court read the Fourth Amendment to protect the "sanctity" of home and the "privacies of life" from government intrusion.[19] The Court cited with approval a legal scholar who believed: 1) that the main reason why people decided to leave the state of nature and create government was to protect their property; and 2) that the British fight against general warrants inspired the Fourth Amendment provision requiring special warrants.[20] This version of the social contract theory of government appeals to the conservative justices who are reviving the trespass reading of the Fourth Amendment.

The place-based reading of the Fourth Amendment treats privacy rights in the home (and the curtilage, or area surrounding a home) different than claims made in other places. The place-based reading is still relevant to current analysis of search and seizure, but more emphasis is now placed on the privacy reading as the better way to ensure limited government. Consequently, science and technology have revived the old debates about what those privacies of life are in a digital age where almost everything can be monitored, and the public expects that they will be monitored. Justice Scalia describes originalism as an alternative to the idea of a living constitution that is continually being interpreted to keep up with contemporary societal expectations. He downplayed the reasonable expectation of privacy test that the Court adopted in *Katz*. And in *Kyllo v. U.S.*

(2001),[21] he wrote for a majority that struck down the warrantless use of a thermal-imaging device to detect heat radiating from inside a home that the police suspected of being used to grow marijuana. According to Scalia, the police use of a sense-enhancing device that was not in general public use to get information that would previously have been unknowable without a physical search was a search that ordinarily required a warrant. To rule otherwise, he explained, would allow the police to evade the Fourth Amendment's intended protection of the interior of a private home from warrantless search.

Special and General Warrants

Reading Fourth Amendment opinions provide insights into why there is skepticism about the legality and wisdom of using big data searches for preventive policing. The Court has accepted the argument that the Fourth Amendment was a response to the colonists' experiences with officers of the British Crown who used general warrants (known as writs of assistance) to gather evidence of violations of tax laws. In *Stanford v. Texas* (1965),[22] the Court noted that these writs "bedeviled the colonists," who described these general warrants as instruments of arbitrary power, threats to liberty, and violations of fundamental principles of law. The *Stanford* case arose when the police got a warrant to search Stanford's home under a 1951 Texas statute that outlawed the Communist Party and authorized searches for and seizure of any documents related to the Communist Party in Texas. The police seized around 2,000 documents, none of which were related to the Communist Party. The Supreme Court ruled that general warrants such as the one authorized under the Texas statute were unconstitutional, because the Fourth Amendment was adopted specifically to protect the people of the new nation from the abuses they experienced under the old form of government when officers of the British Crown used general search warrants without meaningful legal limits on their power.

The *Stanford* case arose during the Cold War, but it raised one concern that is again being raised by counterterrorism policy—specifically, the concern that threats to national security are being used to justify weakening the Fourth Amendment prohibition against general search warrants. Images of government officials rummaging around in an attic or garage looking for dirt on someone have been replaced with images of government officials rummaging through a person's electronic files. The concerns began with reports of secret National Security Agency surveillance programs that accessed the files of telecommunication service providers. Congressional hearings in 2013 confirmed the existence of the programs and broadened the focus to include intelligence gathering programs in law enforcement agencies. In national security, as in policing, prevention programs gather large amounts of data (creating a haystack), which can subsequently be queried for suspicious individuals or actions (searching for the needle). The National Security Agency and the Federal Bureau of Investigation have intelligence programs that are designed to prevent attacks by using search warrants, subpoenas, and national security letters to gather data on individuals who are not suspected of any illegal activity. This kind of data gathering calls to mind the general warrants that the colonists detested and that the Fourth Amendment was intended to prohibit by requiring special warrants. The government initially claimed that the general warrants applied only to the communications metadata, the information (telephone numbers, dates, times, and e-mail addresses) about the information (the actual content of the communications) and that search warrants were still required to target individuals or access the contents of their communications. But subsequent revelations about the scope of the programs and mistakes in the handling of data revived debates about whether claims regarding national security and public safety were being used to trump privacy rights. The Supreme Court's Fourth Amendment privacy cases have primarily been criminal cases, but recent developments in electronic surveillance have reintroduced national security to debates about informational privacy.

An Alternative Approach to Search and Seizure

The justices' current dissatisfaction with the law of search and seizure is primarily based on the belief that the old doctrines are showing their age and showing that they are not a good fit for the digital age. This belief is not limited to one end of the ideological spectrum, which is why Fourth Amendment privacy cases do not always produce the ideological voting patterns that are typical of criminal law and national security. For example, in *U.S. v. Jones* (2012),[23] the Court struck down the warrantless attachment of a GPS device to a car that tracked the vehicle's movement for 28 days. Justice Scalia's opinion for a unanimous Court—but one where a plurality of the justices, including Scalia, relied on 18th-century common law of trespass, and another plurality, including the liberal Sotomayor and the conservative Alito, relied on the right to privacy.

Justice Alito's concurring opinion acknowledged that the Court should strive to assure the degree of privacy against government that existed when the Fourth Amendment was adopted, but he described Scalia's reliance on the 18th-century trespassory doctrine as "unwise," straining the language of the Fourth Amendment, artificial, and with little support in current case law.[24] According to Alito, "[d]isharmony with the substantial body of existing case law" was not the only problem with the trespassory approach.[25] Property rights also vary by state. More importantly, dramatic technological change creates flux in popular expectations of privacy: "New technology may provide increased convenience or security at the expense of privacy, and many people may find the tradeoff worthwhile."[26] Or they may think the tradeoff is inevitable and will ultimately result in "the end of privacy."[27] According to Alito, it is especially during times of changes in technology and attitudes that legislative bodies are better designed than courts to "gauge changing public attitudes, to draw detailed lines, and to balance privacy and public safety in a comprehensive way."[28] But he was not confident that Congress was up to the task. Justice Alito's views on technology and privacy are particularly noteworthy because he has had a long-standing interest in the topic. In 1972, he wrote the *Final Report on the Conference on the Boundaries of Privacy in American Society*, which highlighted the importance of privacy and the threats to it presented by technological developments.[29]

Justice Sotomayor's concurrence acknowledged the inadequacy of both the current search-and-seizure doctrines and the property-based trespassory approach that Scalia has supported as an alternative. She considered both approaches inadequate for life in a digital world, where the government can easily and cheaply use technology to electronically search without physically trespassing. According to Sotomayor, the common law concept of trespassing can readily be applied to cases where the police do not get a warrant before physically attaching a GPS device to a person's car, but the trespassory approach could not be readily applied to electronic searches of databases. Furthermore, electronic searches are used more often than physical searches, and they have a broader scope and longer duration.

Justice Sotomayor thinks that the solution to the digital-privacy problem is a new approach to search-and-seizure law. She thinks that the solution requires reconsidering the current rule of Fourth Amendment law that a person has no reasonable expectation of privacy in information that he or she has "voluntarily disclosed to third parties."[30] This is the third-party doctrine. The third-party doctrine is the legal policy that a person does not have a privacy right to any communications or documents that have been willingly shared with another person. One problem with this rule of law is that it will continually tip the scales of justice toward government power to search and seize as police and national security agencies increase their use of the new and cheaper technologies to conduct large-scale electronic searches rather than the targeted physical searches. These digital technologies reveal the limits of the trespassory or property approach to search and seizure even if, as conservatives claim, the original purpose of the Fourth Amendment was to protect property rights.

Justice Sotomayor's analysis is also valuable because it brings important social and political issues to the legal debates about Fourth Amendment doctrines. Fourth Amendment issues are usually debated almost exclusively as differences of opinion about the best way to fight crime. Advocates of due process emphasize the importance of protecting the civil liberties and advocates of crime control emphasize empowering police to effectively fight crime. Sotomayor's analysis of search and seizure and the right to privacy adds a civics dimension that is an important aspect of the scholarly analysis of the meaning and value of privacy. Privacy cases are typically described as conflicts between an individual and the government. The individual is claiming an exemption from the societal values expressed in law, and the government is claiming the power to enforce the values. The conventional understanding of these conflicts is that an individual is claiming a privacy right to be exempt from the community's norms of behavior. An alternative approach, which is especially relevant to Fourth Amendment privacy rights in national security settings, emphasizes the sociality of privacy, not the individuality of privacy. This social or civic approach treats privacy as a social value, not merely a social cost.[31]

According to the third-party doctrine, the Fourth Amendment does not even apply to voluntarily shared information such as telephony data, e-mails, and other electronic communication. As a result, more information as well as more of the ordinary activity of everyday life is exposed to warrantless government searches. Sotomayor worried that the increasing exposure will have a "chilling effect" on civic engagement because people will be wary of any political participation or of taking any public or private action, knowing that such action is incidentally swept up by the government as it monitors human interactions under programs designed to prevent crime or national security threats.

The civics (or social or political) approach to Fourth Amendment privacy issues can be traced to the nation's founders. Is the fact that so many provisions of the Bill of Rights protect the rights of suspects and convicted offenders evidence that the Founding Fathers were soft on crime? No. They had experienced the political use of the criminal law powers during the colonial era, when colonial governors used their substantial criminal law powers to investigate critics or opponents, and therefore wrote the Fourth Amendment to prevent such abuses of power. This history with political criminal justice is worth remembering when thinking about granting government broad power to eavesdrop at home and abroad.

Fitting Technology into the Fourth Amendment

Technological innovations have historically required fitting the Fourth Amendment to contemporary circumstances. The law has adapted to automobiles, telephones, airplanes, computers, and the Internet, but the traditional Fourth Amendment doctrines do not protect informational privacy in an age of smart electronic devices, location tracking, and DNA analysis.

The Court has allowed warrantless aerial surveillance in drug investigations because aerial surveillance is not ordinarily considered a search, and there is no reasonable expectation of privacy from above.[32] The use of electronic tracking devices also raises search-and-seizure questions. The Court has allowed the police to, without a warrant, put tracking beepers in chemicals and equipment that are the kinds of materials used to make controlled substances, follow the beeps to a ranch where, after peering through and climbing over fences, they determined that there was an illegal drug lab in a barn on the property. This warrantless search was constitutional because the protection against unreasonable search of a home and the curtilage does not extend to a barn located some distance from the home.[33] More recently, however, Scalia's majority opinions in *U.S. v. Jones* (2012) and *Florida v. Jardines* (2013)[34] reflect the property rights reading of the Fourth Amendment. In *Jardines*, the Court held that the police use of trained drug dogs to sniff the front porch of a single-family home was a physical search for the purposes of the Fourth Amendment, and it required a warrant. The Court's precedents clearly recognize that a person has a reasonable expectation of privacy in the home and its curtilage. The legal question in *Jardines* was whether a sidewalk leading to a porch and the porch itself were protected areas. The majority held that they were protected areas and that the warrantless dog sniff for drugs was unconstitutional.

In *Jardines*, Scalia emphasizes the difference between a physical search and seizure, which the Fourth Amendment specifically addresses, and an electronic search, which it does not. However, the property-rights approach could protect against physical or electronic searches and thereby become an effective alternative to the privacy reading of the Fourth Amendment that liberals have supported. The conservative justices may prefer the place-based and property-rights-based analysis of search and seizure to the privacy analysis because conservatives are generally more supportive of property rights than privacy rights.

This may explain recent rulings in cases involving strip searches of prisoners and DNA swabs of people who have been arrested. The conservative majority, in an ideologically divided 5–4 ruling, upheld suspicionless searches of prisoners prior to their entering the prison population.[35] And in another

5–4 ruling that included four of the five conservative justices, the Court held that conducting a DNA swab as part of arrest procedures did not violate the Fourth Amendment because: 1) the government has a legitimate interest in determining an arrestee's identity and criminal history; and 2) the swab, which it considered comparable to fingerprinting, is not so invasive that it requires a warrant.[36] Justice Scalia's dissent did not compare the invasiveness of fingerprinting and DNA swabs. He compared the DNA swab policy to the British general warrants that the colonists hated, and he concluded that both policies were the kind of suspicionless searches for information that the Fourth Amendment was intended to prohibit.

These cases illustrate how science and technology can raise questions about exactly what constitutes a search for the purposes of the Fourth Amendment. When does using a search aid constitute a Fourth Amendment search? The police and other government officials use low-tech aids such as binoculars and high-tech aids such as thermal-imaging devices. The Court has held that the police use of a search aid that is normally available to civilians is not a Fourth Amendment search. What about using trained dogs to smell drugs or explosives? It depends on where and how the trained dogs are used. Trained dogs are a search aid because of their ability to sniff out 19,000 explosive scents, which is one of the reasons why the CIA's K-9 dog teams compete in the United States Police Canine Association Certification and Trial events.[37]

The general-public-use rule illustrates why technology makes search and seizure such a dynamic area of criminal law. As technology that was developed for the military, national security agencies, or the police becomes widely available to civilians, it can be used for warrantless searches. For example, imaging devices are now widely available for businesses such as heating, ventilation, air-conditioning, and insulating.

Changing the Third-Party Doctrine?

Technology has changed methods of communication. The telephone changed communications, and wiretap technology followed shortly thereafter. Telecommunications were eventually incorporated into Fourth Amendment law. Computers also changed communication. Congress has struggled to keep federal privacy laws in step with the pace and scope of change in digital technology that created the Internet, social networking, and mobile communications devices. The Electronic Communications Privacy Act of 1986[38] regulates access to e-mail. It provides that the government must get a search warrant to obtain newer e-mails, those that are less than 180 days old, but not older e-mails. The government can also access older e-mails without getting a search warrant by getting a court to issue a subpoena. Electronic communications have changed a great deal since 1986 and now include much more than e-mail. The law currently treats print and aural communication different than electronic communication. A person who stores print documents in a file cabinet in the home or office has greater Fourth Amendment protection than when telecommunication service providers hold digital communications.

The third-party doctrine effectively exempts electronic communications from the provisions of the Fourth Amendment. The third party could be the store where a consumer used a credit card or check to buy a good or service or app, the bank that processed the payment, or the telephone company or Internet service provider whose telecommunication services are used by a consumer. Congressional efforts to amend the 1986 law to treat electronic communications like other communications are complicated by the familiar debates about where to strike the balance between privacy rights and government power to access information for criminal or national security purposes.

In one sense, shared information is, by definition, public information. But as it becomes harder to function in the political, economic, social, and professional worlds without sharing electronic data, the current legal regime means that the Fourth Amendment covers less and less. The fact that the government does not need a search warrant to follow the digital information trails we leave behind with every keystroke is one reason for the widespread dissatisfaction with the law of search and seizure. The search for alternative approaches that will ensure that the Fourth Amendment remains relevant for informational privacy in the digital age requires thinking about expectations of privacy. Technology changes expectations of privacy, and changes in expectations of privacy change search-and-seizure case law. The legal regime can be adapted to digital technology precisely because the regime is based on the reasonable expectation of privacy. However, adapting it requires developing a new consensus on where to strike the balance between security and privacy for electronic information. The justices have asked Congress to decide the question, but congressional inaction presents the Supreme Court with an opportunity to establish a new policy. Cases challenging the power of the police to conduct warrantless searches of a person's cell phone or smartphone are working their way through the federal court system.

THE EXCLUSIONARY RULE

The second rule of Fourth Amendment law is the exclusionary rule. The Fourth Amendment declares a right to be free from unreasonable search and seizure, but it does not provide a remedy for violations of the right. The Court created the exclusionary rule as a remedy for violations of constitutional rights. The exclusionary rule provides that illegally obtained evidence cannot be used in court to obtain a conviction.

Conservatives have been especially vocal critics of the exclusionary rule since the Warren Court applied it to state courts in Mapp v. Ohio (1961). They describe it as a judge-created policy that is not required by the Constitution and that sometimes allows a guilty person to go free on a legal technicality—to literally get away with murder. In 1969, when he was an assistant attorney general in the Nixon administration, William Rehnquist wrote a memo that described the exclusionary rule as a court-created evidentiary rule, not a constitutional principle. The memo also stated the crime-control belief that the exclusionary rule would result in fewer confessions because the Court's Miranda ruling required police to tell suspects that they had a right to remain silent prior to questioning them.[39] Congress had recently tried to overturn the Miranda decision. A provision of a major federal statute, the Omnibus Crime Control and Safe Streets Act of 1968, directed judges to take into consideration the totality of the circumstances surrounding questioning when considering how much weight to give a confession. Suspects did not have to be read their rights—confessions were admissible if they were voluntary. Ironically, in 2000, Chief Justice Rehnquist wrote for a majority ruling that Congress could not legislatively overturn the Miranda holding that governed the admissibility of custodial statements.[40]

The Burger, Rehnquist, and Roberts Courts have fundamentally changed the exclusionary rule by applying a cost-benefit analysis that heavily weighs the societal costs of the rule, creating numerous exceptions to the rule and limiting its purpose to deterrence. The exclusionary rule was originally described as serving two purposes: deterring police misconduct by suppressing illegally obtained evidence and protecting judicial integrity by eliminating the possibility that courts would preside over trials where illegally obtained evidence was used to obtain a conviction. Since then, conservative justices have supported the crime-control model of justice, which places greater value on effective crime fighting than on due-process protection of individual rights. The result is a balance that weighs the high costs that the exclusionary rule imposes on society when the guilty go free because evidence was suppressed. On July 6, 1976, the Court announced four decisions that used the balancing test to determine whether evidence was admissible.[41] Balancing one individual's rights against society's collective interest in public safety will likely result in more convictions, because the heavier weight is on the side of the greater good. Balancing also affects the assumption of innocence and the belief that it is better to allow 100 guilty to go free than convict 1 innocent person.

In 1977, Robertson advised the Court to take a "bold new approach" to the exclusionary rule, which he called an "archaic judicial approach" to providing a remedy for violations of Fourth Amendment rights, by limiting the rule to those instances where the police are guilty of a bad faith violation of the law."[42] The Court has done so. It also created the following exceptions to the exclusionary rule:

- **Grand Juries**
- **Harmless Error**: Evidence may be admitted if it is obtained by a harmless error.
- **Civil Trials**
- **Good Faith**: Evidence may be admitted if it is obtained illegally but by good faith actions.
- **Independent Source**: Evidence can be admitted if one of its sources was constitutional.
- **Inevitable Discovery**: Evidence can be admitted if it would have eventually been discovered anyway.
- **Public Safety**: Evidence is admissible if it was obtained in an emergency or an exigent circumstance such as a threat to public safety.
- **Preventive Detention**
- **Parole Revocation**
- **Prisoners**: The rule does not apply to prison disciplinary hearings.
- **Impeaching a Witness**: The rule does not apply to a defendant's own testimony or testimony of an accomplice.
- **Physical Evidence**: A possible new exception is for physical evidence, such as a gun or bullet. In contrast to testimonial evidence, whose validity can be affected by the way it was obtained (e.g., a confession obtained by the third degree or torture), the validity of physical evidence is often not affected by how it was obtained.

The Court has created so many exceptions to the exclusionary rule that it is misleading to say that evidence obtained in violation of the Constitution is inadmissible in court. Have the exceptions become the rule? It may now be more accurate to say that evidence obtained in violation of the Constitution will be admitted unless there are special circumstances, such as a pattern of extreme police misconduct, that justify keeping it out of court.

The conservative justices certainly have changed the way the exclusionary rule works. They have limited its scope, created numerous exceptions to it, restated its purpose, and transformed it from a rule of law into a regulatory policy consideration. The conservative working majority on the current Court may ultimately abolish the rule altogether. Conservatives have two major criticisms of the exclusionary rule. The first is that it is a judge-created policy rather than an explicit provision of the Constitution; the second is that it functions as the death penalty for evidence. Chief Justice Burger's dissent in Bivens v. Six Unknown Federal Narcotics Agents (1971)[43] called it "capital punishment" for evidence.

Justice Scalia's opinion in Hudson v. Michigan (2006)[44] reflects both criticisms. The police had a search warrant but violated the knock-and-announce rule that required waiting 20–30 seconds before entering a home. Hudson moved to exclude the evidence found during the search. Scalia wrote for a conservative majority, holding that the exclusionary rule did not apply to such cases. He also questioned the continued need for the rule in an era when the legal regime governing policing includes civil remedies for violations of Fourth Amendment rights, police professionalism, internal disciplinary procedures and citizen review processes, and increased police accountability. Scalia is saying, in effect, that times change. And he questions the continued need for a judicial policy that was created to solve a problem that existed in the old legal regime of policing but which no longer exists, or is a problem for which there are less drastic remedies.

Another case that illustrates how the conservative majority has substantially limited the exclusionary rule is Herring v. U.S. (2009),[45] where the Court said that the rule applied only to the deterrence of culpable behavior that was sufficiently serious to outweigh the costs of excluding the evidence obtained. This reasoning has fundamentally changed the way the exclusionary rule works. The exclusionary rule is no longer a rule in the sense that it operates the way a rule of law operates. A rule of law is one that is applied to relevant cases. If the exclusionary rule prohibited illegally obtained evidence from being used to obtain a conviction, then it would be operating as a rule of law. But it no longer works this way. It is now just one of the factors to be taken into consideration when weighing evidence. This change reflects the conservative belief that the exclusionary rule should not operate as a rule of law and that a judge or jury should use a balancing approach, introducing proportionality to the weighing of evidence. Proportionality is an ancient principle of justice that is captured by aphorisms such as "Let the punishment fit the crime" and "an eye for an eye." The conservative justices consider proportionality an especially appropriate principle for applying the exclusionary rule because they have held that its purpose is punishment (to deter police misconduct), and proportionality is an ancient principle used to determine whether a rule (or policy) is just.

The Court has, for decades, considered the exclusionary rule part of the judicial regulation of police conduct that is intended to give effect to Fourth Amendment rights.[46] Today, the rule is of declining importance as a matter of legal policy, and the possibility that it may be abolished has raised the question whether the exclusionary rule is essential for the fair administration of criminal justice. The study of comparative law suggests that the exclusionary rule is not essential, because most other countries, including Western-style democracies with legal systems that protect individual rights, do not use the exclusionary rule. However, eliminating one of the main deterrents to police misconduct weakens judicial implementation of Fourth Amendment privacy rights.

INFORMATIONAL PRIVACY AND THE SECURITY STATE

This section of the chapter examines two developments in law enforcement that raise Fourth Amendment privacy issues: the trends toward (1) intelligence-based policing to prevent crime and other threats to public safety and (2) integrating national security surveillance and domestic law enforcement. It is important to examine these developments, because one theme of the Supreme Court's crime-control approach during the last several decades is judicial deference to law enforcement officials. Judicial deference has increased executive discretion. Law enforcement officials have considerable leeway to decide how to implement court rulings and federal statutes related to search and seizure, surveillance, and privacy rights. As a result, law enforcement and surveillance agencies play a more important role in developing the legal regime for Fourth Amendment privacy rights. These agencies include local, state, and national law enforcement. The main federal agencies are the Federal Bureau of Investigation, the National Security Agency and Central Security Service, the Customs and Border Protection (which is part of the Department of Homeland Security), and even the Internal Revenue Service—which has a criminal tax division that investigates suspected criminal activity by obtaining tax-related information from paper documents and electronic communications.

Intelligence-Based Policing

The general worry about intelligence-based policing is that it is one more face of big government. The specific worry is that using big data to prevent crimes or other threats is similar to the general search warrants that the colonists detested so much that they wrote the Fourth Amendment to prohibit them. Traditional policing is usually described as reactive, responsive, or investigatory policing. Intelligence-based policing is usually described as proactive or preventive. But traditional policing also used information to prevent crime. For instance, the Uniform Crime Reporting Program was conceived in 1929 by the International Association of Chiefs of Police,[47] and CompStat is a data-based management system that police departments have used for years to strategically allocate resources.[48]

Stop-and-frisk programs are another example of traditional policing to prevent crime. Justice Scalia has had a major impact on this area of Fourth Amendment law. The Court reads the Fourth Amendment to protect against unreasonable search or unreasonable seizure. In *Hodari D.* (1991),[49] Scalia wrote for a majority that narrowed the definition of a seizure to allow more police interactions on the street that do not trigger a Fourth Amendment seizure. He held that a seizure does not begin with first police contact. It begins when a police officer applies physical force or when a person submits to a show of police authority. A cop on the beat can say, "Come here," "Hold it a minute," or "Show your hands," without triggering a seizure if the person feels free to end the encounter with the officer. And in *Whren v. U.S.* (1996),[50] Scalia wrote for a unanimous Court in a vehicular stop-and-frisk case. The Court upheld a police department practice of patrolling high-crime neighborhoods in unmarked cars, stopping drivers for minor traffic infractions, and then looking for evidence of other crimes. These stops were challenged as pretextual stops—vehicle stops that were supposedly for traffic enforcement but were actually for drugs. Scalia explained that for purposes of the Fourth Amendment, the relevant question is whether the stop was objectively reasonable; a police officer's intentions or ulterior motives are not relevant because they are subjective. These two rulings allowed police departments to adopt stop-and-frisk policies without worrying very much about the Fourth Amendment.

Big Data and Predictive Analytics: Working the Right End of the Problem

The phrase "big data" refers to the ability to gather, maintain, and use very large databases. Life in the digital information age makes information commercially and politically valuable to individuals, businesses, organizations, and governments who want to know more about consumer behavior, telecommunications, driver's licenses, passport and immigration documents, and records (e.g., medical, educational, financial, criminal, and social security). These information management systems provide valuable data that can be mined in search of patterns of activity, whether behavior or communication, that arouse suspicion. The suspicion may then trigger searches for evidence to prosecute or perhaps prevent crime. The ability to crunch big numbers has upset the balance between government power and individual rights.

Digital technology has created new tools that enable governments to do what they do better, but it is also a game changer, because it is transforming the missions of law enforcement agencies by driving the trend toward intelligence-based preventive policing. Prevention is sometimes called working the right end of the crime problem. Traditional law enforcement primarily responds to crimes by conducting after-the-fact investigations; preventive policing is proactive. Prevention seems logical and commonsensical, but it does raise important political and legal questions. Prevention invariably leads law enforcement and surveillance agencies to look into areas of people's lives that are considered private and beyond the reach of the long arm of the law. Furthermore, the legal regime that governs search and seizure was based on the assumption that policing was primarily responsive, not preventive.

Digital technology is likely to continue to drive policy without much legal direction because the law struggles to keep up with the pace of scientific and technological change. The ability to economically gather, curate, and use data is changing policing priorities and shifting resources from investigating crimes to intelligence gathering and predictive analytics. Closed-circuit television cameras, red-light cameras and license-plate readers, and crowd-sourced photos increase the ability to conduct surveillance and to investigate crimes. Algorithmic criminology and predictive policing use databases to detect patterns of human activity, whether communication or behavior, and create actionable intelligence. Predictive analytics is an increasingly important aspect of how police and national security officials do their jobs. Given its size, its location, and the fact that it has been the target of major terrorist attacks, it is not surprising that the New York City Police Department is leading the way toward combining crime, terrorism, and intelligence gathering.[51]

One way to predict developments in policing is to read what the leadership of law enforcement and surveillance agencies think about recent trends and future directions in law enforcement. An article on predictive policing in *The Police Chief*, a publication for law enforcement professionals, describes three modern eras of policing:[52]

- Professionalism: This era, which began in the 1960s, emphasized police training and systems management.
- Community Policing: This era, which began in the 1990s, stressed reconnecting police to the communities they served.
- Intelligence-Based Policing: This era, which began with the terrorist attacks on 9/11, promotes preventive policing. Police departments see themselves as on the front lines in the war on terror and align themselves with the intelligence and prevention priorities of national security agencies.

The current era of intelligence-based policing emphasizes using digital technology to perform law enforcement missions that integrate domestic policing and national security. The new methods and missions present a challenge for changing while remaining true to the original purposes of the Fourth Amendment.

Legal Fiction and Scientific Fact

Criminal justice is a familiar subject of fiction. Steven Spielberg's science fiction film *Minority Report* (2002)[53] describes a precrime squad whose homicide detectives have been so successful using facial recognition systems to track down people who are about to commit murder that they have eliminated murder within their jurisdiction. In one scene, the Tom Cruise character walks through a mall as software identifies his face and presents digital ads targeted for his demographic. This fictional storyline seems plausible because facial recognition software is not merely the stuff of science fiction. Facebook has a biometric database with billions of photos. Companies can equip mannequins with cameras for eyes that use face recognition software to target ads to potential customers as they walk through the store's door.

In *The Anatomy of Violence: The Biological Roots of Crime* (2013),[54] Adrian Raine argues that public policy should use scientific knowledge to prevent crime. Raine calls his preventive policing program, which resembles the one described in the film *Minority Report*, "Legal Offensive on Murder: Brain Research Operation for the Screening of Offenders" or LOMBROSO. LOMBROSO is a reference to the famous 19th-century criminologist Cesare Lombroso, who believed that criminal behavior had biological origins and that a person's innate criminality was revealed in certain physical features such head shape or eye spacing. Twenty-first-century science can now look inside the body for evidence of future criminality. Raine's LOMBROSO program would subject all men 18 years old and over to brain scans and DNA testing. Those whose tests indicated future criminality would be monitored and even detained as at-risk individuals. Scientific innovations such as genomic testing can now be used to identify individuals who are at risk for certain diseases. Should public policy use medical testing for evidence of criminal predisposition? Would a genomic test search be a reasonable Fourth Amendment search? This is an area where science and technology affect Fourth Amendment privacy rights.

THE FEDERAL BUREAU OF INVESTIGATION

The FBI was created to investigate federal crimes, but its mission expanded over time. The FBI's website describes itself as a "proactive, threat-driven security agency" that helps protect the country from all of "the most dangerous threats" facing the nation, from crime and corruption to international and domestic terrorists and spies on U.S. soil.[55] The FBI's Biometric Center of Excellence[56] is developing a facial-recognition database with the photos of individuals who have been arrested on the assumption that face prints will replace fingerprints and may eventually make it unnecessary for the police to physically stop someone and ask, "Can I see some identification?" The Biometric Center's mission is not limited to domestic crime, because the FBI works for crime control around the world. Its goal is to predict and prevent crime or other threats by taking preemptive action. Today's law enforcement agencies want what the Central Intelligence Agency was created to provide: actionable intelligence. In the good old days, the police got actionable intelligence by nailing "Wanted: Dead or Alive" posters to trees, hanging pictures on post office walls, or walking the streets to talk to people in the community. Today, the FBI hires intelligence analysts with skill sets that are a far cry from those used by the traditional cop on the beat.[57] In the past, federal agents or police detectives investigating a bank robbery might begin by rounding up the usual suspects—bringing the locals whose modus operandi fits the crime down to the police station for questioning. This investigative approach is no longer as effective because the "usual suspects" in a cybercrime bank robbery could be anyone, anywhere in the world, with a laptop. E-commerce, which allows rapidly moving money around the world with electronic transfers, requires developing global policing to detect and investigate crime. Intelligence can be used for crime in the streets, identifying a high-crime neighborhood, and for crime in the suites, nonviolent white-collar crimes for financial gain.

The FBI maintains that its effort to develop the Next Generation Identification Program (NGI) is driven by technology, customer (that is, police department) requirements, and growing demand for Integrated Automated Fingerprint Identification System (IAFS) services. The FBI describes the NGI as a collaborative effort by the bureau, the Criminal Justice Information Services Advisory Board, and members of the Compact Council—which is made up of local, state, federal, and international representatives.[58] Current identification systems use palm prints, passport pictures, mug shots, and driver's licenses. The Department of Homeland Security's Science and Technology Directorate is testing a Biometric Optical Surveillance System (BOSS) for scanning crowds. The rather ominous-sounding BOSS began as a military project to identify terrorists and suicide bombers overseas, but in 2010 it was transferred to the DHS for police use.[59]

The use of information systems to detect patterns of criminal activity, allocate resources, and prevent crime has become one measure of professionalism in law enforcement agencies. Databases are created for many purposes and often used for others. DNA databases are useful for medical and nonmedical purposes such as establishing insurance policies, conducting criminal investigations, and preventive policing. Facial-recognition software is being developed into an investigative technique that may replace lineups, fingerprints, and palm prints as ways to identify arrestees. But it can also be used for other purposes. The FBI describes its Criminal Justice Information Services Division as a place where "statisticians are compiling vast amounts of data" from law enforcement agencies that allow investigators and police professionals to use state-of-the-art technologies "to catch crooks and terrorists" by searching the "world's largest repository of criminal fingerprints and history records in a flash."[60]

Government agencies want authority to gather more data, greater access to private-sector databases, and databases that are integrated. With statutory authority and sometimes with judicial supervision, agencies get information using search warrants approved by judges, grand jury subpoenas, and national security letters (NSLs). The FBI's use of NSLs has increased because the USA PATRIOT Act expanded its authority to use them to demand that companies produce data. Furthermore, the NSL demands are secret—the target cannot disclose the demand to produce data; they do not require judicial approval; and the standard for issuance is that the information must be relevant to a counterterrorism investigation.[61]

Going Dark

The FBI has provided congressional testimony supporting legislation to require wire and electronic communications providers to provide government access to their databases. The bureau maintains that the legislation is necessary because its capacity to conduct court-approved eavesdropping by intercepting electronic communication is "going dark" as more communications become electronic and the service providers are not required to ensure access to records.[62] The FBI's general counsel elaborated on the problem in an address to the American Bar Association's Standing Committee on Law and National Security. He explained that Congress had to update the government's authority to conduct surveillance of telecommunications in order to prevent police and national security intelligence from going dark. He described "going dark" as the government losing its ability to actually conduct court-ordered searches of telecommunications, because the 1994 law that requires companies to cooperate with government interception of communications for law enforcement or other purposes, the Communication Assistance for Law Enforcement Act, requires telephone companies to provide technical assistance with things such as attaching an alligator clip to a phone line, but not other kinds of assistance. Digital technology is making those other kinds of assistance, such as requiring companies to build interception capabilities into their database systems, including those systems that use the Internet to converse, crucial in an era where more communication occurs in forms not covered by the 1994 law. The general counsel warned that the outdated law was making court-ordered surveillance obsolete because the new telecommunications companies do not always comply with current law, which instructs companies that receive orders to provide technical assistance to law enforcement officials who request information. The general counsel described proposed legislation to strengthen wiretap orders issued by judges as a solution to the problem of applying old law to new technology.[63]

The General Counsel also called attention to a practical law enforcement problem created by divided or unclear Supreme Court rulings in search-and-seizure cases involving technology. It is hard for criminal justice officials to implement such rulings by training law enforcement officers. The *Jones* ruling on the warrantless attachment of a GPS device to a car, for instance, raised almost as many questions as it settled, because the justices were divided into two camps—those who supported the trespass approach and those who supported the reasonable expectation of privacy approach. The Court has been searching for a limiting principle that could be applied to digital technology in order to protect privacy in the age of preventive analytics, but it has failed to settle on a bright-line rule. The Court is clearly dissatisfied with the current legal regime but is not clear about an alternative approach. Congress remains divided. And executive branch officials are uncertain about how to comply with Court rulings. Under these circumstances, developments in the legal regime governing Fourth Amendment privacy rights are likely to include continued criticism.

Getting Used to the New

New law enforcement tools, policies, and programs are sometimes used first against organized crime, drug dealers, and immigrants or to prevent especially serious and heinous crimes such as terrorism, child pornography, pedophilia, or murder. Limiting the use of new government powers for special needs or populations eases the general public's concerns about privacy rights and increases acceptance of using the powers for broader purposes. For example, using preventive or administrative detention for sexual predators increases the likelihood that preventive detention will be considered a legal approach to crime or national security generally. The biometric information systems that are proposed to increase border control and keep track of immigrants are likely to be used for citizens for special purposes such as monitoring sex-offender registries or abusers of drug and alcohol. The U.S. Customs and Border Protection Agency (CBP), which uses unarmed Predator drones to protect public safety along the Mexican and Canadian borders, states that it does so in compliance with current laws governing the collection and storage of photos, video, and other data. These laws include privacy laws that require the CBP to limit access to information to law enforcement personnel with "an official need-to-know in a law enforcement capacity." In recognition of the privacy dimensions of its special law enforcement function, the CBP also writes privacy impact assessments.

Digital surveillance technology was less controversial when it targeted national security threats, foreigners, and the borders. Now that the technology is being used for domestic policing and the missions of law enforcement agencies have been expanded to include national security, more attention is being paid to public policies governing digital surveillance. Agencies and organizations are aware of public concerns. The Police Executive Research Forum acknowledges that intelligence-led and predictive policing methods that use cameras and computer-based predictive analytics raise public fears of Big Brother that need to be addressed as technology transforms policing missions.[64]

THE NATIONAL SECURITY MODEL OF JUSTICE

The initial legislative response to the failure to prevent the terrorist attacks on 9/11 was the passage of the USA PATRIOT Act in 2001. The PATRIOT Act increased government power to conduct searches and to gather intelligence. It also increased the coordination of agency missions by removing the wall of separation between criminal investigations and intelligence gathering. Both the increase in government power and the organizational changes reflected conservative thinking that law, rights, and courts had made it harder and harder to protect national security. The sources of this conservative thinking about law are the Cold War and the war on crime. During the Cold War, conservatives worried that the American commitment to democratic values and the rule of law meant that the United States would only fight a limited war against the Soviet Union, which was not constrained by such values. And domestically, conservatives had for decades described standards for due process of law as unduly tying the hands of police officers responsible for fighting street crime.

The Bush administration brought this conservative thinking about crime to thinking about national security. It chose the war model rather than the crime model for counterterrorism policy because it did not want to apply the extensive body of criminal law, including the law of search and seizure, to counterterrorism policy. The administration worried that laws would unduly tie the hands of those who were responsible for protecting national security while also allowing the nation's enemies to use laws and rights against the United States. This is the conservative fear that terrorists, who recognized that they could not win conventional warfare with the United States, would instead wage law-fare against the United States.

The PATRIOT Act prompted changes in the missions of criminal and national security agencies. According to the Department of Justice, the this act was intended to increase the coordination of law enforcement, intelligence, and national defense communities—all the different agencies with some responsibility for public safety—by removing the legal wall that separated criminal investigations (a primarily domestic activity) from intelligence gathering (focused primarily on foreign agents). By doing so, the PATRIOT Act modernized government to more effectively wage digital-age battles rather than leaving it to fight with "antique weapons—legal authorities leftover from the era of rotary telephones."[65] Given the general public's skepticism of big government, it is politically astute for government agencies to describe their plans to prevent crime, protect national security, gather intelligence, and secure the borders as modernization that is intended to create smarter, not bigger government.

Congress created the Privacy and Civil Liberties Oversight Board in 2004 (PCLOB) to review laws, regulations, and policies and to make recommendations about where to strike the balance between security and privacy in an era where public policy officially promoted integrating policing and national security. The disclosure of secret FBI and National Security Agency (NSA) surveillance programs revived the debate about how best to protect informational privacy and national security. The government claimed the surveillance programs were authorized by section 215 of the PATRIOT Act, which authorized the collection of business records of telecommunications companies, and the Foreign Intelligence Surveillance Act, which authorized certain access to telecommunications data for foreign intelligence investigations.

At a PCLOB workshop[66] to discuss legal, technical, and policy perspectives on NSA surveillance, the participants who defended the intelligence programs claimed that the programs were nothing new—they merely used new technology to do what existing legal frameworks authorized. The critics claimed that the programs were actually a new kind of surveillance. Both the defenders and the critics compared the intelligence programs to the way the law treats mail sent through the post office. The government can monitor letters by looking at addresses and stamps on an envelope, but it cannot access the contents of a letter without a search warrant. The defenders of the intelligence program said that the NSA's gathering of metadata on telephone calls (e.g., the numbers, time, and length of calls) was like the government keeping track of the addresses on the outside of an envelope sent through the mail. The problem with this analogy is that it does not adequately account for the fact that technology has changed the relative value of the content and the metadata. The content of communication (the transactional information) used to be more valuable than the metadata (the information on the outside of an envelope, the telephony data). But digital technology makes the metadata more valuable than the content for the purpose of intelligence analysis. This is an unsatisfactory state of affairs, because it means that current Fourth Amendment doctrines protect less valuable information but not more valuable information.

The PCLOB workshop participants explained that the first step in searching for a needle in a haystack is to create a haystack (of information). They also acknowledged that this reverses the regular law of search and seizure, which begins with individualized suspicion: the police convince a magistrate that there is probable cause to believe that searching a particular person or place will produce evidence of a particular crime. Intelligence analytics assumes that all information is relevant to a search in the sense that it is necessary to have all the information (e.g., every phone record or e-mail or electronic funds transfer) in order to perform any subsequent analysis or query of the data targeting suspicious individuals.

Some evidence of the national security or military model is readily apparent. Police use hardware including armored vehicles, assault rifles, bulletproof vests, and drones and tactics such as SWAT teams. Other evidence is less visible but nevertheless very important because it includes methods, specifically data-driven policing analytics, that have transformed the missions of law enforcement agencies at all levels of government. Domestic law enforcement agencies are becoming more militarized and more national security–oriented as their missions are broadened to include counterterrorism. The integration of national security and policing is a manifest goal of national counter-terrorism policy. Data-driven policing is not new, but the emphasis on using intelligence for preventive policing or predictive analytics is new, and the legal framework for determining its limits is uncertain.

The general public's reaction to worries that broad, suspicionless intelligence gathering by the FBI or the NSA weakens Fourth Amendment privacy rights is typically, "What do you have to hide? If you are not doing anything wrong, you should not be concerned about the government monitoring what you are doing." Perhaps Justices Alito and Sotomayor are right to worry about the civic cost of widespread government surveillance. The civic or social costs include the chilling effect that ubiquitous surveillance has on freedom of expression in a legal regime where privacy is becoming an archaic concept.

The National Security Agency

The National Security Agency (NSA) was created in 1952. The NSA's mission statement on its website describes the agency as leading the government in cryptology, signals intelligence, information assurances products and services, and computer network operations. The mission statement reveals why there are concerns about the NSA's broad powers. Although the NSA targets "foreign signals intelligence information," its mission is to give the United States and its allies decision-making advantages "under all circumstances" in an ever-changing global environment. Its computer network operations specifically enable network warfare operations to defeat terrorists and their organizations at home and abroad.[67] The exposure of secret surveillance programs that monitored telecommunications under general warrants authorizing data gathering raised privacy concerns because they are so broad that they resemble fishing expeditions, a term that is used to describe inappropriate government efforts to search until they catch someone doing something illegal.

The National Security Court System

The Foreign Intelligence Surveillance Act of 1978 (FISA) created a special legislative court system (the Foreign Intelligence Surveillance Courts) to review applications to gather intelligence on foreign powers or agents of foreign powers. One of Congress's roles is legislative oversight of the executive branch. Congressional hearings on warrantless domestic and foreign surveillance programs resulted in passage of the FISA. The act authorized foreign surveillance, provided for judicial supervision of it, and required annual reports to Congress. The legal regime for national security makes it easier to get a search warrant for national security purposes than for criminal investigations. The government does not have to show probable cause to get the FISA court to approve an application for foreign surveillance. The government only has to show that the surveillance is relevant to a valid investigation.[68]

One of the lessons of the failure to prevent the terrorist attacks on 9/11 is the need to coordinate missions and connect databases. The integration of databases is a force multiplier, because connectivity increases the effectiveness of individual agency missions. Integration also makes informational privacy more vulnerable. Connectivity and mobility have made it surprisingly easy to determine with a high degree of confidence a person's identity, simply by examining a small number of patterns of anonymized mobility datasets that provide geospatial and temporal data but not a name, home address, telephone number, or other specific identifiers. Combining this data with medical records and voter records, for instance, allows detailed tracking of people without having to get a search warrant.[69]

Legal Globalization

A final development that merits some attention is legal globalization. Economic globalization has promoted international trade by integrating the national legal regimes governing trade. Legal globalization promotes international perspectives on law enforcement problems such as trafficking in drugs and other illegal goods and services; terrorism, including providing financial and other material support organizations that support terrorism; and compliance with tax, environmental, and human rights laws. Legal globalization is likely to further integrate U.S. domestic criminal law and international criminal law relating to search and seizure and the privacy protections for personal data. Globalization has affected economics, politics, and law by developing a body of international law and by coordinating national policies. Economic globalization promotes international trade. Political globalization has produced a body of international human rights and other law. Legal globalization promotes the development of international criminal law and the coordination of national efforts targeting transnational and international crime. The Department of State has an Office of Global Criminal Justice that deals with certain kinds of violent crime, and the FBI shares the Combined DNA Index System (CODIS) with domestic and "certain international crime laboratories."[70] The investigation of white-collar crimes presents special challenges for the protection of informational privacy. The ease with which electronic funds can move assets across national borders has prompted intergovernmental efforts to increase tax compliance and reduce tax avoidance by moving financial assets to countries with reputations as tax havens. These government efforts include coordinating national tax laws and monitoring financial transactions to detect and investigate potential criminal activity. Treaty-based law enforcement is part of the international web of laws.

Two developments in legal globalization of criminal justice are directly related to the Fourth Amendment: the first is increased coordination of U.S. criminal laws with the criminal laws of other countries; the second is the continued development of the body of international criminal law. The coordination of national laws will be less controversial than the development of international law. The coordination is more informal and less visible, and it can occur through nongovernmental organizations or intergovernmental organizations. Nongovernmental organizations, such as the International Association of Chiefs of Police, work for criminal and national security cooperation to bring drug kingpins or terrorists to justice. Intergovernmental organizations, such as The International Criminal Police Organization (INTERPOL) and The United Nations, the largest intergovernmental organization, work for international cooperation.

When police and surveillance agencies integrate their software systems and countries integrate their systems, the capacity for global eavesdropping will be greatly enhanced. But there is resistance to global law enforcement. The development of a body of international criminal law is politically controversial for some of the same reasons that the nationalization of the Bill of Rights—the process by which the provisions of the Bill of Rights were applied to the states—was controversial. Some states resisted the nationalization of crime policy. Today, critics of globalization worry that it will erode U.S. national sovereignty. Treaty-based law enforcement is more controversial than multinational or intergovernmental accords for international trade policy because the American states defend their sovereignty over criminal justice policy. Therefore, the future development of an international legal regime for criminal justice that includes international bodies with authority to make crime policy is less certain than is the continued development of an international legal regime for trade policy. Conservatives have been critical of the trend toward comparative and international law, particularly when judges cite or rely on foreign sources of law in cases dealing with issues such as the death penalty, privacy rights, sexual behavior, marriage, or the rights of children. On the other hand, advocates of informational privacy rights consider the further development of transnational legal processes as a way to bring U.S. standards up to the standards of other countries or the European Union.

INTEREST GROUPS AND INFORMATIONAL PRIVACY

The websites of all government agencies with law enforcement and national security missions officially state that they protect informational privacy rights. However, interest groups play an important role in putting informational privacy on the government's agenda. The American Civil Liberties Union advocates for the development of a legal regime for drones before they are widely used domestically. Without a legal framework or public policy addressing issues such as when warrants are required, image retention, and the use of armed drones, the individual police departments, border security agencies, and surveillance agencies will decide for themselves what practices or policies are acceptable without civilian input or oversight.

The Center for Democracy and Technology opposed the legislation that the FBI supported as a way to solving the going dark problem of going dark by requiring telecommunication service providers to ensure government access to data, including encrypted data. The center worried that legislation requiring telecommunication service providers to allow government access to communications records, by—among other things—decrypting encrypted files, would make Internet communications less secure.[71] The Future of Privacy Forum advocates for responsible data practices in the private and public sectors because the government gathers its own data (arrest records, fingerprint files, terrorist organizations) and gets data from the private sector. This means that the scope and quality of private-sector data is relevant to public policies about government access to private data, particularly as the Internet expands to include connections with smart devices ranging from phones and computers to cars, implanted medical devices, electric meters that alert the police about a home with unusually high electrical usage rates typical of marijuana grow houses, or even guns that provide a data stream every time the weapon is fired.

The website of the International Association of Privacy Professionals describes the association as committed to developing good data policies in the same way that organizations worked to develop industry standards for manufacturing throughout the world. Big data is currently used to analyze automobile- and airplane-repair patterns to make predictions about part failures, to make stock trading decisions, to make pricing predictions, and to make product recommendations for customers based on past purchase patterns.[72] The Electronic Privacy Information Center (EPIC) and The Electronic Frontier Foundation (EFF) provide information about and advocate for privacy in the digital age. The EFF's Surveillance Self-Defense Project provides information about state privacy laws and the Fourth Amendment.

CONCLUSIONS

Conservative dissatisfaction with the rules for search warrants and the exclusionary rule changed the Supreme Court's case law to reflect the values and policies of the crime-control model of justice. The current dissatisfaction with the state of search-and-seizure law focuses on two developments. The first is scientific and technological advances that have greatly increased the government's power to search without having to get warrants, primarily because the third-party doctrine leaves electronic communications outside the Fourth Amendment's system protecting informational privacy.

The second development is the application of the national security model of justice to domestic law enforcement. These two developments present a familiar challenge: how to accommodate change while remaining true to the original purpose of the Fourth Amendment. The participants in the debates about where to strike the balance between government power and individual rights include the government officials responsible for implementing Fourth Amendment privacy rights (the Court, Congress, and various executive branch agencies), interest groups, and the general public whose expectations of privacy play an important role in determining the direction of this very dynamic area of law. The search for a limiting principle that can clearly establish the boundaries of government power in the digital era illustrates how challenges create demand for change that maintains continuity with foundational values, thereby bringing life to the law.

NOTES

1. Akhil Reed Amar, "Fourth Amendment First Principles," *Harvard Law Review* 107 (1994): 757–759.

2. Norman M. Robertson, "Reason and the Fourth Amendment—the Burger Court and the Exclusionary Rule," *Fordham Law Review* 46 (1977):174.

3. Morgan Cloud, "Rube Goldberg Meets the Constitution: The Supreme Court, Technology and the Fourth Amendment," *Mississippi Law Journal* 72 (2002): 28–29.

4. In *Coolidge v. New Hampshire*, 403 U.S. 443, 490 (1971), Justice Harlan (concurring) noted that it was "apparent" that the law of search and seizure was "due for an overhauling."

5. "Fourth Amendment—Search and GPS Surveillance: *United States v. Jones*," *Harvard Law Review* 126 (2012): 226, 231.

6. Herbert L. Packer, *The Limits of the Criminal Sanction* (Stanford, CA: Stanford University Press, 1968).

7. James A. Morone, *Hellfire Nation* (New Haven: Yale University Press, 2003), 41. Louis Hartz, *The Liberal Tradition in America* (New York: Harcourt, Brace & World, 1955), 9, described the American "cult of constitution worship." In "The Living Constitution" (2010), David Strauss describes allegiance to the Constitution as central to what it means to be an American.

8. Donald P. Kommers and John E. Finn, *American Constitutional Law: Essays, Cases, and Comparative Notes* (Belmont, CA: Wadsworth Publishing Company, 1998), 3.

9. *Brigham City, Utah v. Stuart*, 547 U.S. 398, 403 (2006).

10. Helen A. S. Popkin commenting on proposed changes in Facebook's privacy policy, "You Have Until Friday to Comment on Facebook Changes," *NBC News*, March 19, 2012, accessed May 30, 2013. http://www.nbcnews.com/technology/you-have-until-friday-comment-facebook-changes-487695.

11. Ralph A. Rossum and G. Alan Tarr, *American Constitutional Law*. 3rd edition (New York: St. Martin's Press, 1991), 475–76.

12. *Olmstead v. U.S.*, 277 U.S. 438, 478 (1928).

13. Antonin Scalia and Bryan A. Garner, *Reading Law: The Interpretation of Legal Texts* (St. Paul: West Thomson, 2012).

14. Thomas K. Clancy, "The Irrelevancy of the Fourth Amendment in the Roberts Court," *Chicago-Kent Law Review* 84 (2010): 191–208.

15. Clancy, "The Irrelevancy of the Fourth Amendment in the Roberts Court," 193.

16. Ibid., 195–96.

17. *Terry v. Ohio*, 392 U.S. 1 (1968).

18. The special needs doctrine was established in *Skinner v. Railway Labor Executives' Association*, 489 U.S. 602 (1989 and *National Treasury Employees Union v. Raab*, 489 U.S. 656 (1989).

19. *Boyd v. U.S.*, 116 U.S. 616, 630 (1886).

20. *Boyd*, 116 U.S. at 627–28.

21. *Kyllo v. U.S.*, 533 U.S. 27 (2001).

22. *Stanford v. Texas*, 379 U.S. 476 (1965).

23. *U.S. v. Jones*, 565 U.S. ___ (2012).

24. *U.S. v. Jones*, 565 U.S. ___, 2–3 (2012).

25. *Jones*, 565 U.S. at 7.

26. *Jones*, 565 U.S. at 10.

27. Alito citing "The End of Privacy," NPR, accessed August 15, 2013. http://www.npr.org/series/114250076/endofprivacy.

28. *Jones*, 565 U.S. at 13.

29. Samuel Alito, "Final Report on the Conference on the Boundaries of Privacy in American Society," Princeton University Woodrow Wilson School of Public and International Affairs, Princeton, NJ. January 4, 1972, accessed August 19, 2013, http://epic.org/privacy/justices/alito/princeton/3.pdf.

30. *Jones*, 565 U.S. at 957.

31. Daniel J. Solove, *The Digital Person: Technology and Privacy in the Information Age* (New York: New York University Press, 2004); and *Nothing to Hide: The False Tradeoff between Privacy and Security* (New Haven: Yale University Press, 2011).

32. *California v. Ciraolo*, 476 U.S. 207 (1986); *Florida v. Riley*, 488 U.S. 445 (1989).

33. *U.S. v. Dunn*, 480 U.S. 294 (1987).

34. *U.S. v. Jones*, 565 U.S. ___ (2012); *Florida v. Jardines*, 569 U.S. ___ (2013).

35. *Florence v. Board of Chosen Freeholders of the County of Burlington*, 566 U.S. ___ (2012).

36. *Maryland v. King*, 569 U.S. ___ (2013).

37. Accessed June 13, 2013. https://www.cia.gov/news-information/featured-story-archive/2013-featured-story-archive/agency-k-9-unit-places-first-in-competition.html.

38. http://www.justice.gov/jmd/ls/legislative_histories/pl99-508/pl99-508.html.

39. Yale Kamisar, "How Rehnquist Spared the Landmark Confession Case, But Weakened Its Impact," *ABA Journal*, June 23, 2006, accessed June 30, 2013. http://www.abajournal.com/magazine/article/mirandas_reprieve/.
40. *Dickerson v. U.S.*, 530 U.S. 428 (2000).
41. Robertson, "Reason and the Fourth Amendment—the Burger Court and the Exclusionary Rule," 158–9.
42. Ibid., 174–5.
43. *Bivens v. Six Unknown Federal Narcotics Agents*, 403 U.S. 388, 419 (1971).
44. *Hudson v. Michigan*, 547 U.S. 586, 597 (2006).
45. *Herring v. U.S.*, 555 U.S. 135 (2009).
46. *Stone v. Powell*, 428 U.S. 465 (1976).
47. See http://www.fbi.gov/about-us/cjis/ucr/ucr.
48. *The Police Chief*, accessed May 22, 2013, http://www.policechiefmagazine.org/magazine/index.cfm?fuseaction=display&article_id=998&issue_id=92006.
49. *California v. Hodari D.*, 499 U.S. 621 (1991).
50. *Whren v. U.S.*, 517 U.S. 806 (1996).
51. The NYPD Intelligence Division & Counter-terrorism Bureau, accessed July 25, 2013, http://www.nypdintelligence.com/.
52. Charlie Beck and Colleen McCue, "Predictive Policing: What We Can Learn from Wal-Mart about Fighting Crime in a Recession," *The Police Chief*, accessed August 20, 2013. http://www.policechiefmagazine.org/magazine/index.cfm?fuseaction=display_arch&article_id=1942&issue_id=112009.
53. *Minority Report*, DVD, directed by Steven Spielberg (Los Angeles, CA: 20th Century Fox and Dreamworks Pictures, 2002).
54. Raine, *The Anatomy of Violence: The Biological Roots of Crime*.
55. FBI website accessed June 25, 2013, http://www.fbi.gov/about-us.
56. http://www.biometriccoe.gov/.
57. https://www.fbijobs.gov/121.asp.
58. http://www.fbi.gov/about-us/cjis/fingerprints_biometrics/ngi.
59. Charlie Savage, "Facial Scanning Is Making Gains in Surveillance," *The New York Times*, August 21, 2013, accessed August 21, 2013, http://www.nytimes.com/2013/08/21/us/facial-scanning-is-making-gains-in-surveillance.html?pagewanted=all.
60. FBI Criminal Justice Information Services Division, http://www.fbi.gov/about-us/cjis.
61. For information about National Security Letters, see the Electronic Privacy Information Center website: http://epic.org/privacy/nsl/.
62. Website accessed May 7, 2013. http://www.fbi.gov/news/testimony/going-dark-lawful-electronic-surveillance-in-the-face-of-new-technologies
63. Andrew Weissmann, "New Technology, National Security, and the Law," Address to the American Bar Association, Standing Committee on Law and National Security, May 20, 2013, accessed August 19, 2013. http://www.c-spanvideo.org/program/NewTechnolo.

64. The Police Executive Research Forum, "How Are Innovations in Technology Transforming Policing?" January 2012, accessed July 30, 2013, http://policeforum.org/library/critical-issues-in-policing-series/Technology_web2.pdf.

65. The text of the PATRIOT Act and the Department of Justice description of it are available at the DOJ website, Accessed June 5, 2013. http://www.justice.gov/archive/ll/highlights.htm.

66. A recording of the July 9, 2013, PCLOB workshop is available at http://www.c-spanvideo.org/program/BoardMo.

67. Accessed September 21, 2013, http://www.nsa.gov/about/mission/index.shtml.

68. http://uscode.house.gov/view.xhtml?path=/prelim@title50/chapter36&edition=prelim.

69. Yves-Alexandre de Montjoye, Cesar A. Hidalgo, Michael Verleysen, and Vincent D. Blondel, "Unique in the Crowd: The Privacy Bounds of Human Mobility," *Nature*, March 25, 2013, accessed August 21, 2013, http://www.nature.com/srep/2013/130325/srep01376/full/srep01376.html.

70. FBI website, accessed August 22, 2013, http://www.fbi.gov/about-us/lab/biometric-analysis/codis.

71. "FBI Seeks New Mandates on communications Technology," accessed September 18, 2013, https://www.cdt.org/policy/fbi-seeks-new-mandates-communications-technologies.

72. Justin Brookman, "Privacy in a World of Persistent Surveillance," *Privacy Perspectives International Association of Privacy Professionals*, June 19, 2013, accessed August 15, 2013, https://www.privacyassociation.org/privacy_perspectives/post/privacy_in_a_world_of_persistent_surveillance.

13

The Changing Expectations of Privacy in the Digital Age

Meghan E. Leonard

INTRODUCTION

The history of the right to privacy surely dates back to the founders, in their room in Philadelphia in 1787. To be certain, the founders were concerned with how individual citizens might be protected from a large national government. When the Bill of Rights was added to the Constitution, underlying at least the First, Third, Fourth, Fifth, and Ninth Amendments was the right of citizens to some level of privacy. Though this was the context, the founders did not go so far as to grant an explicit right to privacy in the Constitution; they merely hinted at it. Thus, over time, the federal Supreme Courts, Congress, and the citizens have had to determine just how much privacy to grant and to what extent the government might be able to intrude on that privacy.

This intellectual debate over the right to privacy of the citizens has played out over time, but complicating the matter are the ever-changing ways in which someone's privacy can be intruded upon by the government or other institutions. As technology has developed from postal mail, to the telephone, computers, the Internet, GPS tracking, and even smartphones, information about individual citizens is both ubiquitous and easier to obtain. This matters because of the ease in which information can be collected. As Justice Alito puts it:

> In the pre-computer age, the greatest protections of privacy were neither constitutional nor statutory, but practical. Traditional surveillance for any extended period of time was difficult and costly and therefore rarely undertaken. The surveillance at issue in this case—constant monitoring of the location of a vehicle for four weeks—would have

required a large team of agents, multiple vehicles, and perhaps aerial assistance. Only an investigation of unusual importance could have justified such an expenditure of law enforcement resources. Devices like the one used in the present case, however, make long-term monitoring relatively easy and cheap . . . (concurring in *United States v. Jones* 565 U.S. __ (2012))

In what contexts should these issues be of concern to American citizens? If information gathering is easier for the government and law enforcement, is that not better for solving crimes or predicting terrorist plots? Isn't easy a good thing in this sense? As the debate over the collection of mass amounts of information by the National Security Administration was reignited in 2013, many people argued that citizens who have not done anything wrong, who have not violated any laws, should have nothing to worry about. That the ability to collect this information in a digital context exists is a good thing for the ease of investigation. Indeed, examples of beneficial uses of these technologies are certain. For example, investigators can track missing persons by the location features on their cell phones. But should people, even those who are not doing anything wrong or are not breaking the law, be concerned? That is an important debate underlying this entire chapter.

While the ability to violate one's privacy is certainly easier, is it not also possible that citizens are more willing to give away their privacy? Many people post personal information on sites like Facebook, Twitter, and LinkedIn daily. How much information should be available, and what are citizens' expectations about this information? How have these debates changed in this current and ever-changing digital age? In this chapter, we will address these questions. To do so, we will first trace our historic expectations of privacy as they relate to changing technologies. We will then address how the expectations of privacy have changed in relation to the Internet and other digital technologies including social networks and smartphones. In the final section, we will review how these changes affect the use of these tools by law enforcement and the federal government to collect information. We will address social networks, smartphones, and digital evidence as examples.

THE BEGINNINGS OF THE AMERICAN RIGHT TO PRIVACY

While the technologies we are concerned with in this chapter might be new, the debate over an individual's right to privacy is not. The Fourth Amendment states that citizens have a right "to be secure in their persons, houses, papers, and effects, against unreasonable searches and seizures."

The Fourth Amendment's relation to the right to privacy, which was not codified until much later, was first outlined in a 1890 article in the *Harvard Law Review*, where Louis Brandeis and Samuel Warren argue that a person has the "right to be let alone." Seemingly foreshadowing current debates,[1] Warren and Brandeis argue, "instantaneous photographs and newspaper enterprise have invaded the scared precincts of private and domestic life; and numerous mechanical devices threaten to make good the prediction that 'what is whispered in the closet shall be proclaimed from the house-tops.'"[2] While we still grapple with these topics, one must wonder what this means in a society where some people might want their information to stay in the proverbial closet, but others seemingly shout everything from the rooftops of Facebook and Twitter.

Initially, the Supreme Court would not agree with Brandeis and Warren. Over the then Justice Brandeis's dissent, the Court ruled in *Olmstead v. United States* (1928) that the Fourth or Fifth amendments did not protect citizens from wiretapping by law enforcement, because there was no physical trespass on a person's property.[3] While one's expectation of privacy would not be adopted by the court for a number of years, as Lepore argues, Brandeis continued his arguments about privacy and his foreshadowing in his dissent.[4] He wrote, "Ways may someday be developed by which the government, without removing papers from secret drawers, can reproduce them in court, and by which it will be enabled to expose a jury to the most intimate occurrences of the home."[5] Indeed, 85 years before the country would engage in a national debate over the National Security Agency's data-gathering efforts and whether or not we should consider Edward Snowden a national hero or traitor, questions related to these debates were raised but left to the future to answer. In the rest of this chapter, we will examine how the Supreme Court and others have or have not answered these questions.

One point it might be helpful to make is to answer the question proposed by Alva Noe: "[w]hat's the big deal about privacy?"[6] Noe continues, "Privacy is the state of being unobserved."[7] Privacy is the idea that we can live autonomously and anonymously. It is the idea that we can exist in a way in which others do not know what we are doing. Surely, when Brandeis and Warren were writing their famous article, this was in some ways possible. An individual could commute from one place to another without being caught on camera. A person could communicate via letter without concern that the government would store the information in a massive database. Surely, technology has made our lives better. It has improved medicine, travel, education, and communication. But, has it eliminated our privacy? As Noe continues, "[t]hese days we keep our diaries in public. We've replaced 'Dear Diary' with 'hey, fb friends!' We're less Anne Frank than we are PT Barnum, presenting our lives online and in real-time. Each of us runs a media empire devoted to our own exhibition. Millions of us, at minimum, are the authors of fan magazines devoted to ourselves. And that powerful, vain impulse to broadcast ourselves is just the tip of the iceberg. We use phones that literally map our every move. Our credit cards leave a permanent and transmittable record of our every purchase. And you can't walk down the street, or drive anywhere, without being photographically recorded."[8] So, what does this mean for privacy? Can it exist anymore? Surely it is still possible. Indeed, not everyone has an online presence. But what does it mean when people argue that privacy is necessary for democracy? We will address this big question as we examine privacy's relation to our constitutional rights. We will do so keeping in mind, as Noe continues, "[i]f we really value privacy—if, for example, we really believe that being unobserved is necessary to securing our freedom in a democracy—then why are so few of us bothering to pull down the shades and lock the door?"[9]

EXPECTATIONS OF PRIVACY AND THE FOURTH AMENDMENT

Though the beginnings of the digital age would not be upon us for at least another decade, the central tenants to how we should think about our constitutional rights to privacy were set out in the 1960s. Driven in part by the appointment of Earl Warren to Chief Justice, the Court would come to take on a much more expansive view of the rights of individuals, especially those accused of crimes. Though not a criminal case, the Supreme Court first found a formal right to privacy in the penumbras of the Constitution in 1965 in *Griswold v. Connecticut*.[10] Yet, two years later, the Court would make their landmark decision in how privacy should be balanced with law enforcement's need to investigate a crime. Like *Olmstead*, *Katz v. United States*[11] involved the use of warrantless wiretapping by law enforcement. This time, the Court found that an individual's privacy was violated by this device. In *Katz*, the majority outlined a two-part test for determining if a person has a reasonable expectation of privacy: "(1) a person must have an actual, subjective expectation of privacy, and (2) the expectation of privacy must be 'one that society is prepared to recognize as 'reasonable' given societal expectations."[12] *Katz* is first important for its outline of an individual's expectation of privacy. It is secondly important because it protects not just a physical person, but an individual even if there was no "physical penetration."[13] Thus, if a person has a reasonable expectation of privacy, the ability for law enforcement to search becomes more difficult and requires a warrant. However, the obvious question *Katz* would pose is what a reasonable expectation of privacy might be.

Is something no longer private once you have shared it with another person? What about if you searched for it in Google? There is no definitive answer to these questions, but we will explore how they have been addressed under certain circumstances and what is still left to be answered.

Let's take two more examples dealt with by the Supreme Court, and then we will apply those principles to more complex digital or informational technologies. The first example comes from a 2001 case, *Kyllo v. United States*.[14] In this case, investigators used a thermal-imaging device to determine the level of heat loss occurring in a home. They believed one of the occupants was growing marijuana, a process that requires significant heat lamps. The thermal scan was done off the property of Kyllo. Significant heat loss was detected and used to gather other evidence. Kyllo was arrested. Applying *Katz*, the Court found that even without going into the home, the officers conducted a formal search of the home and therefore needed a warrant to conduct such a search. To expect privacy inside one's home, Justice Scalia wrote, was most important and absolutely fell within any reasonable expectations. But, one's home is the area in which we most expect privacy. What about a car? Does the same reasonable expectation of privacy extend to one's vehicle?

In the case of *United States v. Jones*, officers attached a (technically) warrantless GPS device to the car of Antoine Jones.[15] The police had requested and were granted a warrant to place it on the car of Jones during an investigation for drug trafficking. However, the warrant included a 10-day period in which the GPS device could be placed. It was placed on Jones's car on the 11th day. As a result, the search was outside the scope of the warrant but continued. The GPS was on the car for 28 days and tracked the movements of Jones, ultimately gaining as many as 2,000 pages of data on Jones's whereabouts. Combined with other evidence, Jones was found guilty and sentenced to life in prison for drug and other charges. The concurring opinions of Justice Sotomayor and Alito in the Jones case made it very clear that in applying the *Katz* standard, Jones's expectation of privacy had been violated.

What can be learned from the *Katz*, *Kyllo*, and *Jones* cases? One, related to law enforcement's ability to search, the physical penetration or trespass rule from *Olmstead* no longer stands. Technological changes made it so that there was no need to go into a person's house or follow an individual's car to engage in a search. Thus, the Court has made it clear that no formal trespass may be necessary to alter one's expectation of privacy. These cases have set the standard for how an individual's expectation of privacy has changed in the eyes of Constitutional law from 1928 through 2012. Yet, the technologies at issue in these cases, wiretapping, thermal imaging, and GPS, are relatively simple in their use as law enforcement tools. The Supreme Court has not answered questions about our expectations of privacy as related to digital technologies.

When they ultimately do address questions related to privacy and digital technologies, it may be one of the most historically significant decisions they ever make. In the next sections, we will address how digital technology has changed and what this might mean for our future Fourth Amendment rights and our expectations of privacy.

There is, of course, one other common thread to the *Katz*, *Kyllo*, and *Jones* cases—all of the searches in question were conducted without a warrant. So, in dealing with questions related to the Fourth Amendment and searches of new technologies, the questions go beyond what and how law enforcement or the government can search to what a warrant might cover or require. One such complication is the plain-view doctrine. The best way to describe this doctrine is through a hypothetical scenario. Say law enforcement has a warrant to conduct a search of your house for illegally owned guns. But you leave drug paraphernalia out on your coffee table, in plain view, so that any reasonable person walking through your home might see that. While the drug evidence was not related to the warrant, it is argued that it can be collected and used in a court because it was in plain view to law enforcement. But the translation of this scenario to a search of a computer is not so clear. If there is a warrant to search one's computer, and files are found that are not within the scope of the warrant, are these files permissible as evidence? Are they equivalent to the drug paraphernalia in plain view? Some lower federal courts have equated this evidence with the plain-view doctrine; others have been more hesitant to do so.[16] Thus, in understanding the complexities of our expectations of privacy as related to digital technologies, we must consider not only warrantless searches and their legality but also searches conducted with a warrant and how much information might be collected in these searches.

CHANGING TECHNOLOGY AND THE LAW

For today's students, it may be hard if not impossible to remember life before the Internet. The beginning of the transitions from the industrial to the digital or informational age occurred most formally in the 1970s and 1980s. Although the origins of what would become the Internet date back much earlier, we will focus here on the availability of these technologies to a majority of Americans.[17] The linking of computers by network began as early as the 1950s, when defense researchers invented ways to send messages from one computer to another. From there, this networking evolved to sending messages to computers connected via wires, to wireless connections. When it was clear that computers could share information, the bones of what we now know as e-mail were developed. Soon, it was determined that many computers could be connected, and without wires. Ultimately, the development of the Internet took place, with people across the world being able to communicate with one another in an instant.

By the 1990s, the Internet would be commonplace, with people accessing information via the web and sending information via e-mail. Soon after, people were posting on blogs and Myspace and then Facebook and Twitter. Individuals were connected to one another like never before, but information was being stored and collected at the same rapid pace. What was not moving at this same rapid pace was the law. While there are necessarily laws behind the technologies, what is surprising is just how few laws have been passed to deal with this rapidly changing technology and individual's rights.

Electronic Communications Privacy Act

The central statute that deals with these digital-age complications is the Electronic Communications Privacy Act (ECPA) of 1986. When the ECPA was written and passed, it helped protect individuals by developing standards for the monitoring of cell phone conversations and Internet communications. The Center for Democracy and Technology provides a very helpful brief history on the ECPA and what led to its adoption.[18] After the Katz decision, Congress passed a crime bill that made it a crime to intercept phone calls without a warrant, except for under very specific circumstances. But by the 1980s, communications were more likely to be wireless, via cell phones or e-mail. Because previous law covered only wired communications, it was unclear how these new technologies were to be dealt with. The ECPA added both wireless and digital communications to the protected categories.

One significant problem that arose in updating the law to deal with technologies was how to deal with the information that was stored as a result of things like e-mail. In a compromise, and given the then limitations on data storage, Congress decided that after 180 days, e-mail would no longer be protected by the same warrant standards and could be accessed with a subpoena without judicial approval.[19] It goes without saying that much has changed in the world of technology since 1986. Without listing all of the changes, one that is seemingly most important is the storage of data. Terabytes of information can be stored in devices smaller than a cell phone. A terabyte worth of information equates to millions of document pages, thousands of songs, and hundreds of video files. In today's context, searching one's digital storage is like searching that person's "photo albums, stereos, telephones, desktops, file cabinets, waste paper baskets, televisions . . . postal services, playgrounds, jukeboxes, dating services, movie theaters, daily planners, shopping mall, personal secretaries, virtual diaries, and more."[20] Yet laws protecting this information have not been updated since the last time the New York Mets won the World Series and *Alf* was one of the most popular shows on television.[21]

Even though questions still remain as to how Congress might go about updating the ECPA, individuals and groups across the political spectrum agree that something must be done, and soon. Groups such as the American Civil Liberties Union (ACLU)[22] and Digital Due Process have outlined sets of principles that should guide Congress in dealing with the ever-changing technologies. For example, while digital storage has increased, many people are now moving toward cloud storage and computing. In other words, changes are necessary but must be flexible to new technologies.[23] Making changes and updating the law might be possible, but given that it necessitates some prediction of the technological future, this task will be very challenging.

We know where the statutes on digital technologies stand. But we need to know, given specific types of information technology, what is being done, how courts are ruling, and what discussions are being had. In the next section we will find out where our expectations of privacy stand as related to the Internet and social networking as well as smartphones and location based applications. Finally, we will discuss digital storage and the expectation of privacy.

THE INTERNET AND SOCIAL NETWORKING

In her law-review note, Lindsay Feuer provides a helpful historical overview of social networking as it relates to the Fourth Amendment and privacy.[24] Many students will be familiar with social networking sites and tools, and most will even use them in their everyday life. The most common of these cited by the literature are Facebook, Twitter, and Foursquare; though, to be sure, these may even be out of date by the time you are reading this chapter. Social networks have been defined by boyd and Ellison as "services that allow individuals to 1) construct a public or semi-public profile within a bounded system, 2) articulate a list of other users with whom they share a connection, and 3) view and traverse their list of connections and those made by others within the system."[25] This technology allows individuals to connect with other people and express their views. More lofty explanations of these sites focus on their ability to connect people and to efficiently share information. These websites launched in the early 2000s. For example, MySpace launched in 2003, LinkedIn in 2003, Facebook in 2004, and Twitter in 2006.[26]

Facebook and the New Social Norm of Sharing

As most students will be passingly familiar with this technology, I will provide only a brief introduction to how these technologies are used as they relate to an individual's expectation of privacy. We will focus on the most popular of these sites, Facebook, although the generalities are applicable to other similar sites. Individuals join a site, and after reading privacy policies,[27] they develop a profile for others to see. One can adjust his or her profile, adding as little or as much information as that person sees fit. Users can communicate with others through a variety of methods, such as status updates, information on their profile, chat features, direct messages, "poking," posting pictures or videos, joining or creating groups, playing games, "liking" pages, and others.[28] The content of this information ranges from the mundane, to political opinions, personal events, and even sometimes illegal activity. Indeed, in 2013, a man confessed to murdering his wife on Facebook and even posted a picture of her deceased body.[29] The information shared on social networking sites is varied, and individuals can post as much or as little information as they prefer.

Posting information about one's personal life on social networking sites is part of an ever-changing social atmosphere, where sharing this information has become a norm. Feuer summarizes the view of the founder of Facebook, Mark Zuckerberg:

> [A]ccording to Zuckerberg, "people have really gotten comfortable not only sharing more information and different kinds, but more openly and with more people. The social norm is just something that has evolved over time." Like a noisy cafeteria at lunchtime, or a playground at recess, "Facebook is a place of indiscriminate musings and minutiae, where people report their every thought, mood, hiccup, cappuccino, increased reps at the gym or switch to a new brand of toothpaste." This "social norm" demonstrates the difficulties in maintaining any expectation of privacy.[30]

This norm of sharing a lot of personal information has changed how we interact with people and the outside world. No longer does law enforcement need a GPS, wiretap, or thermal imaging to know what is going on in many people's lives. Facebook and other social-networking sites make it easy for an individual to share as much or as little information about themselves as they might like. This information might be shared with just a certain group of people, all of the users' friends, or the entire Facebook community, and because some Facebook information is searchable through search engines, this information could be made available to anyone with access to the Internet. This then "demonstrates the difficulty of ensuring that anything is truly 'private' when posted on Facebook."[31]

The first response a person might have to this discussion is about the privacy settings available on Facebook and other social-networking sites. To be certain, there are changeable settings, where users can limit who sees what information they post. But, these are also complicated. Knowing exactly which settings to choose and how to best protect your privacy on Facebook is difficult for even the most adept of users. For the everyday or less-aware user, adjusting the privacy settings to one in which the user is most comfortable might be difficult. In addition, the privacy setting options change[32] frequently, as does the Facebook interface.[33] The newest update to Facebook that has raised concerns about privacy is the graph search function. "Facebook's Graph Search function is kind of like a regular search function, only more complicated. But the bottom line is that it indexes everyone's public posts, likes, photos, interests, etc. to make them as easy as possible for everyone else—from friends to exes to cops to advertisers to your boss—to find."[34] In other words, Facebook is moving toward getting the information a person posts pushed out to more people, rather than fewer.

Facebook and other social-networking sites remind users of the privacy risks when creating an account. Not only does Facebook (and other sites) warn that this information is public, if an individual is concerned about it, he or she can deactivate their account; within their privacy policy statements, these sites make clear the other ways in which one's information is being used. Facebook itself collects and stores all of the information, pictures, status updates, GPS locations, and other information posted by users. Feuer continues, "Specifically, Facebook warns its users to always think before they post because just like anything else on the Internet, information shared on Facebook may be shared with others."[35] But how can this be reconciled with the sharing of information as the "new norm," as the Facebook founder himself pointed out? In other words, given this norm and the privacy warnings provided by the sites themselves, do we have a reasonable expectation of privacy under the *Katz* decision in our social-network communications?

Knowingly Exposed, Plain-View Doctrine, and Social Networks

The first obvious consideration when dealing with whether information a person posts on social networks is subject to Fourth Amendment protections is whether the person "knowingly exposed" this information to the public. Writing for the majority in *Katz*, Justice Stewart states, "What a person knowingly exposes to the public, even in his own home or office, is not a subject of Fourth Amendment protection."[36] The question, not yet answered by the Courts, is whether simply posting on a social-network website meets this standard. Schmidt continues:

> Because of the privacy guidelines disclosed on each social networking website, location check-ins may be considered knowingly exposed to the public; to join a social network, the user must respond to the standard privacy policy and accept the terms-of-use agreement. A user may argue that her social networking use is not aimed toward the public dissemination of personal information, but that user posts with the hope that the community will see the information. A user posts with the understanding that the information put on social networking websites will be broadcast to the world and, thus, knowingly exposed. Therefore, under the Katz framework, it seems that the average social networking website user will not receive the benefit of the Fourth Amendment's protections for her online social networking activity.[37]

Schmidt clearly argues that social-network postings or location check-ins would fail the knowingly exposed test and therefore would not be subject to Fourth Amendment protections. Indeed, she argues, it is the very nature of social media posting that one is exposing this information to the world. People post because they want others to read the information.

One of the few cases in which the Supreme Court did address newer technologies and privacy was *Ontario v. Quon*,[38] as Schmidt highlights. In this case, text messages sent from government-issued pagers were searched when supervisors suspected the pagers were being used inappropriately. The owners of the pagers were warned that their communication would be monitored, and it was made explicitly clear to them that their usage would be logged. As a result, the Court did not agree that the officers had an expectation of privacy, and therefore their Fourth Amendment protections did not apply. Thus, the ruling was simplified. While the court made no overarching ruling, they were well aware of the implications of new technologies on individual expectations of privacy. As Schmidt notes, "The changing role of technology and communication implies the need for a reevaluation of policy and a continuing vigilance on the part of social networking users."[39] In addition, it is clear that employers must be explicit in their policies about which employee electronic communication is monitored and stored. In any event, this case will be relevant as the Supreme Court takes on these issues when the policy stated by the employer is not as explicit or not stated at all.

A close parallel to the knowingly exposed doctrine is the plain-view doctrine mentioned earlier. Let's compare two cases about growing drugs to see how they might parallel to posting on social-networking sites. Recall the case of *Kyllo* described earlier. Investigators in this case used thermal imaging to determine if a house was giving off an unusual amount of heat. This increased use of heat would be one indicator that the resident was growing marijuana. The Supreme Court determined this to be a search under the Fourth Amendment, one that would require a warrant. In our home, we have an expectation of privacy, and therefore it cannot be entered—in this case, through the images—without a warrant or consent. Eleven years earlier, the Supreme Court took up the case *Florida v. Riley*.[40] In this case, there was a flyover of Riley's property by a sheriff investigating based on a tip that Riley was growing marijuana. As in *Kyllo*, there was no search warrant. The officer, from 400 feet in the air, found drugs. The majority found that this was not a search, and unlike in the *Kyllo* case, did not require a warrant. As Schmidt describes, "[a]ccording to the Court, flying at such an altitude was not contrary to any law, so any member of the public also could have potentially flown that close to Riley's greenhouse and observed his drug operation."[41] Note the difference here. Any person could have seen the drugs growing in Riley's yard, while reasonable people cannot see through the walls of Kyllo's home, and therefore could not see the drugs in that case. As a result, Kyllo was protected under the Fourth Amendment, while Riley was not.

The question that will ultimately be posed to the Courts will be whether social-network postings, check-ins, or location indicators are more similar to the case of *Kyllo* or *Riley*. While it's quite likely that any case will not be phrased in that way, the parallels are worth considering. When a person posts online, is he or she they keeping that information in their house, unexposed to any reasonable member of the public? Or are they metaphorically growing that information in their backyard, where any reasonable member of the public would be able to access the information with a little effort. No one knows how any court will rule on this issue. But many argue that while the *Kyllo* parallel might be most beneficial to the public, the *Riley* argument seemingly is more reasonable, as individuals are warned by the privacy policies and that what they post will be stored by these sites and could possibly be found in an Internet search. The motivations of a social-network post are to expose others to this information. We take up this discussion in the next section. What protection does an individual have once the information they possess is given to another person?

The Third-Party Doctrine and Social Networks

In an article in the Boston College Law Review, Monu Bedi explains in great detail how the third-party doctrine might affect communications and the expectation of privacy in social networks or Internet communications more broadly.[42] The Supreme Court outlined the third party doctrine in *United States v. Miller*.[43] In this case, the Bureau of Alcohol, Tobacco, and Firearms issued subpoenas for Miller's banks when he had been charged with carrying alcohol-distilling equipment on which no liquor tax had been paid. The banks complied with the requests and provided the information to the ATF. After Miller was convicted, he argued that the information received from the banks was in violation of his Fourth Amendment rights. The Supreme Court disagreed with Miller. They argued that because he shared the information with the bank, a third party, he no longer had an expectation of privacy related to that information.[44] The third-party doctrine, thus, refers to the fact that once you share information with a third party, you no longer have a reasonable expectation of privacy over that information, and it is no longer subject to Fourth Amendment protections.

In 1979, the Supreme Court expanded the third-party doctrine to information disclosed to automated machines. In *Smith v. Maryland*, the Court found that this information, in this case telephone numbers recorded by a pen register, was not subject to Fourth Amendment protections because one no longer has a reasonable expectation of privacy over this information.[45] As Bedi argues, this is the closest the Supreme Court has come to indicating whether information transmitted over the Internet would be subject to Fourth Amendment protections. It is argued that because Internet communications are disclosed through ISPs, similar to the storing of phone numbers in *Smith*, this communication would not be protected, due to the third-party doctrine.[46] If the information stored online through an ISP is not subject to Fourth Amendment protections, what does that mean for social networking? Are all social-networking posts automatically denied Fourth Amendment protections? What about the fact that so much communication between people happens over the Internet? Could this possibly mean that no expectation of privacy exists in any Internet communication? This seems like an extreme possibility, and while we wait for a precedential decision regarding such from the courts, scholars will debate these issues. Indeed, one possibility highlights parallels between Facebook communication and communication between individuals within a relationship context.

Monu Bedi links Facebook communication not to that of the technological cases and Fourth Amendment expectations of privacy but to interpersonal privacy. In 1965, the Supreme Court found a right to privacy in the Constitution in *Griswold*. They later upheld this right in the abortion cases of *Roe* and *Casey v. Planned Parenthood*. In 2003, this interpersonal privacy right was further expanded to protect intimate relationships from government intrusion. Of course, this raises the question of how, under the concept of interpersonal privacy, relationships can be protected, but under the third-party doctrine, communication within these relationships is not guaranteed Fourth Amendment protections?[47] Bedi argues that the ISP is simply providing the means for conducting interpersonal relationships. "These service providers facilitate communications between individuals much like landlords or other providers, but there is an important difference. Landlords, maids, and other service individuals play a role—albeit a small one—in the relationship between the two individuals for whom they facilitate communication. They may not have a stake or preferred outcome in the matter. Nevertheless, they are a part of the resulting relationship in that they form a mini-relationship (perhaps not to the same depth or quality) with the sender and/or the recipient. In other words, they are not *merely* conduits that transmit information. This is very different from ISPs, which play no comparable role."[48] Bedi therefore concludes that if we look to case law related to interpersonal privacy, rather than related to the third-party doctrine, we might find a way to grant Fourth Amendment protections to Facebook and other social-networking communications.

To be certain, it is unclear what expectation of privacy one might have in his or her social-networking communications. There is no way to tell how the Supreme Court will eventually balance the plain-view doctrine, the third-party doctrine, and the concept of interpersonal privacy. In addition, given the new norm of sharing personal information without regard to who might see it, there will be significant legal debate on how and when each of these concepts might apply. Further, there is seemingly no time in the very near future in which the Supreme Court might take up a case relevant to these issues. Congress may be able to make changes sooner than a case would reach the court, deciding if and when social-networking interactions might be protected. But, beyond Facebook and Twitter, there are other issues related to technology and our expectations of privacy that we can examine to gain further insight. In the next two sections, we address these.

SMARTPHONES AND LOCATION-BASED APPLICATIONS

If Facebook is the most ubiquitous technological innovation challenging our expectations of privacy, the smartphone is following this same path. Smartphones have more significant capabilities than a traditional cellular phone. These smartphones have operating systems not unlike those of a computer. Their capabilities vary, but many can connect to the Internet, and they provide applications, or apps, that allow users to play games, download music, find out the current weather, or access their banking information, for example. As of 2012, there were 172 million smartphone users in the United

States, making up more than half of all cell phone users.[49] Almost all smartphones and many regular cell phones come equipped with GPS technology. This allows the phone to be tracked and located using the GPS satellites. Tracking using GPS can be done in many different ways. We will discuss two and what they mean for our expectation of privacy. The first is when law enforcement or the government uses their technological capabilities, such as stingrays, to track individuals. Secondly, we will discuss when individuals themselves post location information, such as through sites like Foursquare or through checking in to a location on Facebook.

Stingray Devices and GPS Tracking

The Supreme Court took up the issue of GPS tracking in the *Jones* case mentioned above. Recall that the Court decided that a warrantless GPS device used to follow a suspect was a violation of Jones's Fourth Amendment rights. The obvious intrusion here was placing a device on the car of Jones. But what if law enforcement had the capability to track individuals through the GPS locators in their cell phones or smartphones? Brittany Hampton, in an article in the *University of Louisville Law Review*, takes up this question. She contemplates the stingray. "Stingrays trick cell phones into connecting to a fictitious tower in order to locate the person using the cell phone. This provides law enforcement with an efficient means of obtaining valuable information regarding the location of an individual."[50] More specific information about the inner workings of the stingray technology is not well known outside of the government. However, there are generally three component parts: "an antenna, a computer with mapping software, and a special device. This device mimics a cell phone tower, and tricks the phone into connecting to it."[51] When the phone connects to the device, the device is able to locate the phone, even when it's not in use, and view hardware numbers associated with the phone. This can be done either by aiming the antenna at a certain place to get information on the phones located there or by using numbers associated with a phone to locate the phone specifically.[52]

These stingray devices are most often used without a warrant. Though there are cases pending in the lower courts regarding individuals' Fourth Amendment protections and expectations of privacy as related to stingrays, there is as of yet no precedent on these types of devices. While *Jones* will certainly be an important case in determining how law enforcement must use stingrays, there are two other cases in the *Katz* vein that might be helpful in parsing out the Fourth Amendment protections as related to stingray and stingray-type GPS searches. As Hampton notes, *United States v. Knotts*[53] and *United States*

v. Karo[54] are particularly relevant here.[55] In Knotts, law enforcement officials suspected that individuals were using chloroform in the processing of methamphetamines. They worked with the manufacturer to add a locator beacon to the can, which they used to follow the individual who purchased the can. Using the beeper and following Knotts, they were able to locate him in a cabin. The Court ruled that Knotts did not have a reasonable expectation of privacy here. Knotts was traveling on public roads, and law enforcement could have followed the can without the assistance of the tracking device. In other words, normal visual surveillance would have been sufficient, as the person could be tracked through plain view. As Hampton suggests, this is similar to the Court's decision that one's garbage on the curb in front of that person's house is not protected by the Fourth Amendment, because there is no expectation of privacy once one places trash on the curb.[56]

On the other hand, the Court did find Fourth Amendment protections applied in the Karo case. In the Karo case, law enforcement placed a beeper in a 50-gallon drum of ether, which they believed was going to be used to extract cocaine from clothing. Karo took the ether and brought it back to and inside his home until it was moved to other locations. Using the information from the beeper, law enforcement executed a warrant to search a home, finding evidence to indict those involved. The Court ruled, in a continuance of their decision in Knotts, that the use of the beeper was not a violation of the Fourth Amendment. However, the Court did rule that unlike in Knotts, where following the beeper could have been done without the beeper because of the use of public roads, this same protection did not apply to the inside of a home. Therefore, if an agent had entered the house in following the beeper, that search would have been a violation of the Fourth Amendment, as our expectation of privacy is the greatest in our homes. This is why in the Kyllo case, the Court found a violation of the Fourth Amendment rights because, using the technology, they penetrated the walls of the home, the place where one's expectation of privacy is greatest. In addition, the Court has focused on the idea of an "extrasensory device," or a device that provides information that law enforcement could not obtain through traditional surveillance. Thus, the search in Knotts was not subject to Fourth Amendment protections, but the searches in Karo (if they entered the home) and Kyllo would be protected because the information provided by the device could not be obtained through traditional search methods.[57]

Beyond stingrays, cell phones can be tracked in more traditional ways. Every seven seconds, a cell phone connects to a cellular tower to maintain a signal. This process is called registration and cannot be stopped unless a user turns off his or her cell phone. The cell phone companies who own the towers store this information. In order for law enforcement to obtain this information, they must get a court order. Yet, in order to obtain information from the stingray devices, there is no court order needed, because the government owns the device. Even though the information that law enforcement can obtain is similar, they only need court orders to get the information from the cell phone companies. In early 2013, a federal district court judge heard arguments on a case that asks whether individuals are protected against the use of these devices by the Fourth Amendment and their right to privacy.[58] In this case, the Federal Bureau of Investigation used a court order to obtain information from Verizon about the individual's air card, which is used to connect to the Internet. With this information, the FBI was able to track his location via his air card, using a stingray device.[59] The outcome of this case is not yet known, but it will certainly be relevant as it works its way through the federal courts and they consider the use of these stingray devices.

Stingray devices and other technologies that allow law enforcement to track an individual are only going to become more common as the use of cell phones and other GPS technologies expands even further. Courts will have to deal with questions of how and when law enforcement can use these devices and how and when our Fourth Amendment rights might protect us from these devices. Seemingly, the central question will be whether they equate the device to the one in *Jones* and *Kyllo* or to the locator beacons in *Knotts*. Will it be the case that they will rely on the fact that these are extrasensory devices and therefore require Fourth Amendment protections? Or might a court decide that because one such search could have been done in public spaces with traditional surveillance methods, then our expectation of privacy is not relevant? It is difficult to determine what the outcomes might be, but the courts will likely have something to say sooner rather than later. How our expectation of privacy as it relates to our location might be interpreted will be determined by the outcome of these cases.

Location-Based Applications and Voluntary GPS Location Information

While some GPS location information is involuntary or done by another party to determine where a person might be, other information is provided freely by an individual. Location-based mobile applications include any application that records an individual's location and provides information based on that location. "LBS offer a wide array of services, navigation tools help users reach their destination (e.g., Google Maps); local search applications to help users find and review nearby businesses (e.g., Yelp); location sharing applications that allow users to check in to their location and share it with their friends (e.g., Foursquare, Facebook Places); social networking applications that allow users to geotag content such as photos and posts and share it with their friends (e.g., Facebook, Twitter); ambient social networking applications that run in the background on a smartphone and enhance serendipity by alerting users in real-time to nearby friends or individuals with whom they have affinity (e.g., Google Latitude, Highlight, Sonar); and dating applications that allow users to find romantic and sexual partners nearby (e.g., OkCupid, Grindr)."[60] In these applications, the user's location information is central to the service provided. Though the users opt in to these applications and participation in these applications in voluntary, like the situation with social networking described above, it is not clear to what degree any expectation of privacy exists.

Yet, this information is also used by the companies who create the applications. These companies store data about individuals—data they then sell to other companies. Thus, as more and more people use smartphones and these applications, more companies will have access to sensitive data about individuals. This leads to serious questions about individual privacy, as Horwath outlines.[61] She continues, "In his concurrence in *United States v. Jones*, Justice Samuel Alito suggests that while new technologies 'may provide increased convenience or security at the expense of privacy, many people may find the tradeoff worthwhile.' LBS users consent to the collection and mining of their user data in exchange for free use of the app or service."[62] Indeed, while Facebook and social-networking sites are optional, and individuals post information with some knowledge that other people will see that information, it is unclear if people understand the privacy tradeoffs associated with LBS (location-based services). Indeed, the convenience of searching for directions on one's phone is easier than the more private use of a paper map, and some might argue that the use of some of these LBS are necessary to participate in modern life.

Yet again, we approach the same crossroads, where the third-party doctrine applies, as people share information with companies. Under current case law, this would no longer be protected by the Fourth Amendment. Interestingly, what Horwath argues for is a more nuanced approach. "Since our location information can be ascertained from uses like these, some LBS usage can almost be said to involve involuntary disclosures of our location information. For these reasons, I believe that a better standard than the reasonable expectation of privacy test would be whether a person intends to limit access to his location information in a way that society recognizes as reasonable. The fact that a person uses LBS in furtherance of criminal activity should not matter—'numerous courts have held that privacy expectations are not diminished by the criminality of a defendant's activities.'"[63]

In other words, she argues that rather than being a bright-line third-party doctrine ruling, we should think about the intent of a person in sharing the information with others. In that, if it is clear the individual was trying to limit who knew the information, then that person should have an expectation of privacy, and that individual's Fourth Amendment rights apply. For example, an individual who tweets his or her location, asking for someone to bring marijuana to sell, would have no reasonable expectation of privacy, because the person voluntarily disclosed the location.[64] Although this is only a suggestion and not law, it is likely that some compromise or nuanced position will be necessary to deal with expectations of privacy on social-networking and location-based application posts. One final type of information is subject to similar types of questions. However, this is not information that one shares with others, but it is subject to the same arcane ECPA that is too outdated to deal with this. In the next section, we deal with the questions of digital evidence.

DIGITAL EVIDENCE

As previously stated, today a person can store increasingly large amounts of information in increasingly small physical spaces. Indeed, given newer cloud storage developments, one need not even be in possession of physical space to hold large amounts of digital data. The question that arises as related to digital searches concerns what might be covered by a warrant. Generally, courts have agreed that searches on digital data, especially on an individual's personal computer, place "an extreme burden on both the individual's privacy, as well as on police resources."[65] As a result, courts have a challenge in applying the Fourth Amendment to these searches. The struggle is centered on whether digital searches should be conducted in the same way as traditional searches of physical space and evidence are. There are two sides to the argument. One side suggests that a computer is simply a container and that evidence contained within it should be treated as such. The other side argues that this approach is not fully appropriate, and a new set of specialized rules related to searching of digital information is needed.[66] The problem related to searching digital evidence arises even after a warrant has been secured. Because of the way in which computers are searched, virtually every document on a computer, even those not covered by a warrant, are in plain view. Thus, a warrant for even part of a computer's information is almost always going to cover that entire computer. As with the sections above, those who write about these issues have come to a single conclusion: New laws relating to privacy and digital technology must be passed.

CONCLUSION

In this chapter, we have reviewed the Fourth Amendment and our expectation of privacy as it pertains to the ever-changing digital world. First, we saw how the expectation of privacy has been derived from Fourth Amendment cases at the Supreme Court. Arguing that privacy is one of the most fundamental rights, the Court decided in *Katz* that individuals have a reasonable expectation of privacy in searches conducted by law enforcement. While applying these principles to other cases, it became clear that certain places were more protected. Indeed, we have the greatest expectations of privacy in our homes and a lesser expectation in places where our actions can be clearly observed. Overall, the Court has applied this principle to different types of searches, finding that those that are most invasive into our privacy require Fourth Amendment protection.

However, it may be the case that individuals' expectations of personal privacy have changed as a result of technologies that limit our privacy. Social-networking sites allow users to connect instantly with people all over the world. In doing so, individuals can share information with their friends, acquaintances, and even strangers with the click of a button. While social-networking sites like Facebook or Twitter have privacy policies, it is unclear whether people really understand that what they post could very easily be available for anyone to see, even if they have strict privacy settings. Beyond social networks, individuals check in at places through location-based apps such as Foursquare. In this context, the location of a person might be available to law enforcement or everybody. Certainly, with the technologically advanced smartphones and cell phones today, we can be tracked by the GPS location systems within these technologies even without voluntarily posting this information. So, what does this all mean for our expectation of privacy? As Brandon Crowther highlights, there are four central factors to this gap between our expectation of privacy and our actual digital privacy that I believe sum up the issues at hand. These are "1) the increased gap between subjective and objective expectations in digital contexts, 2) contractual arrangements with Internet service providers, 3) storage of information on third-party servers, and 4) judges' technological inexperience."[67] To deal with these, it is clear that legal changes must be made.

Almost all the authors and scholars cited within this chapter agree that new law or laws governing our expectation of digital privacy must be passed and that this must happen soon. The central federal law related to digital evidence and digital privacy was passed in 1986. Even when the federal courts try to interpret this law, it is almost impossible. The exact changes that need to be made will be debated. However, without change, we will be stuck in the current situation, where citizens have very little certain information on what their current expectation of privacy might be.

GLOSSARY

Fourth Amendment—an amendment to the Bill of Rights that protects citizens against unreasonable search and seizure.

Electronic Communications Privacy Act of 1986—the primary federal statute governing online privacy. It includes three sections: the Wiretap Act, the Stored Communications Act, and the Pen Register Act. The goal of the act was to balance law enforcement's need to collect information and an individual's need for privacy.

Expectation of Privacy—a doctrine in *Katz v. United States* (1967) that outlines to scope of the protection a citizen might expect from their protection against unreasonable search and seizure.

Plain-View Doctrine—When conducting a search, law enforcement officers may collect and use in court what is not covered by the warrant but is in plain view of the officers conducting the search

The Digital or Information Age—a period in human history that covers the time after the Industrial Revolution and industrial age. This period is characterized by the use of computers and the microminiturization of computer technology. This period includes the rise of the Internet.

Third-Party Doctrine—a doctrine that states a person no longer has a reasonable expectation of privacy in any communication voluntarily disclosed to another person or entity.[68]

NOTES

1. Jill Lepore, "The Prism: Privacy in the Age of Publicity," *New Yorker*, June 24, 2013.
2. Louis Brandeis, and Samuel Warren, "The Right to Privacy," *Harvard Law Review* 4 (1890): 193.
3. *Olmstead v. United States*, 277 U.S. 438 (1928).
4. Jill Lepore, "The Prism: Privacy in the Age of Publicity."
5. *Olmstead v. United States*, 277 U.S. 438 (1928, Brandeis, J. dissenting), as quoted in Lepore, "The Prism: Privacy in the Age of Publicity."
6. Alva Noe, "What's the Big Deal about Privacy?" National Public Radio website, August 16, 2013, Accessed August 16, 2013, http://www.npr.org/blogs/13.7/2013/08/16/212546316/whats-the-big-deal-about-privacy.
7. Ibid.
8. Ibid.
9. Ibid.

10. 381 U.S. 479 (1965).
11. 389 U.S. 347 (1967).
12. Brittany Hampton, "From Smartphones to Stingrays: Can the Fourth Amendment Keep Up with the Twenty-First Century," *University of Louisville Law Review* 51 (2012): 159–176. Quoting from *Katz* (361, Harlan, J concurring).
13. Lee Epstein, and Thomas G. Walker, *Rights Liberties and Justice: Constitutional Law for a Changing America* (Thousand Oaks: C.Q. Press, 2013).
14. 533 U.S. 27 (2001).
15. 565 U.S. __ (2012).
16. Scott D. Blake, "Let's Be Reasonable: Fourth Amendment Principles in the Digital Age," *Seventh Circuit Review* 5 (2010): 491–531.
17. Kate Hafner and Matthew Lyon, *Where Wizards Stay Up Late: The Origins of the Internet* (New York: Simon & Schuster, 1998).
18. See Center for Democracy and Technology, "Security and Surveillance," last modified 2013, https://www.cdt.org/issue/wiretap-ecpa.
19. Center for Democracy and Technology, "Security and Surveillance."
20. Lily R. Robinton, "Courting Chaos: Conflicting Guidance from Courts Highlights the Need for Clearer Rules to Govern the Search and Seizure of Digital Evidence," *Yale Journal of Law and Technology* 12 (2010): 311–347.
21. To be certain, the passage of the Patriot Act in 2001 had some effect on these topics. However, the relation of this information to the government and their intelligence gather as related to terrorism is beyond the scope of this chapter.
22. American Civil Liberties Union, "Modernizing the Electronic Communications Privacy Act (ECPA)," last modified 2013, http://www.aclu.org/technology-and-liberty/modernizing-electronic-communications-privacy-act-ecpa.
23. Digital Due Process: About the Issue, "ECPA Reform: Why Now?," http://www.digitaldueprocess.org/index.cfm?objectid=37940370-2551-11DF-8E02000C296BA163.
24. Lindsay S. Feuer, "Who is Poking Around Your Facebook Profile?: The Need to Reform the Stored Communications Act to Reflect a Lack of Privacy on Social Networking Websites," *Hofstra Law Reivew* 40 (2011): 473–515.
25. Dana M. Boyd and Nicole B. Ellison, "Social Network Sites: Definition, History, and Scholarship," *Journal of Computer-Mediated Communication* 13, no. 1 (2007).
26. Feuer, "Who Is Poking Around Your Facebook Profile?: The Need to Reform the Stored Communications Act to Reflect a Lack of Privacy on Social Networking Websites," 478–9.
27. Lisa A. Schmidt, "Social Networking and the Fourth Amendment: Location Tracking on Facebook, Twitter, and Foursquare," *Cornell Journal of Law and Public Policy* 22 (2012): 515–536.
28. Feuer, "Who Is Poking Around Your Facebook Profile?: The Need to Reform the Stored Communications Act to Reflect a Lack of Privacy on Social Networking Websites," 483.
29. Dave Alsup and Ben Brumfield, "Florida man allegedly kills wife, posts confession, photo of body on Facebook," CNN.com, August 11, 2013, Accessed August 12, 2013, http://www.cnn.com/2013/08/09/us/florida-facebook-confession.

30. Feuer, "Who Is Poking Around Your Facebook Profile?: The Need to Reform the Stored Communications Act to Reflect a Lack of Privacy on Social Networking Websites," 482.

31. Ibid., 485.

32. Whitson Gordon, "Facebook Introduces New, Uber-Simple Privacy Controls," *Lifehacker*, December 21, 2012, accessed August 14, 2013, http://lifehacker.com/5970442/facebook-introduces-new-uber+simple-privacy-controls.

33. Will Oremus, "If You've Ever Posted Anything Embarrassing on Facebook, Now Is the Time to Hide It," *Slate*, July 8, 2013, Accessed August 14, 2013, http://www.slate.com/blogs/future_tense/2013/07/08/facebook_graph_search_privacy_nightmare_is_preventable_if_you_change_your.html.

34. Ibid.

35. Feurer, "Who Is Poking Around Your Facebook Profile?: The Need to Reform the Stored Communications Act to Reflect a Lack of Privacy on Social Networking Websites," 486.

36. Schmidt, "Social Networking and the Fourth Amendment: Location Tracking on Facebook, Twitter, and Foursquare," 520.

37. Ibid., 520.

38. 560 U.S. ___ (2010)

39. Schmidt, "Social Networking and the Fourth Amendment: Location Tracking on Facebook, Twitter, and Foursquare," 529.

40. 488 U.S. 445 (1989).

41. Schmidt, "Social Networking and the Fourth Amendment: Location Tracking on Facebook, Twitter, and Foursquare," 530.

42. Monu Bedi, "Facebook and Interpersonal Privacy: Why the Third Party Doctrine Should Not Apply," *Boston College Law Review* 54 (2013): 1–71.

43. 425 U.S. 435 (1976).

44. Oyez.org "*United States v. Miller*," accessed August 16 2013, http://www.oyez.org/cases/1970-1979/1975/1975_74_1179/#sort=ideology.

45. Bedi, "Facebook and Interpersonal Privacy: Why the Third Party Doctrine Should Not Apply," 3.

46. Ibid., 3.

47. Ibid., 44–50.

48. Ibid., 62.

49. Kathryn Nobuko Horwath, "A Check-in on Privacy after *United States v. Jones*: Current Fourth Amendment Jurisprudence in the Context of Location-Based Applications and Services," *Hastings Constitutional Law Quarterly* 40 (2013): 925–966.

50. Hampton, "From Smartphones to Stingrays: Can the Fourth Amendment Keep Up with the Twenty-First Century," 159.

51. Ibid., 171.

52. Ibid., 171.

53. 460 U.S. 276 (1983).

54. 468 U.S. 705 (1984).

55. Hampton, "From Smartphones to Stingrays: Can the Fourth Amendment Keep Up with the Twenty-First Century," 164–5.

56. Ibid., 164.

57. Ibid., 170.

58. Ibid., 172–3.

59. Kim Zetter, "Government Fights for Use of Spy Tool That Spoofs Cell Towers," *Wired*, March 29, 2013, accessed August 18, 2013, http://www.wired.com/threatlevel/2013/03/gov-fights-stingray-case/all/.

60. Howarth, "A Check-in on Privacy after *United States v. Jones*: Current Fourth Amendment Jurisprudence in the Context of Location-Based Applications and Services," 929–30.

61. Ibid., 930.

62. Ibid., 933.

63. Ibid., 960. Quoting *United States v. Skinner*, 690 F.3d 772,785 (6th Cir. 2012). (Donald, J., concurring in part and concurring in judgment).

64. Michelle McQuigge, "Mr. Lube worker fired after tweeting he was looking to buy some pot at his workplace," *The Vancouver Sun*, August 15, 2013, accessed August 18 2013, http://www.vancouversun.com/business/Lube+worker+fired+after+tweeting+Twitter+looking+some+workplace/8792945/story.html.

65. Robinton, "Courting Chaos: Conflicting Guidance from Courts Highlights the Need for Clearer Rules to Govern the Search and Seizure of Digital Evidence," 325.

66. Ibid., 327–8.

67. Brandon T. Crowther, "(Un)Reasonable Expectation of Digital Privacy," *Brigham Young University Law Review* (2012): 343–69.

68. Bedi, "Facebook and Interpersonal Privacy: Why the Third Party Doctrine Should Not Apply," 2.